Fractional Differential Equations

This is Volume 198 in
MATHEMATICS IN SCIENCE AND ENGINEERING

By Igor Podlubny, Technical University of Kosice, Slovak Republic

A list of recent titles in this series appears on page v of this volume.

FRACTIONAL DIFFERENTIAL EQUATIONS

An Introduction to Fractional Derivatives,
Fractional Differential Equations,
to Methods of their Solution
and some of their Applications

by

Igor Podlubny

Technical University of Kosice, Slovak Republic

ACADEMIC PRESS
San Diego • Boston • New York • London • Sydney • Tokyo • Toronto

This book is printed on acid-free paper.

ACADEMIC PRESS
525B Street
Suite 1900
San Diego, California 92101-4495, USA
http://www.apnet.com

ACADEMIC PRESS
24–28 Oval Road
LONDON
NW1 7DX, UK
http://www.hbuk.co.uk/ap/

A catalogue record for this book is available from the British Library

ISBN 0–12–558840–2

99 00 01 02 03 MP 9 8 7 6 5 4 3 2 1

Mathematics in Science and Engineering
Edited by William F. Ames, Georgia Institute of Technology

Recent titles
T.A. Burton, *Volterra Integral and Differential Equations*
Ram P. Kanwal, *Generalized Functions: Theory and Technique*
Marc Mangel, *Decision and Control in Uncertain Resource Systems*
John L. Casti, *Nonlinear System Theory*
Yoshikazu Sawaragi, Hirotaka Nakayama, and Tetsuzo Tanino, *Theory of Multi-objective Optimization*
Edward J. Haug, Kyung K. Choi, and Vadim Komkov, *Design Sensitivity Analysis of Structural Systems*
Yaakov Bar-Shalom and Thomas E. Fortmann, *Tracking and Data Association*
V.B. Kolmanovskii and V.R. Nosov, *Stability of Functional Differential Equations*
V. Lakshmikantham and D. Trigiante, *Theory of Difference Equations: Applications to Numerical Analysis*
B.D. Vujanovic and S.E. Jones, *Variational Methods in Nonconservative Phenomena*
C. Rogers and W.F. Ames, *Nonlinear Boundary Value Problems in Science and Engineering*
W.F. Ames and C. Rogers, *Nonlinear Equations in the Applied Sciences*
Josip E. Pecaric, Frank Proschan, and Y.L. Tong, *Convex Functions, Partial Orderings, and Statistical Applications*
E.N. Chukwu, *Stability and Time-Optimal Control of Hereditary Systems*
Viorel Barbu, *Analysis and Control of Nonlinear Infinite Dimensional Systems*
Yang Kuang, *Delay Differential Equations: With Applications in Population Dynamics*
K.A. Ames and B. Straughan, *Non-Standard and Improperly Posed Problems*
Z. Gajic and M.T.J. Qureshi, *The Lyapunov Matrix Equation in System Stability and Control*
D. Straub, *Alternative Mathematical Theory of Non-equilibrium Phenomena*
B.G. Pachpatte, *Inequalities for Differential and Integral Equations*
I. Podlubny, *Fractional Differential Equations*

To my parents

Contents

List of Figures

Preface

This book is devoted to integrals and derivatives of arbitrary real order, methods of solution of differential equations of arbitrary real order, and applications of the described methods in various fields.

The theory of derivatives of non-integer order goes back to the Leibniz's note in his list to L'Hospital [123], dated 30 September 1695, in which the meaning of the derivative of order one half is discussed.

Leibniz's note led to the appearance of the theory of derivatives and integrals of arbitrary order, which by the end of the XIX century took more or less finished form due primarily to Liouville, Grünwald, Letnikov, and Riemann. Surveys of the history of the theory of fractional derivatives can be found in [44, 153, 179, 226, 232].

For three centuries the theory of fractional derivatives developed mainly as a pure theoretical field of mathematics useful only for mathematicians.

However, in the last few decades many authors pointed out that derivatives and integrals of non-integer order are very suitable for the description of properties of various real materials, e.g. polymers. It has been shown that new fractional-order models are more adequate than previously used integer-order models. Fundamental physical considerations in favour of the use of models based on derivatives of non-integer order are given in [30, 254].

Fractional derivatives provide an excellent instrument for the description of memory and hereditary properties of various materials and processes. This is the main advantage of fractional derivatives in comparison with classical integer-order models, in which such effects are in fact neglected. The advantages of fractional derivatives become apparent in modelling mechanical and electrical properties of real materials, as well as in the description of rheological properties of rocks, and in many other fields.

The other large field which requires the use of derivatives of non-integer order is the recently elaborated theory of fractals [142]. The development of the theory of fractals has opened further perspectives for the theory of fractional derivatives, especially in modelling dynamical processes in self-similar and porous structures.

Fractional integrals and derivatives also appear in the theory of control of dynamical systems, when the controlled system or/and the controller is described by a fractional differential equation.

The mathematical modelling and simulation of systems and processes, based on the description of their properties in terms of fractional derivatives, naturally leads to differential equations of fractional order and to the necessity to solve such equations. However, effective general methods for solving them cannot be found even in the most useful works on fractional derivatives and integrals.

It should be mentioned that from the viewpoint of applications in physics, chemistry and engineering it was undoubtedly the book written by K. B. Oldham and J. Spanier [179] which played an outstanding role in the development of the subject which can be called *applied fractional calculus.* Moreover, it was the first book which was entirely devoted to a systematic presentation of the ideas, methods, and applications of the fractional calculus.

Later there appeared several fundamental works on various aspects of the fractional calculus, including the encyclopedic monograph by S. Samko, A. Kilbas, and O. Marichev [232], books by R. Gorenflo and S. Vessella [90], V. Kiryakova [116], A. C. McBride [148] K. S. Miller and B. Ross [153], K. Nishimoto [167], B. Rubin [230], lecture notes by F. Mainardi and R. Gorenflo [83, 88, 138] in the book [35], and an extensive survey by Yu. Rossikhin and M. Shitikova [228].

The book by M. Caputo [24], published in 1969, in which he systematically used his original definition of fractional differentiation for formulating and solving problems of viscoelasticity, and his lectures on seismology [28] must also be added to this gallery, as well as a series of A. Oustaloup's books on applications of fractional derivatives in control theory [183, 185, 186, 187].

However, numerous references to the books by Oldham and Spanier [179] and by Miller and Ross [153] show that applied scientists need first of all an easy introduction to the theory of fractional derivatives and fractional differential equations, which could help them in their initial steps to adopting the fractional calculus as a method of research.

The main objective of the present book is to provide such an overview of the basic theory of fractional differentiation, fractional-order differential equations, methods of their solution and applications. Taking into account the needs of the audience to which this book is addressed, namely applied scientists in all branches of science, special attention was paid to providing easy-to-follow illustrative examples. For the same reason only those approaches to fractional differentiation are considered which are related to real applications. The language and the general style were influenced by the author's wish to make the methods of the fractional calculus available to the widest possible group of potential users of this nice and efficient theory.

The book consists of ten chapters.

In *Chapter 1* an introduction to the theory of special functions (the gamma and beta function, the Mittag-Leffler function, and the Wright function) is given. These functions play the most important role in the theory of fractional derivatives and fractional differential equations.

In *Chapter 2* some approaches to generalizations of the notion of differentiation and integration are considered. In each case, we start with integer-order derivatives and integrals and show how these notions are generalized using some selected approaches. We consider there the Grünwald-Letnikov, the Riemann–Liouville, and the Caputo fractional derivative, and also the so-called sequential fractional derivatives. The approach using generalized functions is also discussed, as well as the notion of left and right fractional derivatives. Properties of the considered fractional derivatives are introduced, including composition rules, the links between these approaches, and the use of integral transforms (Laplace, Fourier, Mellin).

Chapters 3–8 are devoted to methods of the treatment of fractional differential equations.

In *Chapter 3* some useful existence and uniqueness theorems for initial problems for fractional differential equations are given. Examples are given of the use of the existence and uniqueness theorem as a method of solution of fractional differential equations. We also study the dependence of the solution on initial conditions.

In *Chapter 4* the Laplace transform method for solving linear fractional differential equations is described and illustrated by examples. Special attention is paid to the difference between fractional differential equations containing "standard" and "sequential" fractional derivatives. There are also examples of the use of the Laplace transform for solving partial differential equations.

In *Chapter 5* the definition and some properties of the fractional Green's function are given, and the explicit expression for the Green's functions for a general ordinary linear fractional differential equation is obtained. There are also given its particular cases for the one-term, two-term, three-term, and four-term fractional differential equations. Combining the methods of Chapters 4 and 5, it is possible to easily obtain closed-form solutions of initial-value problems for ordinary linear fractional differential equations.

In *Chapter 6* some other analytical methods are described, namely the Mellin transform method, the power series method, and Yu. I. Babenko's symbolic method. We also include the method of orthogonal polynomials for the solution of integral equations of fractional order, and give a collection of so-called spectral relationships for various types of kernels. All the methods described in this chapter are also illustrated by examples.

Chapters 7 and 8 deal with numerical methods.

In *Chapter 7* we describe the fractional difference approach to numerical evaluation of fractional derivatives, and discuss the order of approximation. We also describe the "short-memory" principle, which allows faster evaluation of fractional derivatives. The use of the fractional difference method and the "short-memory" principle is illustrated by an example of their application to the calculation of heat load intensity change in blast furnace walls. Additionally, there is an example of the use of the fractional difference approximation of fractional derivatives for the numerical evaluation of finite parts of divergent integrals, which often arise in many fields, especially in fracture mechanics.

In *Chapter 8* the fractional difference method is used for the numerical solution of initial-value problems for ordinary fractional differential equations. Again, the use of the method and the "short-memory" principle is illustrated by several examples.

Chapters 9 and 10 are devoted to applications of the fractional calculus, and provide illustrations of the use of the methods described in the other chapters.

In *Chapter 9* fractional-order dynamical systems and controllers are considered. In fact, this chapter is an extensive demonstration of the use of the methods described in the previous chapters.

In *Chapter 10* we give a survey of various fields of application of fractional derivatives. Some of these are already well established, and some have just started their development in the framework of applied fractional calculus. Where possible, we try to apply the methods described in other chapters. Since it often happens that different objects or processes in different branches of science are described by the same equations, this chapter may provide a certain inspiration for applying fractional calculus in further fields.

The bibliography consists of 259 entries, published up to 1997. However, it cannot be considered as a complete bibliography, and interested reader may find many additional references in the monographs which are mentioned above, especially in the fractional calculus encyclopedia by S. Samko, A. Kilbas, and O. Marichev [232].

Acknowledgements

There are many people to whom I am very obliged for their support.

I would like to express my gratitude to Professor Francesco Mainardi (University of Bologna, Italy), Professor Michele Caputo (University "La Sapienza", Rome, Italy), Professor Rudolf Gorenflo (Free University of Berlin, Germany), Professor Virginia Kiryakova (Institute of Mathematics, Bulgarian Academy of Sciences, Sofia, Bulgaria), Professor Hari M. Srivastava (University of Victoria, Canada), Dr Siegmar Kempfle (Universität der Bundeswehr, Hamburg, Germany), Professor Svante Westerlund (University of Kalmar, Sweden), Professor Denis Matignon (Ecole Nationale Supérieure des Télécommunications, France), Dr Ahmed El-Sayed (University of Alexandria, Alexandria, Egypt), Professor Imrich Koštial (Technical University of Košice), and Dr Ľubomír Dorčák (Technical University of Košice), for the exchange of information and valuable discussions.

I am deeply grateful to Professor W. F. Ames (Georgia Institute of Technology) for his suggestion to submit the manuscript, and to Professor Vadim Komkov for encouragement. I am very obliged to the staff at the Academic Press, Inc., especially to Anne Gillaume and Linda Ratts Engelman, for taking care of the preparation of this book for the press.

I would like to acknowledge the help of Mrs V. Juricová, the librarian of the Central Library of the Technical University of Košice, who aided me in collecting the copies of many articles.

I also express my gratitude to Mrs Jean Hopson, Ms Serena Yeo (British Council, Kosice), and Dr Ladislav Pivka (Technical University of Kosice), for their help with improving the language and the style of different parts of this work.

I am grateful to the Open Society Fund (Bratislava, Slovak Republic), to the Charter 77 Foundation (Bratislava), to Professor Karol Flórián, the Rector of the Technical University of Košice, and to Professor Dušan Malindžák, the Dean of the B.E.R.G. Faculty, who arranged for the financial support for my work at different stages.

I am thankful to people from Mathworks, Inc., for MATLAB, which I used for computations and the creation of some plots.

I typeset this book using the LaTeX typesetting system, so I am also thankful to Donald Knuth for inventing TeX and to Leslie Lamport for creating LaTeX.

I am also thankful to my wife Katarina Kassayová and to our sons Igor and Martin for their everyday understanding, love and support.

Chapter 1

Special Functions of the Fractional Calculus

In this chapter some basic theory of the special functions which are used in the other chapters is given. We give here some information on the gamma and beta functions, the Mittag-Leffler functions, and the Wright function; these functions play the most important role in the theory of differentiation of arbitrary order and in the theory of fractional differential equations.

1.1 Gamma Function

Undoubtedly, one of the basic functions of the fractional calculus is Euler's gamma function $\Gamma(z)$, which generalizes the factorial $n!$ and allows n to take also non-integer and even complex values.

We will recall in this section some results on the gamma function which are important for other parts of this work.

1.1.1 Definition of the Gamma Function

The gamma function $\Gamma(z)$ is defined by the integral

$$\Gamma(z) = \int_0^\infty e^{-t} t^{z-1} dt, \tag{1.1}$$

which converges in the right half of the complex plane $Re(z) > 0$. Indeed, we have

$$\Gamma(x+iy) = \int_0^\infty e^{-t} t^{x-1+iy} dt = \int_0^\infty e^{-t} t^{x-1} e^{iy\log(t)} dt$$
$$= \int_0^\infty e^{-t} t^{x-1} [\cos(y\log(t)) + i\sin(y\log(t))] dt. \qquad (1.2)$$

The expression in the square brackets in (1.2) is bounded for all t; convergence at infinity is provided by e^{-t}, and for the convergence at $t = 0$ we must have $x = Re(z) > 1$.

1.1.2 Some Properties of the Gamma Function

One of the basic properties of the gamma function is that it satisfies the following functional equation:

$$\Gamma(z+1) = z\Gamma(z), \qquad (1.3)$$

which can be easily proved by integrating by parts:

$$\Gamma(z+1) = \int_0^\infty e^{-t} t^z dt = \left[-e^{-t} t^z\right]_{t=0}^{t=\infty} + z \int_0^\infty e^{-t} t^{z-1} dt = z\Gamma(z).$$

Obviously, $\Gamma(1) = 1$, and using (1.3) we obtain for $z = 1, 2, 3, \ldots$:

$$\begin{aligned}
\Gamma(2) &= 1 \cdot \Gamma(1) = 1 = 1!, \\
\Gamma(3) &= 2 \cdot \Gamma(2) = 2 \cdot 1! = 2!, \\
\Gamma(4) &= 3 \cdot \Gamma(3) = 3 \cdot 2! = 3!, \\
\ldots &\quad \ldots \quad \ldots \quad \ldots \\
\Gamma(n+1) &= n \cdot \Gamma(n) = n \cdot (n-1)! = n!
\end{aligned}$$

Another important property of the gamma function is that it has simple poles at the points $z = -n$, $(n = 0, 1, 2, \ldots)$. To demonstrate this, let us write the definition (1.1) in the form:

$$\Gamma(z) = \int_0^1 e^{-t} t^{z-1} dt + \int_1^\infty e^{-t} t^{z-1} dt. \qquad (1.4)$$

The first integral in (1.4) can be evaluated by using the series expansion for the exponential function. If $Re(z) = x > 0$ (i.e. z is in the right half-plane), then $Re(z+k) = x + n > 0$ and $t^{z+k}\Big|_{t=0} = 0$. Therefore,

$$\begin{aligned}\int_0^1 e^{-t} t^{z-1} dt &= \int_0^1 \sum_{k=0}^{\infty} \frac{(-t)^k}{k!} t^{z-1} dt \\ &= \sum_{k=0}^{\infty} \frac{(-1)^k}{k!} \int_0^1 t^{k+z-1} dt = \sum_{k=0}^{\infty} \frac{(-1)^k}{k!(k+z)}.\end{aligned}$$

The second integral defines an entire function of the complex variable z. Indeed, let us write

$$\varphi(z) = \int_1^{\infty} e^{-t} t^{z-1} dt = \int_1^{\infty} e^{(z-1)\log(t)-t} dt. \tag{1.5}$$

The function $e^{(z-1)\log(t)-t}$ is a continuous function of z and t for arbitrary z and $t \geq 1$. Moreover, if $t \geq 1$ (and therefore $\log(t) \geq 0$), then it is an entire function of z. Let us consider an arbitrary bounded closed domain D in the complex plane ($z = x + iy$) and denote $x_0 = \max_{z \in D} Re(z)$. Then we have:

$$\begin{aligned}\left|e^{-t} t^{z-1}\right| &= \left|e^{(z-1)\log(t)-t}\right| = \left|e^{(x-1)\log(t)-t}\right| \cdot \left|e^{iy\log(t)}\right| \\ &= \left|e^{(x-1)\log(t)-t}\right| \leq e^{(x_0-1)\log(t)-t} = e^{-t} t^{x_0-1}.\end{aligned}$$

This means that the integral (1.5) converges uniformly in D and, therefore, the function $\varphi(z)$ is regular in D and differentiation under the integral in (1.5) is allowed. Because the domain D has been chosen arbitrarily, we conclude that the function $\varphi(z)$ has the above properties in the whole complex plane. Therefore, $\varphi(z)$ is an entire function allowing differentiation under the integral.

Bringing together the above considerations, we see that

$$\begin{aligned}\Gamma(z) &= \sum_{k=0}^{\infty} \frac{(-1)^k}{k!} \frac{1}{k+z} + \int_1^{\infty} e^{-t} t^{z-1} dt \\ &= \sum_{k=0}^{\infty} \frac{(-1)^k}{k!} \frac{1}{k+z} + \text{entire function},\end{aligned} \tag{1.6}$$

and, indeed, $\Gamma(z)$ has only simple poles at the points $z = -n$, $n = 0, 1, 2, \ldots$.

1.1.3 Limit Representation of the Gamma Function

The gamma function can be represented also by the limit

$$\Gamma(z) = \lim_{n\to\infty} \frac{n!\, n^z}{z(z+1)\dots(z+n)}, \tag{1.7}$$

where we initially suppose $Re(z) > 0$.

To prove (1.7), let us introduce an auxiliary function

$$f_n(z) = \int_0^n \left(1 - \frac{t}{n}\right)^n t^{z-1} dt. \tag{1.8}$$

Performing the substitution $\tau = \frac{t}{n}$ and then repeating integration by parts we obtain

$$\begin{aligned} f_n(z) &= n^z \int_0^1 (1-\tau)^n \tau^{z-1} d\tau \\ &= \frac{n^z}{z} n \int_0^1 (1-\tau)^{n-1} \tau^z d\tau \\ &= \frac{n^z\, n!}{z(z+1)\dots(z+n-1)} \int_0^1 \tau^{z+n-1} d\tau \\ &= \frac{n^z\, n!}{z(z+1)\dots(z+n-1)(z+n)}. \end{aligned} \tag{1.9}$$

Taking into account the well-known limit

$$\lim_{n\to\infty} \left(1 - \frac{t}{n}\right)^n = e^{-t}$$

we may expect that

$$\lim_{n\to\infty} f_n(z) = \lim_{n\to\infty} \int_0^n \left(1 - \frac{t}{n}\right)^n t^{z-1} dt = \int_0^\infty e^{-t} t^{z-1} dt, \tag{1.10}$$

which ends the proof of the limit representation (1.7) of the gamma function, if the interchange of the limit and the integral in (1.10) is justified. To do this, let us estimate the difference

$$\begin{aligned}\Delta &= \int_0^\infty e^{-t}t^{z-1}dt - f_n(z)\\ &= \int_0^n \left[e^{-t} - \left(1-\frac{t}{n}\right)^n\right] t^{z-1}dt + \int_n^\infty e^{-t}t^{z-1}dt. \qquad (1.11)\end{aligned}$$

Let us take an arbitrary $\epsilon > 0$. Because of the convergence of the integral (1.1) there exists an N such that for $n \geq N$ we have

$$\left|\int_n^\infty e^{-t}t^{z-1}dt\right| \leq \int_n^\infty e^{-t}t^{x-1}dt < \frac{\epsilon}{3}, \qquad (x = Re(z)). \qquad (1.12)$$

Fixing now N and considering $n > N$ we can write Δ as a sum of three integrals:

$$\Delta = \left(\int_0^N + \int_N^n\right) \left[e^{-t} - \left(1-\frac{t}{n}\right)^n\right] t^{z-1}dt + \int_n^\infty e^{-t}t^{z-1}dt. \qquad (1.13)$$

The last term is less then $\frac{\epsilon}{3}$. For the second integral we have:

$$\begin{aligned}\left|\int_N^n \left[e^{-t} - \left(1-\frac{t}{n}\right)^n\right] t^{z-1}dt\right| &\leq \int_N^n \left[e^{-t} - \left(1-\frac{t}{n}\right)^n\right] t^{x-1}dt\\ &< \int_N^\infty e^{-t}t^{x-1}dt < \frac{\epsilon}{3}, \qquad (1.14)\end{aligned}$$

where, as above, $x = Re(z)$.

For the estimation of the first integral in (1.13) we need the following auxiliary inequality:

$$0 < e^{-t} - \left(1-\frac{t}{n}\right)^n < \frac{t^2}{2n}, \qquad (0 < t < n), \qquad (1.15)$$

which follows from the relationships

$$1 - e^t\left(1-\frac{t}{n}\right)^n = \int_0^t e^\tau \left(1-\frac{\tau}{n}\right)^n \frac{\tau}{n} d\tau \qquad (1.16)$$

and

$$0 < \int_0^t e^{\tau}\left(1-\frac{\tau}{n}\right)^n \frac{\tau}{n} d\tau < \int_0^t e^{\tau}\frac{\tau}{n} d\tau = e^t \frac{t^2}{2n}. \tag{1.17}$$

(Relationship (1.16) can be verified by differentiating both sides.)

Using the auxiliary inequality (1.15) we obtain for large n and fixed N:

$$\left|\int_0^N \left[e^{-t} - \left(1-\frac{t}{n}\right)^n\right] t^{z-1} dt\right| < \frac{1}{2n}\int_0^N t^{x+1} dt < \frac{\epsilon}{3}. \tag{1.18}$$

Taking into account inequalities (1.12), (1.14) and (1.18) and the arbitrariness of ϵ we conclude that the interchange of the limit and the integral in (1.10) is justified.

This definitely completes the proof of the formula (1.7) for the limit representation of the gamma function for $Re(z) > 0$.

With the help of (1.3) the condition $Re(z) > 0$ can be weakened to $z \neq 0, -1, -2, \ldots$ in the following manner.

If $-m < Re(z) \leq -m+1$, where m is a positive integer, then

$$\begin{aligned}
\Gamma(z) &= \frac{\Gamma(z+m)}{z(z+1)\ldots(z+m-1)} \\
&= \frac{1}{z(z+1)\ldots(z+m-1)} \lim_{n\to\infty} \frac{n^{z+m} n!}{(z+m)\ldots(z+m+n)} \\
&= \frac{1}{z(z+1)\ldots(z+m-1)} \lim_{n\to\infty} \frac{(n-m)^{z+m}(n-m)!}{(z+m)(z+m+1)\ldots(z+n)} \\
&= \lim_{n\to\infty} \frac{n^z n!}{z(z+1)\ldots(z+n)}.
\end{aligned} \tag{1.19}$$

Therefore, the limit representation (1.7) holds for all z excluding $z \neq 0, -1, -2, \ldots$

1.1.4 Beta Function

In many cases it is more convenient to use the so-called beta function instead of a certain combination of values of the gamma function.

The beta function is usually defined by

$$B(z,w) = \int_0^1 \tau^{z-1}(1-\tau)^{w-1} d\tau, \quad (Re(z) > 0, \quad Re(w) > 0). \tag{1.20}$$

To establish the relationship between the gamma function defined by (1.1) and the beta function (1.20) we will use the Laplace transform.

Let us consider the following integral

$$h_{z,w}(t) = \int_0^t \tau^{z-1}(1-\tau)^{w-1} d\tau. \tag{1.21}$$

Obviously, $h_{z,w}(t)$ is a convolution of the functions t^{z-1} and t^{w-1} and $h_{z,w}(1) = B(z,w)$.

Because the Laplace transform of a convolution of two functions is equal to the product of their Laplace transforms, we obtain:

$$H_{z,w}(s) = \frac{\Gamma(z)}{s^z} \cdot \frac{\Gamma(w)}{s^w} = \frac{\Gamma(z)\Gamma(w)}{s^{z+w}}, \tag{1.22}$$

where $H_{z,w}(s)$ is the Laplace transform of the function $h_{z,w}(t)$.

On the other hand, since $\Gamma(z)\Gamma(w)$ is a constant, it is possible to restore the original function $h_{z,w}(t)$ by the inverse Laplace transform of the right-hand side of (1.22). Due to the uniqueness of the Laplace transform, we therefore obtain:

$$h_{z,w}(t) = \frac{\Gamma(z)\Gamma(w)}{\Gamma(z+w)} t^{z+w-1}, \tag{1.23}$$

and taking $t = 1$ we obtain the following expression for the beta function:

$$B(z,w) = \frac{\Gamma(z)\Gamma(w)}{\Gamma(z+w)}, \tag{1.24}$$

from which it follows that

$$B(z,w) = B(w,z). \tag{1.25}$$

The definition of the beta function (1.20) is valid only for $Re(z) > 0$, $Re(w) > 0$. The relationship (1.24) provides the analytical continuation of the beta function for the entire complex plane, if we have the analytically continued gamma function.

With the help of the beta function we can establish the following two important relationships for the gamma function.

The first one is

$$\Gamma(z)\Gamma(1-z) = \frac{\pi}{\sin(\pi z)}. \tag{1.26}$$

We will obtain the formula (1.26) under the condition $0 < Re(z) < 1$ and then show that it holds for $z \neq 0, \pm 1, \pm 2, \dots$

Using (1.24) and (1.20) we can write

$$\Gamma(z)\Gamma(1-z) = B(z, 1-z) = \int_0^1 \left(\frac{t}{1-t}\right)^{z-1} \frac{dt}{1-t}, \tag{1.27}$$

where the integral converges if $0 < Re(z) < 1$. Performing the substitution $\tau = t/(1-t)$ we obtain

$$\Gamma(z)\Gamma(1-z) = \int_0^\infty \frac{\tau^{z-1}}{1+\tau} d\tau. \tag{1.28}$$

Let us now consider the integral

$$\int_L f(s)ds, \quad f(s) = \frac{s^{z-1}}{1+s}, \tag{1.29}$$

along the contour shown in Fig. 1.1. The complex plane is cut along the real positive semi-axis.

The function $f(\tau)$ has a simple pole at $s = e^{\pi\imath}$. Therefore, for $R > 1$ we have

$$\int_L f(s)ds = 2\pi i \Big[\text{Res} f(s)\Big]_{s=e^{\pi\imath}} = -2\pi i e^{i\pi z}. \tag{1.30}$$

On the other hand, the integrals along the circumferences $|s| = \epsilon$ and $|s| = R$ vanish as $\epsilon \to 0$ and $R \to \infty$, and the integral along the lower cut edge differs from the integral along the upper cut edge by the factor $-e^{2\pi i z}$. Because of this, for $\epsilon \to 0$ and $R \to \infty$ we obtain:

$$\int_L f(s)ds = 2\pi i \Big[\text{Res} f(s)\Big]_{s=e^{\pi\imath}} = -2\pi i e^{i\pi z} = \Gamma(z)\Gamma(1-z)(1-e^{2\pi i z}), \tag{1.31}$$

and

$$\Gamma(z)\Gamma(1-z) = \frac{2\pi i e^{\imath\pi z}}{e^{2\pi\imath z} - 1} = \frac{\pi}{\sin(\pi z)}, \quad (0 < Re(z) < 1). \tag{1.32}$$

If $m < Re(z) < m+1$, then we can put $z = \alpha + m$, where $0 < Re(\alpha) < 1$. Using (1.3) we obtain

$$\begin{aligned}\Gamma(z)\Gamma(1-z) &= (-1)^m \Gamma(\alpha)\Gamma(1-\alpha) \\ &= \frac{(-1)^m \pi}{\sin(\pi\alpha)} = \frac{\pi}{\sin(\pi(\alpha+m))} = \frac{\pi}{\sin(\pi z)},\end{aligned} \tag{1.33}$$

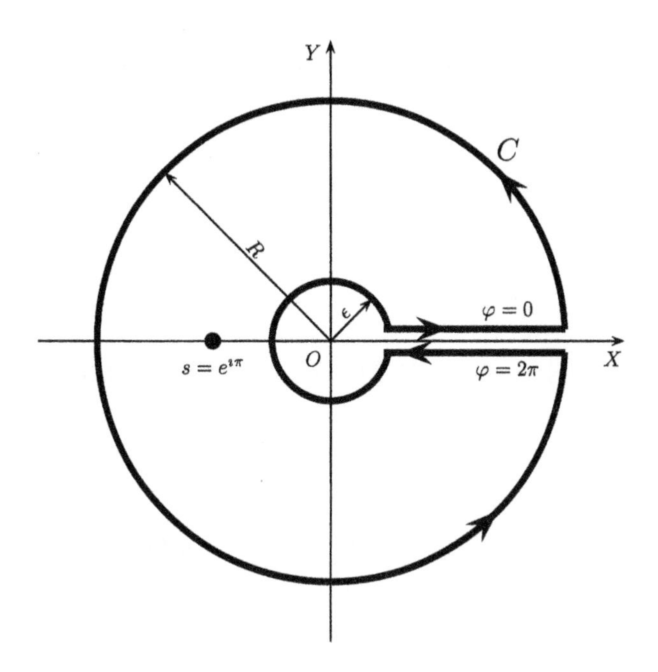

Figure 1.1: *Contour L.*

which shows that the relationship (1.26) holds for $z \neq 0, \pm 1, \pm 2, \ldots$

Taking $z = 1/2$ we obtain from (1.26) a useful particular value of the gamma function:

$$\Gamma\left(\frac{1}{2}\right) = \sqrt{\pi}. \tag{1.34}$$

The second important relationship for the gamma function, easily obtained with the help of the beta function, is the Legendre formula

$$\Gamma(z)\Gamma(z+\frac{1}{2}) = \sqrt{\pi}2^{2z-1}\Gamma(2z), \quad (2z \neq 0, -1, -2, \ldots). \tag{1.35}$$

To prove the relationship (1.35) let us consider

$$B(z,z) = \int_0^1 [\tau(1-\tau)]^{z-1} d\tau, \quad (Re(z) > 0). \tag{1.36}$$

Taking into account the symmetry of the function $y(\tau) = \tau(1-\tau)$

and performing the substitution $s = 4\tau(1-\tau)$ we obtain

$$B(z,z) = 2\int_0^{1/2} [\tau(1-\tau)]^{z-1} d\tau$$

$$= \frac{1}{2^{2z-1}} \int_0^1 s^{z-1}(1-s)^{-1/2} ds = 2^{1-2z} B(z, \frac{1}{2}), \qquad (1.37)$$

and using the relationship (1.24) we obtain from (1.37) the Legendre formula (1.35).

Taking $z = n + \frac{1}{2}$ in (1.35) we obtain a set of particular values of the gamma function

$$\Gamma(n+\frac{1}{2}) = \frac{\sqrt{\pi}\Gamma(2n+1)}{2^{2n}\Gamma(n+1)} = \frac{\sqrt{\pi}(2n)!}{2^{2n}n!} \qquad (1.38)$$

containing also (1.34).

1.1.5 Contour Integral Representation

The integration variable t in the definition of the gamma function (1.1) is real. If t is complex, then the function $e^{(z-1)\log(t)-t}$ has a branch point $t=0$. Cutting the complex plane (t) along the real semi-axis from $t=0$ to $t=+\infty$ makes this function single-valued. Therefore, according to Cauchy's theorem, the integral

$$\int_C e^{-t} t^{z-1} dt = \int_C e^{(z-1)\log(t)-t} dt$$

has the same value for any contour C running around the point $t=0$ with both ends at $+\infty$.

Let us consider the contour C (see Fig. 1.2) consisting of the part of the upper edge $(+\infty, \epsilon)$ of the cut, the circle C_ϵ of radius ϵ with the centre at $t=0$ and the part of the lower cut edge $(\epsilon, +\infty)$.

Taking $\log(t)$ to be real on the upper cut edge, we have

$$t^{z-1} = e^{(z-1)\log(t)}.$$

On the lower cut edge we must replace $\log(t)$ by $\log(t) + 2\pi i$:

$$t^{z-1} = e^{(z-1)(\log(t)+2\pi i)} = e^{(z-1)\log(t)} e^{(z-1)2\pi i} = t^{z-1} e^{2(z-1)\pi i}.$$

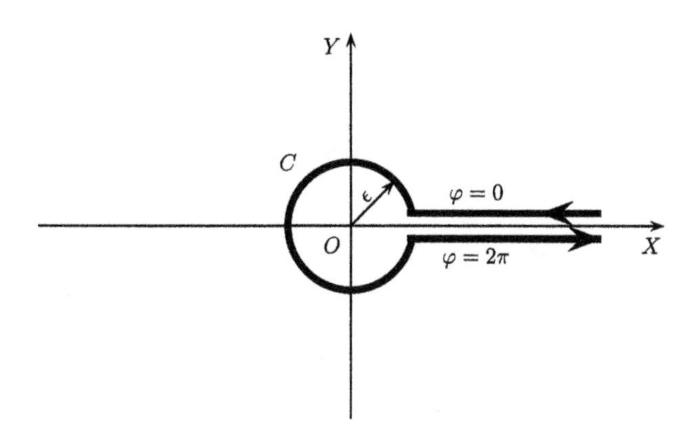

Figure 1.2: *Contour C.*

Therefore,

$$\int_C e^{-t}t^{z-1}dt = \int_{+\infty}^{\epsilon} e^{-t}t^{z-1}dt + \int_{C_\epsilon} e^{-t}t^{z-1}dt + e^{2(z-1)\pi\imath}\int_{\epsilon}^{+\infty} e^{-t}t^{z-1}dt. \tag{1.39}$$

Let us show that the integral along C_ϵ tends to zero as $\epsilon \to 0$. Indeed, taking into account that $|t| = \epsilon$ on C_ϵ and denoting

$$M = \max_{t\in C_\epsilon}\left|e^{-y\arg(t)-t}\right|, \qquad (y = Im(z)),$$

where M is independent of t, we obtain ($z = x + iy$):

$$\begin{aligned}\left|\int_{C_\epsilon} e^{-t}t^{z-1}dt\right| &\leq \int_{C_\epsilon}\left|e^{-t}t^{z-1}\right|dt = \int_{C_\epsilon}\left|t^{x-1}\right|\cdot\left|e^{-y\arg(t)-t}\right|dt \\ &\leq M\epsilon^{x-1}\int_{C_\epsilon}dt = M\epsilon^{x-1}\cdot 2\pi\epsilon = 2\pi M\epsilon^{x},\end{aligned}$$

and therefore

$$\lim_{\epsilon\to 0}\int_{C_\epsilon} e^{-t}t^{z-1}dt = 0 \tag{1.40}$$

and

$$\int_C e^{-t}t^{z-1}dt = \int_{+\infty}^{0} e^{-t}t^{z-1}dt + e^{2(z-1)\pi\imath}\int_0^{+\infty} e^{-t}t^{z-1}dt. \tag{1.41}$$

Using (1.1) we obtain:

$$\Gamma(z) = \frac{1}{e^{2\pi i z} - 1} \int_C e^{-t} t^{z-1} dt. \tag{1.42}$$

The function $e^{2\pi i z} - 1$ has its zeros at the points $z = 0, \pm 1, \pm 2, \ldots$. The points $z = 1, 2, \ldots$ are not the poles of $\Gamma(z)$, because in this case the function $e^{-t} t^{z-1}$ is single-valued and regular in the complex plane (t) and according to Cauchy's theorem

$$\int_C e^{-t} t^{z-1} dt = 0.$$

If $z = 0, -1, -2, \ldots$, then the function $e^{-t} t^{z-1}$ is not an entire function of t and the integral of it along the contour C is not equal to zero. Therefore, the points $z = 0, -1, -2, \ldots$ are the poles of $\Gamma(z)$. According to the principle of analytic continuation, the integral representation (1.42) holds not only for $Re(z) > 0$, as assumed at the beginning, but in the whole complex plane (z).

1.1.6 Contour Integral Representation of $1/\Gamma(z)$

In this section we give formulas for the integral representation of the reciprocal gamma function.

To obtain the simplest integral representation formula for $1/\Gamma(z)$ let us replace z by $1 - z$ in the formula (1.42), which leads to

$$\int_C e^{-t} t^{-z} dt = (e^{-2z\pi i} - 1)\, \Gamma(1 - z), \tag{1.43}$$

and then perform the substitution $t = \tau e^{\pi i} = -\tau$. This substitution will transform (namely, turn it counterclockwise) the complex plane (t) with the cut along the real positive semi-axis into the complex plane (τ) with the cut along the real negative semi-axis. The lower cut edge $\arg(\tau) = -\pi$ in the (τ)-plane will correspond to the upper cut edge $t = 0$ in the (t)-plane. The contour C will be transformed to Hankel's contour Ha shown in Fig. 1.3. Then we have:

$$\int_C e^{-t} t^{-z} dt = -\int_{Ha} e^{\tau} (e^{\pi i} \tau)^{-z} d\tau = -e^{-z\pi i} \int_{Ha} e^{\tau} \tau^{-z} d\tau. \tag{1.44}$$

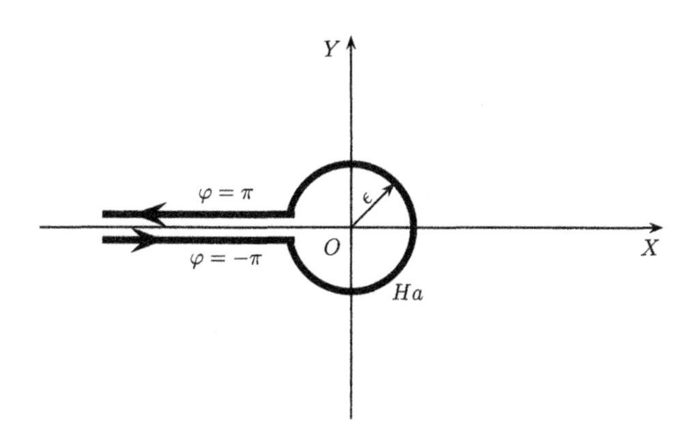

Figure 1.3: *The Hankel contour Ha.*

Taking into account the relationships (1.43) and (1.26) we obtain

$$\int_{Ha} e^{\tau}\tau^{-z}d\tau = (e^{z\pi i}-e^{-z\pi i})\Gamma(1-z) = 2i\sin(\pi z)\Gamma(1-z) = \frac{2\pi i}{\Gamma(z)}. \quad (1.45)$$

Therefore, we have the following integral representation for the reciprocal gamma function:

$$\frac{1}{\Gamma(z)} = \frac{1}{2\pi i}\int_{Ha} e^{\tau}t^{-z}d\tau. \quad (1.46)$$

Let us now denote by $\gamma(\epsilon, \varphi)$ ($\epsilon > 0$, $0 < \varphi \le \pi$) the contour consisting of the following three parts:

1) $\arg\tau = -\varphi, \quad |\tau| \ge \epsilon$;
2) $-\varphi \le \arg\tau \le \varphi$, $|\tau| = \epsilon$;
3) $\arg\tau = -\varphi, \quad |\tau| \ge \epsilon$.

The contour is traced so that $\arg\tau$ is non-decreasing. It is shown in Fig. 1.4.

The contour $\gamma(\epsilon, \varphi)$ divides the complex plane τ into two domains, which we denote by $G^-(\epsilon, \varphi)$ and $G^+(\epsilon, \varphi)$, lying correspondingly on the left and on the right side of the contour $\gamma(\epsilon, \varphi)$ (Fig. 1.4).

If $0 < \varphi < \pi$, then both $G^-(\epsilon, \varphi)$ and $G^+(\epsilon, \varphi)$ are infinite domains.

If $\varphi = \pi$, then $G^-(\epsilon, \varphi)$ becomes a circle $|\tau| < \epsilon$ and $G^+(\epsilon, \varphi)$ becomes a complex plane excluding the circle $|\tau| < \epsilon$ and the line $|\arg\varphi| = \pi$.

Let us show that instead of integrating along Hankel's contour Ha in (1.46) we can integrate along the contour $\gamma(\epsilon, \varphi)$, where $\frac{\pi}{2} < \varphi < \pi$,

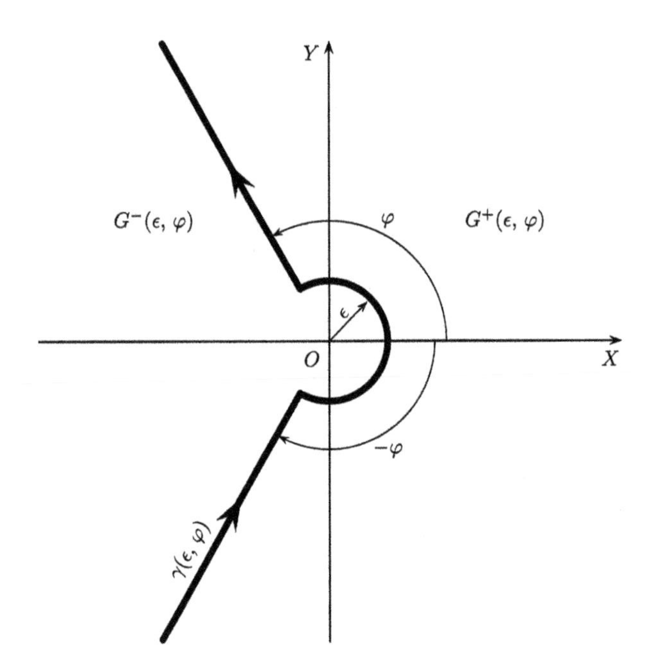

Figure 1.4: *Contour* $\gamma(\epsilon, \varphi)$.

i.e. that

$$\frac{1}{\Gamma(z)} = \frac{1}{2\pi i} \int_{\gamma(\epsilon,\varphi)} e^{\tau}\tau^{-z}d\tau, \qquad (\epsilon > 0, \quad \frac{\pi}{2} < \varphi \leq \pi). \tag{1.47}$$

Let us consider the contour $(A^+B^+C^+D^+)$ shown in Fig. 1.5. Using the Cauchy theorem for the contour gives:

$$0 = \int_{(A^+B^+C^+D^+)} e^{\tau}\tau^{-s}d\tau = \int_{A^+}^{B^+} + \int_{B^+}^{C^+} + \int_{C^+}^{D^+} + \int_{D^+}^{A^+}. \tag{1.48}$$

On the arc (A^+B^+) we have $|\tau| = R$ and

$$\begin{aligned} |e^{\tau}\tau^{-z}| &= e^{R\cos(\arg\tau) - x\log R + y\arg\tau} \\ &\leq e^{-R\cos(\pi-\varphi) - x\log R + 2\pi y}, \end{aligned}$$

from which it follows that

$$\lim_{R\to\infty} \int_{A^+}^{B^+} = 0. \tag{1.49}$$

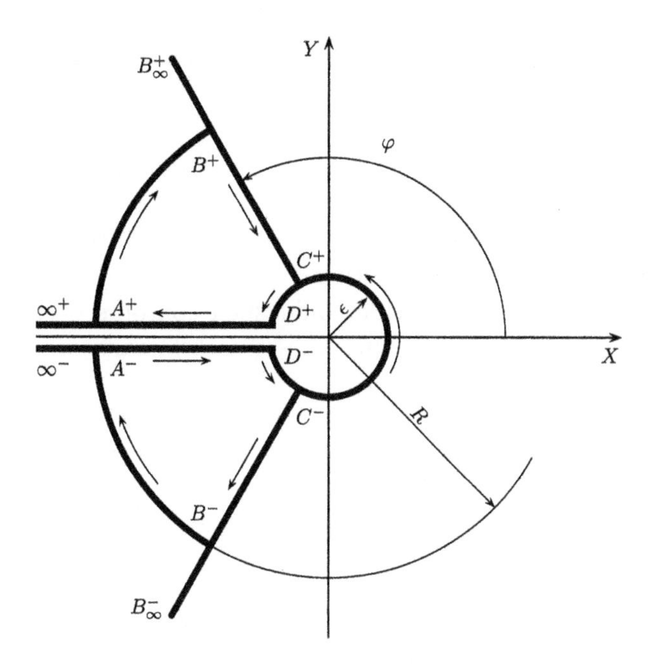

Figure 1.5: *Transformation of the contour Ha to the contour $\gamma(\epsilon, \varphi)$.*

Taking $R \to \infty$ in (1.48) and using (1.49) we obtain:

$$\int_{C^+}^{D^+} + \int_{D^+}^{\infty^+} + \int_{B^+_\infty}^{C^+} = 0$$

or

$$\int_{C^+}^{D^+} + \int_{D^+}^{\infty^+} = \int_{C^+}^{B^+_\infty} . \tag{1.50}$$

Similarly, consideration of the contour $(A^- D^- C^- B^-)$ leads to

$$\int_{\infty^-}^{D^-} + \int_{D^-}^{C^-} = \int_{B^-_\infty}^{C^-} . \tag{1.51}$$

Using (1.50) and (1.51) we see that

$$\int_{Ha} e^\tau t^{-z} d\tau = \left(\int_{B^-_\infty}^{C^-} + \int_{C^-}^{C^+} + \int_{C^+}^{B^+_\infty} \right) e^\tau t^{-z} d\tau = \int_{\gamma(\epsilon,\varphi)} e^\tau t^{-z} d\tau$$

and, indeed, the integral representation (1.47) for the reciprocal gamma function holds for all z.

Now we can obtain the following two integral representations for the reciprocal gamma function.

The first integral representation is obtained for arbitrary complex z.

Let us perform the substitution $\tau = \zeta^{1/\alpha}$, $(\alpha < 2)$ in (1.47) and in the case of $1 \leq \alpha < 2$ consider only such contours $\gamma(\epsilon, \varphi)$ for which $\frac{\pi}{2} < \varphi < \frac{\pi}{\alpha}$. Due to this, since $\epsilon > 0$ is arbitrary, we arrive at the following integral representation

$$\frac{1}{\Gamma(z)} = \frac{1}{2\pi\alpha i} \int\limits_{\gamma(\epsilon,\mu)} \exp(\zeta^{1/\alpha})\zeta^{(1-z-\alpha)/\alpha} d\zeta, \tag{1.52}$$

$$\left(\alpha < 2, \ \frac{\pi\alpha}{2} < \mu < \min\{\pi, \ \pi\alpha\}\right).$$

Another integral representation for $1/\Gamma(z)$ can be obtained if we note that in the case of $Re(z) > 0$ the formula (1.47) holds also for $\alpha = \pi/2$:

$$\frac{1}{\Gamma(z)} = \frac{1}{2\pi i} \int\limits_{\gamma(\epsilon,\frac{\pi}{2})} e^u u^{-z} du, \qquad (\epsilon > 0, \quad Re(z) > 0). \tag{1.53}$$

Performing the substitution $u = \sqrt{\zeta}$ in (1.53), we obtain the integral representation

$$\frac{1}{\Gamma(z)} = \frac{1}{4\pi i} \int\limits_{\gamma(\epsilon,\pi)} \exp(\zeta^{1/2})\zeta^{-(z+1)/2} d\zeta, \qquad (\epsilon > 0, \quad Re(z) > 0). \tag{1.54}$$

We would like to emphasize that the integral representation (1.52) is valid for arbitrary z, whereas the integral representation (1.54) holds only if $Re(z) > 0$.

1.2 Mittag-Leffler Function

The exponential function, e^z, plays a very important role in the theory of integer-order differential equations. Its one-parameter generalization, the function which is now denoted by [65]

$$E_\alpha(z) = \sum_{k=0}^{\infty} \frac{z^k}{\Gamma(\alpha k + 1)}, \tag{1.55}$$

was introduced by G. M. Mittag-Leffler [155, 156, 157] and studied also by A. Wiman [256, 257].

The two-parameter function of the Mittag-Leffler type, which plays a very important role in the fractional calculus, was in fact introduced by Agarwal [3]. A number of relationships for this function were obtained by Humbert and Agarwal [107] using the Laplace transform technique. This function could have been called the Agarwal function. However, Humbert and Agarwal generously left the same notation as for the one-parameter Mittag-Leffler function, and that is the reason that now the two-parameter function is called the Mittag-Leffler function. We will use the name and the notation used in the fundamental handbook on special functions [65]. In spite of using the same notation as Agarwal, the definition given there differs from Agarwal's definition by a non-constant factor. Some parts of this section are based on results by M. M. Dzhrbashyan [45, Chapter III].

Regarding the distribution of zeros, the papers by A. Wiman [257], A. M. Sedletskii [240], R. Gorenflo, Yu. Luchko, and S. Rogosin [87], and the book by M. M. Dzhrbashyan [45, pp. 139–146] must be mentioned; we will not discuss them here.

1.2.1 Definition and Relation to Some Other Functions

A two-parameter function of the Mittag-Leffler type is defined by the series expansion [65]

$$E_{\alpha,\beta}(z) = \sum_{k=0}^{\infty} \frac{z^k}{\Gamma(\alpha k + \beta)}, \qquad (\alpha > 0, \quad \beta > 0). \tag{1.56}$$

It follows from the definition (1.56) that

$$E_{1,1}(z) = \sum_{k=0}^{\infty} \frac{z^k}{\Gamma(k+1)} = \sum_{k=0}^{\infty} \frac{z^k}{k!} = e^z, \tag{1.57}$$

$$E_{1,2}(z) = \sum_{k=0}^{\infty} \frac{z^k}{\Gamma(k+2)} = \sum_{k=0}^{\infty} \frac{z^k}{(k+1)!} = \frac{1}{z}\sum_{k=0}^{\infty} \frac{z^{k+1}}{(k+1)} = \frac{e^z - 1}{z}, \tag{1.58}$$

$$E_{1,3}(z) = \sum_{k=0}^{\infty} \frac{z^k}{\Gamma(k+3)} = \sum_{k=0}^{\infty} \frac{z^k}{(k+2)!} = \frac{1}{z^2}\sum_{k=0}^{\infty} \frac{z^{k+2}}{(k+2)} = \frac{e^z - 1 - z}{z^2}, \tag{1.59}$$

and in general

$$E_{1,m}(z) = \frac{1}{z^{m-1}} \left\{ e^z - \sum_{k=0}^{m-2} \frac{z^k}{k!} \right\}. \tag{1.60}$$

The hyperbolic sine and cosine are also particular cases of the Mittag-Leffler function (1.56):

$$E_{2,1}(z^2) = \sum_{k=0}^{\infty} \frac{z^{2k}}{\Gamma(2k+1)} = \sum_{k=0}^{\infty} \frac{z^{2k}}{(2k)!} = \cosh(z), \tag{1.61}$$

$$E_{2,2}(z^2) = \sum_{k=0}^{\infty} \frac{z^{2k}}{\Gamma(2k+2)} = \frac{1}{z} \sum_{k=0}^{\infty} \frac{z^{2k+1}}{(2k+1)!} = \frac{\sinh(z)}{z}. \tag{1.62}$$

The hyperbolic functions of order n [65], which are generalizations of the hyperbolic sine and cosine, can also be expressed in terms of the Mittag-Leffler function:

$$h_r(z,n) = \sum_{k=0}^{\infty} \frac{z^{nk+r-1}}{(nk+r-1)!} = z^{r-1} E_{n,r}(z^n), \qquad (r = 1,\, 2,\, \ldots,\, n), \tag{1.63}$$

as well as the trigonometric functions of order n, which are generalizations of the sine and cosine functions:

$$k_r(z,n) = \sum_{j=0}^{\infty} \frac{(-1)^j z^{nj+r-1}}{(nj+r-1)!} = z^{r-1} E_{n,r}(-z^n), \qquad (r = 1,\, 2,\, \ldots,\, n). \tag{1.64}$$

Using [2, formulas 7.1.3 and 7.1.8] we obtain

$$E_{1/2,1}(z) = \sum_{k=0}^{\infty} \frac{z^k}{\Gamma(\frac{k}{2}+1)} = e^{z^2} \operatorname{erfc}(-z), \tag{1.65}$$

where $\operatorname{erfc}(z)$ is the error function complement defined by

$$\operatorname{erfc}(z) = \frac{2}{\sqrt{\pi}} \int_z^{\infty} e^{-t^2} dt.$$

For $\beta = 1$ we obtain the Mittag-Leffler function in one parameter:

$$E_{\alpha,1}(z) = \sum_{k=0}^{\infty} \frac{z^k}{\Gamma(\alpha k+1)} \equiv E_{\alpha}(z). \tag{1.66}$$

The function $\mathcal{E}_t(\nu, a)$, introduced in [153] for solving differential equations of rational order, is a particular case of the Mittag-Leffler function (1.56):

$$\mathcal{E}_t(\nu, a) = t^{\nu} \sum_{k=0}^{\infty} \frac{(at)^k}{\Gamma(\nu + k + 1)} = t^{\nu} E_{1,\nu+1}(at). \tag{1.67}$$

Yu. N. Rabotnov's [218] function $\ni_{\alpha}(\beta, t)$ is a particular case of the Mittag-Leffler function (1.56) too:

$$\ni_{\alpha}(\beta, t) = t^{\alpha} \sum_{k=0}^{\infty} \frac{\beta^k t^{k(\alpha+1)}}{\Gamma((k+1)(1+\alpha))} = t^{\alpha} E_{\alpha+1,\alpha+1}(\beta t^{\alpha+1}). \tag{1.68}$$

It follows from the relationships (1.67) and (1.68) that the properties of the Miller–Ross function and Rabotnov's function can be deduced from the properties of the Mittag-Leffler function in two parameters (1.56).

Plotnikov [190, cf. [250]] and Tseytlin [250] used in their investigations two functions $Sc_{\alpha}(z)$ and $Cs_{\alpha}(z)$, which they call the fractional sine and cosine. Those functions are also just particular cases of the Mittag-Leffler function in two parameters:

$$Sc_{\alpha}(z) = \sum_{n=0}^{\infty} \frac{(-1)^n z^{(2-\alpha)n+1}}{\Gamma((2-\alpha)n + 2)} = z E_{2-\alpha,2}(-z^{2-\alpha}), \tag{1.69}$$

$$Cs_{\alpha}(z) = \sum_{n=0}^{\infty} \frac{(-1)^n z^{(2-\alpha)n}}{\Gamma((2-\alpha)n + 1)} = E_{2-\alpha,1}(-z^{2-\alpha}). \tag{1.70}$$

Another "fractionalization" of the sine and cosine functions, which can also be expressed in terms of the Mittag-Leffler function (1.56), has been suggested by Luchko and Srivastava [128]:

$$\sin_{\lambda,\mu}(z) = \sum_{k=0}^{\infty} \frac{(-1)^k z^{2k+1}}{\Gamma(2\mu k + 2\mu - \lambda + 1)} = z E_{2\mu,2\mu-\lambda+1}(-z^2), \tag{1.71}$$

$$\cos_{\lambda,\mu}(z) = \sum_{k=0}^{\infty} \frac{(-1)^k z^{2k}}{\Gamma(2\mu k + \mu - \lambda + 1)} = E_{2\mu,\mu-\lambda+1}(-z^2). \tag{1.72}$$

Of course, the properties of both versions of the fractional sine and cosine follow from the properties of the Mittag-Leffler function (1.56).

Generalizations of the Mittag-Leffler function (1.56) to two variables, suggested by P. Humbert and P. Delerue [108] and by A. M. Chak [36], were further extended by H. M. Srivastava [243] to the following symmetric form:

$$\xi^{\nu,\sigma}_{\alpha,\beta,\lambda,\mu}(x,y) = \sum_{m=0}^{\infty}\sum_{n=0}^{\infty} \frac{x^{m+\frac{\beta(\nu n+1)-1}{\alpha}}\, y^{n+\frac{\mu(\sigma m+1)-1}{\lambda}}}{\Gamma\Big(m\alpha+(\nu n+1)\beta\Big)\,\Gamma\Big(n\lambda+(\sigma m+1)\mu\Big)}. \tag{1.73}$$

An interesting generalization of the Mittag-Leffler function to several variables has been suggested by S. B. Hadid and Yu. Luchko [100], who used it for solving linear fractional differential equations with constant coefficients by the operational method:

$$E_{(\alpha_1,\dots,\alpha_m),\beta}(z_1,\dots,z_m) \tag{1.74}$$

$$= \sum_{k=0}^{\infty} \sum_{\substack{l_1+\ldots+l_m=k\\ l_1>0,\ \dots,\ l_m>0}} \frac{(k;l_1,\dots,l_m)\prod\limits_{i=1}^{m} z_i^{l_i}}{\Gamma(\beta+\sum\limits_{i=1}^{m}\alpha_i l_i)},$$

where $(k;l_1,\dots,l_m)$ are multinomial coefficients [2].

1.2.2 The Laplace Transform of the Mittag-Leffler Function in Two Parameters

As follows from relationship (1.57), the Mittag-Leffler function $E_{\alpha,\beta}(z)$ is a generalization of the exponential function e^z and, therefore, the exponential function is a particular case of the Mittag-Leffler function.

We will outline here the way to obtain the Laplace transform of the Mittag-Leffler function with the help of the analogy between this function and the function e^z. For this purpose, let us obtain the Laplace transform of the function $t^k e^{at}$ in an untraditional way.

First, let us prove that

$$\int_0^{\infty} e^{-t} e^{\pm zt}\,dt = \frac{1}{1\mp z}, \qquad |z|<1. \tag{1.75}$$

Indeed, using the series expansion for e^z, we obtain

$$\int_0^{\infty} e^{-t}e^{zt}\,dt = \frac{1}{1-z} = \sum_{k=0}^{\infty}\frac{(\pm z)^k}{k!}\int_0^{\infty} e^{-t}t^k\,dt = \sum_{k=0}^{\infty}(\pm z)^k = \frac{1}{1\mp z}. \tag{1.76}$$

Second, we differentiate both sides of equation (1.75) with respect to z. The result is

$$\int_0^\infty e^{-t} t^k e^{\pm zt} dt = \frac{k!}{(1-z)^{k+1}}, \quad (|z| < 1), \tag{1.77}$$

and after obvious substitutions we obtain the well-known pair of Laplace transforms of the function $t^k e^{\pm at}$:

$$\int_0^\infty e^{-pt} t^k e^{\pm at} dt = \frac{k!}{(p \mp a)^{k+1}}, \quad (Re(p) > |a|). \tag{1.78}$$

Let us now consider the Mittag-Leffler function (1.56). Substitution of (1.56) in the integral below leads to

$$\int_0^\infty e^{-t} t^{\beta-1} E_{\alpha,\beta}(zt^\alpha) dt = \frac{1}{1-z}, \qquad (|z| < 1), \tag{1.79}$$

and we obtain from (1.79) a pair of Laplace transforms of the function $t^{\alpha k+\beta-1} E^{(k)}_{\alpha,\beta}(\pm z t^\alpha)$, $(E^{(k)}_{\alpha,\beta}(y) \equiv \frac{d^k}{dy^k} E_{\alpha,\beta}(y))$:

$$\int_0^\infty e^{-pt} t^{\alpha k+\beta-1} E^{(k)}_{\alpha,\beta}(\pm a t^\alpha) dt = \frac{k!\, p^{\alpha-\beta}}{(p^\alpha \mp a)^{k+1}}, \qquad (Re(p) > |a|^{1/\alpha}). \tag{1.80}$$

The particular case of (1.80) for $\alpha = \beta = \frac{1}{2}$

$$\int_0^\infty e^{-pt} t^{\frac{k-1}{2}} E^{(k)}_{\frac{1}{2},\frac{1}{2}}(\pm a\sqrt{t}) dt = \frac{k!}{(\sqrt{p} \mp a)^{k+1}}, \qquad (Re(p) > a^2). \tag{1.81}$$

is useful for solving the semidifferential equations considered in [179, 153].

1.2.3 Derivatives of the Mittag-Leffler Function

By the Riemann–Liouville fractional-order differentiation ${}_0D_t^\gamma$ (γ is an arbitrary real number) of series representation (1.56) we obtain

$${}_0D_t^\gamma \left(t^{\alpha k+\beta-1} E^{(k)}_{\alpha,\beta}(\lambda t^\alpha)\right) = t^{\alpha k+\beta-\gamma-1} E^{(k)}_{\alpha,\beta-\gamma}(\lambda t^\alpha). \tag{1.82}$$

The particular case of relationship (1.82) for $k = 0$, $\lambda = 1$ and integer γ is given in [65], equation 18.1(25) and has the form

$$\left(\frac{d}{dt}\right)^m \left(t^{\beta-1}E_{\alpha,\beta}(t^\alpha)\right) = t^{\beta-m-1}E_{\alpha,\beta-m}(t^\alpha), \quad (m = 1, 2, 3, \ldots). \tag{1.83}$$

Formula (1.83) has some interesting consequences. Taking $\alpha = \frac{m}{n}$, where m and n are natural numbers, we obtain

$$\left(\frac{d}{dt}\right)^m \left(t^{\beta-1}E_{m/n,\beta}(t^{m/n})\right) = t^{\beta-1}E_{m/n,\beta}(t^{m/n}) + t^{\beta-1}\sum_{k=1}^{n}\frac{t^{-\frac{m}{n}k}}{\Gamma\left(\beta - \frac{m}{n}k\right)}, \tag{1.84}$$

$$(m, n = 1, 2, 3, \ldots).$$

Setting $n = 1$ and taking into account the well-known property of the gamma function

$$\frac{1}{\Gamma(-\nu)} = 0 \quad (\nu = 0, 1, 2, \ldots),$$

we obtain from (1.84) that

$$\left(\frac{d}{dt}\right)^m \left(t^{\beta-1}E_{m,\beta}(t^m)\right) = t^{\beta-1}E_{m,\beta}(t^m), \tag{1.85}$$

$$(m = 1, 2, 3, \ldots; \quad \beta = 0, 1, 2, \ldots, m).$$

Performing the substitution $t = z^{n/m}$ in (1.84) we obtain

$$\left(\frac{m}{n}z^{1-\frac{n}{m}}\frac{d}{dz}\right)^m \left(z^{(\beta-1)n/m}E_{m/n,\beta}(z)\right)$$

$$= z^{(\beta-1)n/m}E_{m/n,\beta}(z) + t^{(\beta-1)n/m}\sum_{k=1}^{n}\frac{z^{-k}}{\Gamma\left(\beta - \frac{m}{n}k\right)} \tag{1.86}$$

$$(m, n = 1, 2, 3, \ldots).$$

Taking $m = 1$ in (1.86), we obtain the following expression:

$$\frac{1}{n}\frac{d}{dz}\left(z^{(\beta-1)n}E_{1/n,\beta}(z)\right) = z^{\beta n-1}E_{1/n,\beta}(z) + z^{\beta n-1}\sum_{k=1}^{n}\frac{z^{-k}}{\Gamma\left(\beta - \frac{k}{n}\right)}, \tag{1.87}$$

$$(n = 1, 2, 3, \ldots).$$

1.2.4 Differential Equations for the Mittag-Leffler Function

It is worthwhile noting that relationships (1.84)–(1.87) can also be interpreted as differential equations for the Mittag-Leffler function; namely, if we denote

$$\begin{aligned}
y_1(t) &= t^{\beta-1} E_{m/n,\beta}(t^{m/n}), \\
y_2(t) &= t^{\beta-1} E_{m,\beta}(t^{m}), \\
y_3(t) &= t^{(\beta-1)n/m} E_{m/n,\beta}(t), \\
y_4(t) &= t^{(\beta-1)n} E_{1/n,\beta}(t),
\end{aligned}$$

then these functions satisfy the following differential equations respectively:

$$\frac{d^m y_1(t)}{dt^m} - y_1(t) = t^{\beta-1} \sum_{k=1}^{n} \frac{t^{-\frac{m}{n}k}}{\Gamma\left(\beta - \frac{m}{n}k\right)}, \tag{1.88}$$

$$(m, n = 1, 2, 3, \ldots)$$

$$\frac{d^m y_2(t)}{dt^m} - y_2(t) = 0, \tag{1.89}$$

$$(m = 1, 2, 3, \ldots; \quad \beta = 0, 1, 2, \ldots, m)$$

$$\left(\frac{m}{n} t^{1-\frac{n}{m}} \frac{d}{dt}\right)^m y_3(t) - y_3(t) = t^{(\beta-1)n/m} \sum_{k=1}^{n} \frac{t^{-k}}{\Gamma\left(\beta - \frac{m}{n}k\right)} \tag{1.90}$$

$$(m, n = 1, 2, 3, \ldots)$$

$$\frac{1}{n}\frac{dy_4(t)}{dt} y_4(t) - t^{n-1} y_4(t) = t^{\beta n-1} \sum_{k=1}^{n} \frac{t^{-k}}{\Gamma\left(\beta - \frac{k}{n}\right)}, \tag{1.91}$$

$$(n = 1, 2, 3, \ldots).$$

1.2.5 Summation Formulas

Let us start with the obvious relationship

$$\sum_{\nu=0}^{m-1} e^{i2\pi\nu k/m} = \begin{cases} m, & \text{if } k \equiv 0 \pmod{m} \\ 0, & \text{if } k \not\equiv 0 \pmod{m} \end{cases} \tag{1.92}$$

where the notation $k \equiv p \pmod{m}$ means that the remainder of the division of $k - p$ by m is zero (k, p and m are integer numbers).

Combining (1.92) and the definition (1.56) of the Mittag-Leffler function, we obtain

$$\sum_{\nu=0}^{m-1} E_{\alpha,\beta}(ze^{i2\pi\nu/m}) = mE_{m\alpha,\beta}(z^m), \quad (m \geq 1). \tag{1.93}$$

Replacing α with $\frac{\alpha}{m}$ and z with $z^{1/m}$ in (1.93), we arrive at

$$E_{\alpha,\beta}(z) = \frac{1}{m}\sum_{\nu=0}^{m-1} E_{\alpha/m,\beta}(z^{1/m}e^{i2\pi\nu/m}), \quad (m \geq 1). \tag{1.94}$$

The following particular case of formula (1.94) must be mentioned. Taking $m = 2$ and $z = t^2$, we obtain

$$E_{\alpha,\beta}(z) + E_{\alpha,\beta}(-z) = 2E_{\alpha,\beta}(z^2). \tag{1.95}$$

Similarly, starting with the obvious formula

$$\sum_{\nu=-m}^{m} e^{i2\pi\nu k/(2m+1)} = \begin{cases} 2m+1, & \text{if } k \equiv 0 \pmod{2m+1} \\ 0, & \text{if } k \not\equiv 0 \pmod{2m+1}, \end{cases} \tag{1.96}$$

we obtain

$$E_{\alpha,\beta}(z) = \frac{1}{2m+1}\sum_{\nu=-m}^{m-1} E_{\alpha/(2m+1),\beta}(z^{1/(2m+1)}e^{i2\pi\nu/(2m+1)}), \quad (m \geq 0). \tag{1.97}$$

A generalization of the summation formula (1.93) has been obtained by H. M. Srivastava [244]:

$$\sum_{\nu=0}^{m-1} e^{i2\pi\nu(m-n)/m} E_{\alpha,\,\beta}(ze^{i2\pi\nu}/m) = mz^n E_{m\alpha,\,\beta+n\alpha}(z^m). \tag{1.98}$$

Obviously, for $n = 0$ the relationship (1.98) gives the summation formula (1.93).

1.2.6 Integration of the Mittag-Leffler Function

Integrating (1.56) term-by-term, we obtain

$$\int_0^z E_{\alpha,\beta}(\lambda t^\alpha)t^{\beta-1}dt = z^\beta E_{\alpha,\beta+1}(\lambda z^\alpha), \quad (\beta > 0). \tag{1.99}$$

Relationship (1.99) is a particular case of the following more general relationship obtained by the fractional-order term-by-term integration of the series (1.56):

$$\frac{1}{\Gamma(\nu)}\int_0^z (z-t)^{\nu-1} E_{\alpha,\beta}(\lambda t^{\alpha}) t^{\beta-1} dt = z^{\beta+\nu-1} E_{\alpha,\beta+\nu}(\lambda z^{\alpha}), \qquad (1.100)$$

$$(\beta > 0, \quad \nu > 0).$$

From (1.100) and formulas (1.57), (1.61) and (1.62) we obtain:

$$\frac{1}{\Gamma(\alpha)}\int_0^z (z-t)^{\alpha-1} e^{\lambda t} dt = z^{\alpha} E_{1,\alpha+1}(\lambda z), \quad (\alpha > 0), \qquad (1.101)$$

$$\frac{1}{\Gamma(\alpha)}\int_0^z (z-t)^{\alpha-1} \cosh(\sqrt{\lambda} t) dt = z^{\alpha} E_{2,\alpha+1}(\lambda z^2), \quad (\alpha > 0), \qquad (1.102)$$

$$\frac{1}{\Gamma(\alpha)}\int_0^z (z-t)^{\alpha-1} \frac{\sinh(\sqrt{\lambda} t)}{\sqrt{\lambda} t} dt = z^{\alpha+1} E_{2,\alpha+2}(\lambda z^2), \quad (\alpha > 0). \qquad (1.103)$$

Let us also prove the following formula for the fractional integration of the Mittag-Leffler function:

$$\frac{1}{\Gamma(\alpha)}\int_0^z (z-t)^{\alpha-1} E_{2\alpha,\beta}(t^{2\alpha}) t^{\beta-1} dt = -z^{\beta-1} E_{2\alpha,\beta}(z^{2\alpha}) + z^{\beta-1} E_{\alpha,\beta}(z^{\alpha}).$$

$$(1.104)$$

To prove (1.104), let us consider the integral

$$\int_0^z E_{2\alpha,\beta}(t^{2\alpha}) t^{\beta-1} \left\{1 + \frac{(z-t)^{\alpha}}{\Gamma(1+\alpha)}\right\} dt$$

$$= \sum_{k=0}^{\infty} \frac{1}{\Gamma(2k\alpha+\beta)} \int_0^z t^{2k\alpha+\beta-1} \left\{1 + \frac{(z-t)^{\alpha}}{\Gamma(1+\alpha)}\right\} dt$$

$$= z^{\beta} \sum_{k=0}^{\infty} \frac{z^{2k\alpha}}{\Gamma(2k\alpha+\beta+1)} + z^{\beta} \sum_{k=0}^{\infty} \frac{z^{(2k+1)\alpha}}{\Gamma((2k+1)\alpha+\beta+1)}$$

$$= z^{\beta} \sum_{k=0}^{\infty} \frac{z^{k\alpha}}{\Gamma(k\alpha+\beta+1)} = z^{\beta} E_{\alpha,\beta+1}(z^{\alpha}). \qquad (1.105)$$

Comparing (1.105) and (1.99) we have

$$\int_0^z E_{2\alpha,\beta}(t^{2\alpha})t^{\beta-1}\left\{1+\frac{(z-t)^\alpha}{\Gamma(1+\alpha)}\right\}dt = \int_0^z E_{\alpha,\beta}(\lambda t^\alpha)t^{\beta-1}dt, \quad (\beta>0). \tag{1.106}$$

Differentiating (1.106) with respect to z, we obtain (1.104).

There is also an interesting relationship for the Mittag-Leffler function, which is similar to the Cristoffel–Darboux formula for orthogonal polynomials; namely,

$$\int_0^t \tau^{\gamma-1}E_{\alpha,\gamma}(y\tau^\alpha)(t-\tau)^{\beta-1}E_{\alpha,\beta}(z(t-\tau)^\alpha)d\tau$$

$$= \frac{yE_{\alpha,\gamma+\beta}(yt^\alpha)-zE_{\alpha,\gamma+\beta}(zt^\alpha)}{y-z}t^{\gamma+\beta-1}, \quad (\gamma>0, \quad \beta>0), \tag{1.107}$$

where y and z ($y \neq z$) are arbitrary complex numbers.

Indeed, using the definition of the Mittag-Leffler function (1.56) we have:

$$\int_0^t \tau^{\gamma-1}E_{\alpha,\gamma}(y\tau^\alpha)(t-\tau)^{\beta-1}E_{\alpha,\beta}(z(t-\tau)^\alpha)d\tau$$

$$\begin{aligned}
&= \sum_{n=0}^\infty\sum_{m=0}^\infty \frac{y^n z^m}{\Gamma(\alpha n+\gamma)\Gamma(\alpha m+\beta)}\int_0^t \tau^{\alpha n+\gamma-1}(t-\tau)^{\alpha m+\beta-1}d\tau \\
&= \sum_{n=0}^\infty\sum_{m=0}^\infty \frac{y^n z^m t^{\alpha(n+m)+\beta+\gamma-1}}{\Gamma(\alpha(n+m)+\beta+\gamma)} \\
&= t^{\beta+\gamma-1}\sum_{n=0}^\infty\sum_{k=n}^\infty \frac{y^n z^{k-n}t^{\alpha k}}{\Gamma(\alpha k+\beta+\gamma)} \\
&= t^{\beta+\gamma-1}\sum_{k=0}^\infty \frac{z^k t^{\alpha k}}{\Gamma(\alpha k+\beta+\gamma)}\sum_{n=0}^k\left(\frac{y}{z}\right)^n \\
&= \frac{t^{\beta+\gamma-1}}{y-z}\sum_{k=0}^\infty \frac{t^{\alpha k}(y^{k+1}-z^{k+1})}{\Gamma(\alpha k+\beta+\gamma)},
\end{aligned} \tag{1.108}$$

and utilizing the definition (1.56) we obtain (1.107).

Another interesting formula establishes the relationship between the Mittag-Leffler function and the function $e^{-x^2/4t}$. This relationship plays an important role in the solution of the diffusion (heat conduction, mass transfer) equation:

$$\int_0^\infty e^{-x^2/4t} E_{\alpha,\beta}(x^\alpha) x^{\beta-1} dx = \sqrt{\pi} t^{\beta/2} E_{\alpha/2,(\beta+1)/2}(t^{\alpha/2}), \qquad (1.109)$$

$$(\beta > 0, \quad t > 0).$$

To prove formula (1.109) we note that for every fixed value of t the series

$$e^{-x^2/4t} E_{\alpha,\beta}(x^\alpha) x^{\beta-1} = \sum_{k=0}^{\infty} \frac{x^{\alpha k+\beta-1}}{\Gamma(\alpha k+\beta)} e^{-x^2/4t}, \quad (\beta > 0) \qquad (1.110)$$

can be term-by-term integrated from 0 to ∞. Performing the integration we obtain:

$$\begin{aligned} \int_0^\infty e^{-x^2/4t} E_{\alpha,\beta}(x^\alpha) x^{\beta-1} dt &= \int_0^\infty \left(\sum_{k=0}^{\infty} \frac{x^{\alpha k+\beta-1}}{\Gamma(\alpha k+\beta)} e^{-x^2/4t} \right) dx \\ &= \sum_{k=0}^{\infty} \frac{1}{\Gamma(\alpha k+\beta)} \int_0^\infty x^{\alpha k+\beta-1} e^{-x^2/4t} dx \\ &= \sum_{k=0}^{\infty} \frac{\Gamma\left(\frac{\alpha k+\beta}{2}\right)}{2\Gamma(\alpha k+\beta)} (2\sqrt{t})^{\alpha k+\beta}, \qquad (1.111) \end{aligned}$$

and the use of the Legendre formula

$$\Gamma(z)\Gamma(z+\frac{1}{2}) = \sqrt{\pi} 2^{1-2z} \Gamma(2z)$$

gives formula (1.109).

The use of the Laplace transform of the Mittag-Leffler function (1.80) is also a convenient way for obtaining various useful relationships for the Mittag-Leffler function.

For example, it follows from the identity (s denotes the Laplace transform parameter)

$$\frac{1}{s^2} = \frac{s^{\alpha-\beta}}{s^\alpha - 1} \left[s^{\beta-2} - s^{\beta-\alpha-2} \right] \qquad (1.112)$$

and from the known Laplace transform of the function t^ν [62, formula 4.3(1)]

$$L\{t^\nu; s\} = \Gamma(\nu+1)s^{-\nu-1}, \qquad (Re(s) > 0) \tag{1.113}$$

that

$$\int_0^t \tau^{\beta-1} E_{\alpha,\beta}(\tau^\alpha) \left[\frac{(t-\tau)^{1-\beta}}{\Gamma(2-\beta)} - \frac{(t-\tau)^{\alpha-\beta+1}}{\Gamma(\alpha-\beta+2)} \right] d\tau = t, \tag{1.114}$$

$$(0 < \beta < 2, \quad \alpha > 0).$$

The formula for the fractional integration of the Mittag-Leffler function (1.104) can also be obtained immediately by the inverse Laplace transform of the identity

$$\frac{s^{2\alpha-\beta}}{s^{2\alpha}-1} \cdot s^{-\alpha} = -\frac{s^{2\alpha-\beta}}{s^{2\alpha}-1} + \frac{s^{\alpha-\beta}}{s^\alpha - 1}. \tag{1.115}$$

The formula (1.109) can also be obtained with the help of the Laplace transform technique. Indeed, if $F(s)$ denotes the Laplace transform of a function $f(t)$, i.e.

$$F(s) = L\{f(t); s\} = \int_0^\infty e^{-st} f(t) dt,$$

then [62, formula 4.1(33)]

$$L\left\{ \frac{1}{\sqrt{\pi t}} \int_0^\infty e^{-x^2/4t} f(x) dx; s \right\} = s^{-1/2} F(s^{1/2}). \tag{1.116}$$

Let us take in (1.116)

$$f(x) = x^{\beta-1} E_{\alpha,\beta}(x^\alpha) \tag{1.117}$$

According to (1.80) we have

$$F(s) = \frac{s^{\alpha-\beta}}{s^\alpha - 1}$$

and, therefore,

$$s^{-1/2} F(s^{1/2}) = \frac{s^{\alpha/2-(\beta+1)/2}}{s^{\alpha/2}-1} = L\left\{ t^{\frac{\beta+1}{2}-1} E_{\alpha,\beta}(t^{\alpha/2}); s \right\}. \tag{1.118}$$

Comparing (1.116) and (1.118) we arrive at the relationship (1.109).

Similarly, using the Laplace transform of the Mittag-Leffler function (1.80), starting with the identity

$$\frac{s^{\alpha-\beta}}{s^\alpha - a} \cdot \frac{s^{\alpha-\gamma}}{s^\alpha + a} = \frac{s^{2\alpha-(\beta+\gamma)}}{s^{2\alpha} - a^2} \tag{1.119}$$

we obtain the convolution of two Mittag-Leffler functions:

$$\int_0^t \tau^{\beta-1} E_{\alpha,\beta}(a\tau^\alpha)(t-\tau)^{\gamma-1} E_{\alpha,\gamma}(-a(t-\tau)^\alpha)d\tau = t^{\beta+\gamma-1} E_{2\alpha,\beta+\gamma}(a^2 t^{2\alpha}) \tag{1.120}$$

$$(\beta > 0, \quad \gamma > 0).$$

The relationship (1.120) can also be obtained from (1.107), where we can take $z = -y$ and then utilize the relationship (1.95).

1.2.7 Asymptotic Expansions

Integration of the relationship (1.87) gives

$$E_{1/n,\beta}(z) = z^{(1-\beta)n} e^{z^n} \Big\{ z_0^{(1-\beta)n} e^{-z_0^n} E_{1/n,\beta}(z_0) + n \int_{z_0}^{z} e^{-\tau^n} \left(\sum_{k=1}^{n} \frac{\tau^{-k}}{\Gamma\left(\beta - \frac{k}{n}\right)} \tau^{\beta n - 1} \right) d\tau \Big\}, \quad (n \geq 1) \tag{1.121}$$

which is valid for arbitrary $z_0 \neq 0$.

If $\beta = 1$, then $z_0 = 0$ can be taken in (1.121). This gives:

$$E_{1/n,1}(z) = e^{z^n} \left\{ 1 + n \int_0^z e^{-\tau^n} \left(\sum_{k=1}^{n-1} \frac{\tau^{k-1}}{\Gamma\left(\frac{k}{n}\right)} \right) \right\}, \quad (n \geq 2). \tag{1.122}$$

Taking $n = 2$ in (1.122), we obtain the formula

$$E_{1/2,1}(z) = e^{z^2} \left\{ 1 + \frac{2}{\sqrt{\pi}} \int_0^z e^{-\tau^2} d\tau \right\} \tag{1.123}$$

from which the following asymptotic formula follows:

$$E_{1/2,1} \sim 2e^{z^2}, \quad |\arg(z)| < \frac{\pi}{4}, \quad |z| \to \infty. \tag{1.124}$$

General asymptotic formulas for the Mittag-Leffler function $E_{\alpha,\beta}(z)$ are given below in the form of theorems. The contour $\gamma(\epsilon,\varphi)$ and the domains $G^{-}(\epsilon,\varphi)$ $G^{+}(\epsilon,\varphi)$ used below have been defined in Section 1.1.6. The cases $\alpha < 2$, $\alpha = 2$ and $\alpha > 2$ are considered separately.

First let us obtain the corresponding integral representation formulas, which are necessary for obtaining the asymptotic formulas.

THEOREM 1.1 ∘ *Let $0 < \alpha < 2$ and let β be an arbitrary complex number. Then for an arbitrary $\epsilon > 0$ and μ such that*

$$\pi\alpha/2 < \mu \leq \min\{\pi, \pi\alpha\} \tag{1.125}$$

we have

$$E_{\alpha,\beta}(z) = \frac{1}{2\alpha\pi i} \int\limits_{\gamma(\epsilon,\mu)} \frac{\exp(\zeta^{1/\alpha})\zeta^{(1-\beta)/\alpha}}{\zeta - z} d\zeta, \qquad z \in G^{-}(\epsilon,\mu), \tag{1.126}$$

$$E_{\alpha,\beta}(z) = \frac{1}{\alpha} z^{(1-\beta)/\alpha} \exp(z^{1/\alpha}) + \frac{1}{2\alpha\pi i} \int\limits_{\gamma(\epsilon,\mu)} \frac{\exp(\zeta^{1/\alpha})\zeta^{(1-\beta)/\alpha}}{\zeta - z} d\zeta, \tag{1.127}$$

$$z \in G^{+}(\epsilon,\mu) \quad \bullet$$

Let us prove this statement.

If $|z| < \epsilon$, then

$$\left|\frac{z}{\zeta}\right| < 1, \qquad \zeta \in \gamma(\epsilon,\mu). \tag{1.128}$$

Using the definition of the Mittag-Leffler function $E_{\alpha,\beta}(z)$ (1.56) and the integral representation for the function $1/\Gamma(s)$ (1.52) and taking into account the inequality (1.128), we obtain for $\alpha < 2$ and $|z| < \epsilon$ that

$$\begin{aligned} E_{\alpha,\beta}(z) &= \sum_{k=0}^{\infty} \frac{1}{2\alpha\pi i} \left\{ \int\limits_{\gamma(\epsilon,\mu)} \exp(\zeta^{1/\alpha})\zeta^{(1-\beta)/\alpha-k-1} d\zeta \right\} z^k \\ &= \frac{1}{2\alpha\pi i} \int\limits_{\gamma(\epsilon,\mu)} \exp(\zeta^{1/\alpha})\zeta^{(1-\beta)/\alpha-1} \left\{ \sum_{k=0}^{\infty} \left(\frac{z}{\zeta}\right)^k \right\} d\zeta \\ &= \frac{1}{2\alpha\pi i} \int\limits_{\gamma(\epsilon,\mu)} \frac{\exp(\zeta^{1/\alpha})\zeta^{(1-\beta)/\alpha}}{\zeta - z} d\zeta. \end{aligned} \tag{1.129}$$

It follows from the condition (1.125) that this integral is absolutely convergent and defines a function of z, which is analytic in $G^-(\epsilon,\mu)$ and in $G^+(\epsilon,\mu)$. On the other hand, for every $\mu \in (\pi\alpha/2, \min\{\pi, \pi\alpha\})$ the circle $|z| < \epsilon$ lies in $G^-(\epsilon,\mu)$. Therefore, in accordance with the principle of analytic continuation, the integral (1.129) is equal to $E_{\alpha,\beta}(z)$ not only in the circle $|z| < \epsilon$, but in the entire domain $G^-(\epsilon,\mu)$, and we have proved formula (1.126).

Now let us take $z \in G^+(\epsilon,\mu)$. Then for an arbitrary $\epsilon_1 > |z|$ we have $z \in G^-(\epsilon_1,\mu)$, and using the formula (1.126) gives, on the one hand,

$$E_{\alpha,\beta}(z) = \frac{1}{2\alpha\pi i} \int\limits_{\gamma(\epsilon_1,\mu)} \frac{\exp(\zeta^{1/\alpha})\zeta^{(1-\beta)/\alpha}}{\zeta - z} d\zeta. \tag{1.130}$$

On the other hand, if $\epsilon < |z| < \epsilon_1$ and $-\mu < \arg(z) < \mu$, then the use of the Cauchy theorem gives

$$\frac{1}{2\alpha\pi i} \int\limits_{\gamma(\epsilon_1,\mu)-\gamma(\epsilon,\mu)} \frac{\exp(\zeta^{1/\alpha})\zeta^{(1-\beta)/\alpha}}{\zeta - z} d\zeta = \frac{1}{\alpha} z^{(1-\beta)/\alpha} \exp(z^{1/\alpha}), \tag{1.131}$$

and combining (1.130) and (1.131) we obtain the integral representation formula (1.127).

THEOREM 1.2 ∘ *If $Re(\beta) > 0$, then for arbitrary $\epsilon > 0$*

$$E_{2,\beta}(z) = \frac{1}{4\pi i} \int\limits_{\gamma(\epsilon,\pi)} \frac{\exp(\zeta^{1/2})\zeta^{(1-\beta)/2}}{\zeta - z} d\zeta, \qquad z \in G^-(\epsilon,\pi), \tag{1.132}$$

$$E_{2,\beta}(z) = \frac{1}{2} z^{(1-\beta)/2} \exp(z^{1/2}) + \frac{1}{4\pi i} \int\limits_{\gamma(\epsilon,\pi)} \frac{\exp(\zeta^{1/2})\zeta^{(1-\beta)/2}}{\zeta - z} d\zeta, \tag{1.133}$$

$$z \in G^-(\epsilon,\pi). \quad \bullet$$

The proof of this theorem is similar to the previous one. However, instead of the integral representation (1.52) of the function $1/\Gamma(s)$ we must use the formula (1.54) leading to the relationship (1.132). The integral on the right-hand side of equation (1.132) converges for $Re(\beta) > 0$ and converges absolutely for $Re(\beta) > 1$. Taking into account that formula (1.131) holds also for $\alpha = 2$ and $\mu = \pi$, we obtain (1.133).

Now let us use Theorem 1.1 for establishing the following asymptotic formulas.

THEOREM 1.3 ∘ *If* $0 < \alpha < 2$, β *is an arbitrary complex number and* μ *is an arbitrary real number such that*

$$\frac{\pi\alpha}{2} < \mu < \min\{\pi, \pi\alpha\}, \tag{1.134}$$

then for an arbitrary integer $p \geq 1$ *the following expansion holds:*

$$E_{\alpha,\beta}(z) = \frac{1}{\alpha} z^{(1-\beta)/\alpha} \exp(z^{1/\alpha}) - \sum_{k=1}^{p} \frac{z^{-k}}{\Gamma(\beta - \alpha k)} + O\left(|z|^{-1-p}\right), \tag{1.135}$$

$$|z| \to \infty, \qquad |\arg(z)| \leq \mu. \quad \bullet$$

Let us start the proof of formula (1.135) by taking φ satisfying the condition

$$\frac{\pi\alpha}{2} < \mu < \varphi \leq \min\{\pi, \pi\alpha\}. \tag{1.136}$$

Taking now $\epsilon = 1$ and substituting the representation

$$\frac{1}{\zeta - z} = -\sum_{k=1}^{p} \frac{\zeta^{k-1}}{z^k} + \frac{\zeta^p}{z^p(\zeta - z)} \tag{1.137}$$

into equation (1.127) of Theorem 1.1, we obtain the following expression for the Mittag-Leffler function $E_{\alpha,\beta}(z)$ in the domain $G^+(1,\varphi)$ (i.e., on the right side of the contour $\gamma(1,\varphi)$):

$$\begin{aligned} E_{\alpha,\beta}(z) &= \frac{1}{\alpha} z^{(1-\beta)/\alpha} \exp(z^{1/\alpha}) \\ &- \sum_{k=1}^{p} \left(\frac{1}{2\pi\alpha i} \int_{\gamma(1,\varphi)} \exp(\zeta^{1/\alpha}) \zeta^{(1-\beta)/\alpha+k-1} d\zeta \right) z^{-k} \\ &+ \frac{1}{2\pi\alpha i z^p} \int_{\gamma(1,\varphi)} \exp(\zeta^{1/\alpha}) \zeta^{(1-\beta)/\alpha+p} d\zeta. \end{aligned} \tag{1.138}$$

The first integral can be evaluated with the help of formula (1.52):

$$\frac{1}{2\pi\alpha i} \int_{\gamma(1,\varphi)} \exp(\zeta^{1/\alpha}) \zeta^{(1-\beta)/\alpha+k-1} d\zeta = \frac{1}{\Gamma(\beta - \alpha k)}, \qquad (K \geq 1). \tag{1.139}$$

Substituting this expression into equation (1.138) and taking into account the condition (1.136), we obtain:

$$E_{\alpha,\beta}(z) = \frac{1}{\alpha} z^{(1-\beta)/\alpha} \exp(z^{1/\alpha}) - \sum_{k=1}^{p} \frac{z^{-k}}{\Gamma(\beta - \alpha k)}$$

$$+ \frac{1}{2\pi\alpha i z^p} \int\limits_{\gamma(1,\varphi)} \exp(\zeta^{1/\alpha}) \zeta^{(1-\beta)/\alpha + p} d\zeta, \qquad (1.140)$$

$$(|\arg(z)| \leq \mu, |z| > 1).$$

Let us estimate the integral

$$I_p(z) = \frac{1}{2\pi\alpha i z^p} \int\limits_{\gamma(1,\varphi)} \exp(\zeta^{1/\alpha}) \zeta^{(1-\beta)/\alpha + p} d\zeta,$$

for large $|z|$ and $|\arg(z)| \leq \mu$.

For large $|z|$ and $|\arg(z)| \leq \mu$ we have

$$\min_{\zeta \in \gamma(1,\varphi)} |\zeta - z| = |z| \sin(\varphi - \mu),$$

and therefore for large $|z|$ and $|\arg(z)| \leq \mu$ we have

$$|I_p(z)| \leq \frac{|z|^{-1-p}}{2\pi\alpha \sin(\varphi - \mu)} \int\limits_{\gamma(1,\varphi)} \left|\exp(\zeta^{1/\alpha})\right| \left|\zeta^{(1-\beta)+p}\right| d\zeta. \qquad (1.141)$$

The integral on the right-hand side converges, because for ζ such that $\arg(\zeta) = \pm\varphi$ and $|\zeta| \geq 1$ the following holds:

$$\left|\exp(\zeta^{1/\alpha})\right| = \exp\left(|\zeta|^{1/\alpha} \cos\left(\frac{\varphi}{\alpha}\right)\right),$$

where $\cos(\varphi/\alpha) < 0$ due to condition (1.136).

Combining equation (1.140) and the estimate (1.141) we obtain the asymptotic formula (1.135).

THEOREM 1.4 ∘ *If* $0 < \alpha < 2$, β *is an arbitrary complex number and* μ *is an arbitrary real number such that*

$$\frac{\pi\alpha}{2} < \mu < \min\{\pi, \pi\alpha\}, \qquad (1.142)$$

then for an arbitrary integer $p \geq 1$ *the following expansion holds:*

$$E_{\alpha,\beta}(z) = -\sum_{k=1}^{p} \frac{z^{-k}}{\Gamma(\beta - \alpha k)} + O\left(|z|^{-1-p}\right), \tag{1.143}$$

$$|z| \to \infty, \qquad \mu \leq |\arg(z)| \leq \pi. \quad \bullet$$

To prove Theorem 1.4, let us take

$$\frac{\pi\alpha}{2} < \varphi < \mu < \min\{\pi, \pi\alpha\} \tag{1.144}$$

Taking $\epsilon = 1$ in equation (1.126) of Theorem 1.1 and using formula (1.137), we obtain

$$E_{\alpha,\beta}(z) = -\sum_{k=1}^{p} \frac{z^{-k}}{\Gamma(\beta - \alpha k)} + I_p(z), \qquad z \in G^-(1, \varphi), \tag{1.145}$$

where $I_p(z)$ is the same as above.

For large $|z|$, such that $\mu \leq |\arg(z)| \leq \pi$, the following holds:

$$\min_{\zeta \in \gamma(1,\varphi)} |\zeta - z| = |z| \sin(\varphi - \mu).$$

Additionally, the domain $\mu \leq |\arg(z)| \leq \pi$ lies in the domain $G^-(1, \varphi)$, for which equation (1.145) holds. Therefore, for large $|z|$ we have the estimate

$$|I_p(z)| \leq \frac{|z|^{-1-p}}{2\pi\alpha \sin(\varphi - \mu)} \int_{\gamma(1,\varphi)} \left|\exp(\zeta^{1/\alpha})\right| \left|\zeta^{(1-\beta)+p}\right| d\zeta, \tag{1.146}$$

$$(\mu \leq |\arg(z)| \leq \pi).$$

Combining equation (1.145) and the estimate (1.146), we obtain the asymptotic formula (1.143).

The following two theorems, which give estimates of the behaviour of the Mittag-Leffler function $E_{\alpha,\beta}(z)$ in different parts of the complex plane, are obvious consequences of Theorems 1.3 and 1.4:

THEOREM 1.5 ∘ *If $\alpha < 2$, β is an arbitrary real number, μ is such that $\pi\alpha/2 < \mu < \min\{\pi, \pi\alpha\}$ and C_1 and C_2 are real constants, then*

$$|E_{\alpha,\beta}(z)| \le C_1(1+|z|)^{(1-\beta)/\alpha} \exp\left(Re(z^{1/\alpha})\right) + \frac{C_2}{1+|z|}, \qquad (1.147)$$

$$(|\arg(z)| \le \mu), \quad |z| \ge 0. \quad \bullet$$

THEOREM 1.6 ∘ *If $\alpha < 2$, β is an arbitrary real number, μ is such that $\pi\alpha/2 < \mu < \min\{\pi, \pi\alpha\}$ and C is a real constant, then*

$$|E_{\alpha,\beta}(z)| \le \frac{C}{1+|z|}, \qquad (1.148)$$

$$(\mu \le |\arg(z)| \le \pi), \quad |z| \ge 0. \quad \bullet$$

Let us now turn our attention to the case of $\alpha \ge 2$.

THEOREM 1.7 ∘ *If $\alpha \ge 2$ and β is arbitrary, then for an arbitrary integer number $p \ge 1$ the following asymptotic formula holds:*

$$E_{\alpha,\beta}(z) = \frac{1}{\alpha}\sum_{n}\left(z^{1/\alpha}\exp\left(\frac{2\pi n i}{\alpha}\right)\right)^{1-\beta} \exp\left\{\exp\left(\frac{2\pi n i}{\alpha}\right) z^{1/\alpha}\right\}$$
$$- \sum_{k=1}^{p} \frac{z^{-k}}{\Gamma(\beta - \alpha k)} + O\left(|z|^{-1-p}\right), \qquad (1.149)$$

where the sum is taken for integer n satisfying the condition

$$|\arg(z) + 2\pi n| \le \frac{\pi\alpha}{2}. \quad \bullet$$

Let us start the proof by recalling formula (1.97)

$$E_{\alpha,\beta}(z) = \frac{1}{2m+1}\sum_{\nu=-m}^{m-1} E_{\alpha/(2m+1),\beta}(z^{1/(2m+1)}e^{i2\pi\nu/(2m+1)}), \quad (m \ge 0),$$

where $\alpha > 0$. Taking into account that under the conditions of the theorem $\alpha \ge 2$, let us take integer $m \ge 1$ such that $\alpha_1 = \alpha/(2m+1) < 2$.

In such a case we can apply Theorems 1.3 and 1.4 to all terms of the above sum (1.97).

Let us take an arbitrary μ satisfying the inequality

$$\frac{\pi\alpha_1}{2} < \mu < \min\{\pi, \pi\alpha_1\}, \qquad \left(\alpha_1 = \frac{\alpha}{2m+1}\right).$$

Taking an arbitrary integer $q \geq 1$ and using the asymptotic formula (1.135) of Theorem 1.3 and (1.143) of Theorem 1.4, we obtain

$$E_{\alpha,\beta}(z) = \frac{1}{\alpha}\sum\left(z^{1/\alpha}\exp\left(\frac{2\pi ni}{\alpha}\right)\right)^{1-\beta}\exp\left\{\exp\left(\frac{2\pi ni}{\alpha}\right)z^{1/\alpha}\right\}$$

$$-\frac{1}{2m+1}\sum_{n=-m}^{m}\left\{\sum_{k=1}^{q}\frac{z^{-k/(2m+1)}\exp\left(\frac{-2\pi kni}{2m+1}\right)}{\Gamma(\beta-\frac{k\alpha}{2m+1})}+O\left(|z|^{-(q+1)/(2m+1)}\right)\right\}. \tag{1.150}$$

The first sum in (1.150) is taken for integer values of n satisfying the condition

$$\left|\arg\left(z^{1/(2m+1)}\exp\left(\frac{2\pi ni}{2m+1}\right)\right)\right| \leq \mu. \tag{1.151}$$

Obviously, the condition (1.151) is equivalent to the condition

$$|\arg(z) + 2\pi n| \leq (2m+1)\mu. \tag{1.152}$$

Now let us suppose that z is fixed. If we take $\mu_* > \pi\alpha/2$ and μ_* is close enough to $\pi\alpha/2$, then the inequalities

$$|\arg(z) + 2\pi n| \leq \frac{\pi\alpha}{2} \tag{1.153}$$

and

$$|\arg(z) + 2\pi n| \leq \mu_*$$

are satisfied for the same set of values of n.

The number $(2m+1)\mu > \frac{\pi\alpha}{2}$ can be chosen close enough to $\frac{\pi\alpha}{2}$; therefore, the expression (1.150) can be written as

$$E_{\alpha,\beta}(z) = \frac{1}{\alpha}\sum\left(z^{1/\alpha}\exp\left(\frac{2\pi ni}{\alpha}\right)\right)^{1-\beta}\exp\left\{\exp\left(\frac{2\pi ni}{\alpha}\right)z^{1/\alpha}\right\}$$

$$-\frac{1}{2m+1}\sum_{k=1}^{q}\frac{z^{-k/(2m+1)}}{\Gamma\left(\beta-\frac{k\alpha}{2m+1}\right)}\left\{\sum_{n=-m}^{m}\exp\left(-\frac{2\pi kni}{2m+1}\right)\right\}$$

$$+O\left(|z|^{-(q+1)/(2m+1)}\right), \tag{1.154}$$

where the first sum is taken for n satisfying the condition (1.153).

Until now, q was an arbitrary natural number. Now for a given p let us take

$$q = (2m+1)(p+1) - 1.$$

Then, taking into account that

$$\sum_{n=-m}^{m} \exp\left(-\frac{2\pi kni}{2m+1}\right) = \begin{cases} 2m+1 & k \equiv 0(\mathrm{mod}(2m+1)) \\ 0 & k \not\equiv 0(\mathrm{mod}(2m+1)), \end{cases} \tag{1.155}$$

the asymptotic expansion (1.149) follows from (1.154). The proof of Theorem 1.7 is complete.

1.3 Wright Function

The Wright function plays an important role in the solution of linear partial fractional differential equations, e.g. the fractional diffusion–wave equation.

This function, related to the Mittag-Leffler function in two parameters $E_{\alpha,\beta}(z)$, was introduced by Wright [258, cf. [65, 107]]. A number of useful relationships were obtained by Humbert and Agarwal [107] with the help of the Laplace transform.

For convenience we adopt here Mainardi's notation for the Wright function $W(z;\alpha,\beta)$.

1.3.1 Definition

The Wright function is defined as [65, formula 18.1(27)]

$$W(z;\alpha,\beta) = \sum_{k=0}^{\infty} \frac{z^k}{k!\,\Gamma(\alpha k+\beta)}. \tag{1.156}$$

1.3.2 Integral Representation

This function can be represented by the following integral [65, formula 18.1(29)]:

$$W(z;\alpha,\beta) = \frac{1}{2\pi i} \int_{Ha} \tau^{-\beta} e^{\tau + z\tau^{-\alpha}} d\tau \tag{1.157}$$

where Ha denotes Hankel's contour.

To prove (1.157), let us write the integrated function in the form of a power series in z and perform term-by-term integration using the integral representation formula for the reciprocal gamma function (1.46)

$$\frac{1}{\Gamma(z)} = \int_{Ha} e^{\tau} \tau^{-z} d\tau.$$

1.3.3 Relation to Other Functions

It follows from the definition (1.156) that

$$W(z; 0, 1) = e^z \tag{1.158}$$

$$\left(\frac{z}{2}\right)^{\nu} W(\mp\frac{z^2}{4}; 1, \nu+1) = \begin{cases} J_\nu(z) \\ I_\nu(z). \end{cases} \tag{1.159}$$

Taking $\beta = 1 - \alpha$, we obtain Mainardi's function $M(z; \alpha)$:

$$W(-z; -\alpha, 1-\alpha) = M(z; \alpha) = \sum_{k=0}^{\infty} \frac{(-1)^k z^k}{k!\, \Gamma(-\alpha(k+1)+1)}. \tag{1.160}$$

The following particular case of the Wright function was considered by Mainardi [131]:

$$W(-z; -\frac{1}{2}, -\frac{1}{2}) = M(z; \frac{1}{2}) = \frac{1}{\sqrt{\pi}} \exp\left(-\frac{z^2}{4}\right). \tag{1.161}$$

We see that the Wright function is a generalization of the exponential function and the Bessel functions. For $\alpha > 0$ and $\beta > 0$ it is an entire function in z [65].

Recently Mainardi [131] pointed out that $W(z; \alpha, \beta)$ is an entire function in z also for $-1 < \alpha < 0$.

Let us prove this statement. Using the well-known relationship (1.26)

$$\Gamma(y)\Gamma(1-y) = \frac{\pi}{\sin(\pi y)},$$

we can write the Wright function in the form

$$W(z; \alpha, \beta) = \frac{1}{\pi} \sum_{k=0}^{\infty} \frac{z^k \Gamma(1 - \alpha k - \beta) \sin \pi(\alpha k + \beta)}{k!}. \tag{1.162}$$

Let us introduce an auxiliary majorizing series

$$S = \frac{1}{\pi}\sum_{k=0}^{\infty}\left|\frac{\Gamma(1-\alpha k-\beta)}{k!}\right| |z|^k. \tag{1.163}$$

The convergence radius of series (1.163) for $-1 < \alpha < 0$ is infinite:

$$R = \lim_{k\to\infty}\left|\frac{\Gamma(1-\alpha k-\beta)}{k!}\,\frac{(k+1)!}{\Gamma(1-\alpha k-\alpha-\beta)}\right| = \lim_{k\to\infty}\frac{k+1}{|\alpha|^{\alpha}k^{-\alpha}} = \infty. \tag{1.164}$$

(We use here relationship [63, formula 1.18(4)].)

It follows from the comparison of the series (1.156) and (1.163) that for $\alpha > -1$ and arbitrary β the convergence radius of the series representation of the Wright function $W(z;\alpha,\beta)$ is infinite, and the Wright function is an entire function.

There is an interesting link between the Wright function and the Mittag-Leffler function. Namely, the Laplace transform of the Wright function is expressed with the help of the Mittag-Leffler function:

$$\begin{aligned} L\{W(t;\alpha,\beta);\ s\} &= L\left\{\sum_{k=0}^{\infty}\frac{t^k}{k!\,\Gamma(\alpha k+\beta)};\ s\right\} \\ &= \sum_{k=0}^{\infty}\frac{1}{\Gamma(\alpha k+\beta)}\cdot\frac{1}{s^{k+1}} \\ &= s^{-1}E_{\alpha,\beta}(s^{-1}). \end{aligned} \tag{1.165}$$

Chapter 2

Fractional Derivatives and Integrals

In this chapter several approaches to the generalization of the notion of differentiation and integration are considered. The choice has been reduced to those definitions which are related to applications.

2.1 The Name of the Game

Mathematics is the art of giving things misleading names. The beautiful – and at first look mysterious – name *the fractional calculus* is just one of the those misnomers which are the essence of mathematics.

For example, we know such names as *natural numbers* and *real numbers.* We use them very often; let us think for a moment about these names. The notion of a *natural number* is a natural abstraction, but is the *number* itself *natural*?

The notion of a *real number* is a generalization of the notion of a natural number. The word *real* emphasizes that we pretend that they reflect real quantities. The *real numbers* do reflect real quantities, but this cannot change the fact that they do not exist. Everything is in order in mathematical analysis, and the notion of a *real number* makes it easier, but if one wants to compute something, he immediately discovers for himself that there is no place for *real numbers* in the *real world*; nowadays, computations are performed mostly on digital computers, which can work only with finite sets of finite fractions, which serve as approximations to unreal *real numbers.*

Let us now return to the name of the *fractional calculus*. It does not mean the calculus of fractions. Neither does it mean a fraction of any calculus — differential, integral or calculus of variations. *The fractional calculus* is a name for the theory of integrals and derivatives of arbitrary order, which unify and generalize the notions of integer-order differentiation and n-fold integration.

Let us consider the infinite sequence of n-fold integrals and n-fold derivatives:

$$\ldots, \quad \int_a^t d\tau_2 \int_a^{\tau_2} f(\tau_1) d\tau_1, \quad \int_a^t f(\tau_1) d\tau_1, \quad f(t), \quad \frac{df(t)}{dt}, \quad \frac{d^2 f(t)}{dt^2}, \quad \ldots$$

The derivative of arbitrary real order α can be considered as an interpolation of this sequence of operators; we will use for it the notation suggested and used by Davis [39], namely

$$_aD_t^{\alpha} f(t).$$

The short name for derivatives of arbitrary order is *fractional derivatives.*

The subscripts a and t denote the two limits related to the operation of fractional differentiation; following Ross [227] we will call them the *terminals* of fractional differentiation. The appearance of the terminals in the symbol of fractional differentiation is essential. This helps to avoid ambiguities in applications of fractional derivatives to real problems.

The words *fractional integrals* mean in this book integrals of arbitrary order and correspond to negative values of α. We will not use a separate notation for fractional integrals; we will denote the fractional integral of order $\beta > 0$ by

$$_aD_t^{-\beta} f(t).$$

A *fractional differential equation* is an equation which contains fractional derivatives; a *fractional integral equation* is an integral equation containing fractional integrals.

A *fractional-order system* means a system described by a fractional differential equation or a fractional integral equation or by a system of such equations.

2.2 Grünwald–Letnikov Fractional Derivatives

2.2.1 Unification of Integer-order Derivatives and Integrals

In this section we describe an approach to the unification of two notions, which are usually presented separately in classical analysis: derivative of integer order n and n-fold integrals. As will be shown below, these notions are closer to each other than one usually assumes.

Let us consider a continuous function $y = f(t)$. According to the well-known definition, the first-order derivative of the function $f(t)$ is defined by

$$f'(t) = \frac{df}{dt} = \lim_{h\to 0} \frac{f(t) - f(t-h)}{h}. \tag{2.1}$$

Applying this definition twice gives the second-order derivative:

$$\begin{aligned} f''(t) = \frac{d^2 f}{dt^2} &= \lim_{h\to 0} \frac{f'(t) - f'(t-h)}{h} \\ &= \lim_{h\to 0} \frac{1}{h}\left\{ \frac{f(t) - f(t-h)}{h} - \frac{f(t-h) - f(t-2h)}{h} \right\} \\ &= \lim_{h\to 0} \frac{f(t) - 2f(t-h) + f(t-2h)}{h^2}. \end{aligned} \tag{2.2}$$

Using (2.1) and (2.2) we obtain

$$f'''(t) = \frac{d^3 f}{dt^3} = \lim_{h\to 0} \frac{f(t) - 3f(t-h) + 3f(t-2h) - f(t-3h)}{h^3} \tag{2.3}$$

and, by induction,

$$f^{(n)}(t) = \frac{d^n f}{dt^n} = \lim_{h\to 0} \frac{1}{h^n} \sum_{r=0}^{n} (-1)^r \binom{n}{r} f(t - rh), \tag{2.4}$$

where

$$\binom{n}{r} = \frac{n(n-1)(n-2)\dots(n-r+1)}{r!} \tag{2.5}$$

is the usual notation for the binomial coefficients.

Let us now consider the following expression generalizing the fractions in (2.1)–(2.4):

$$f_h^{(p)}(t) = \frac{1}{h^p} \sum_{r=0}^{n} (-1)^k \binom{p}{r} f(t - rh), \tag{2.6}$$

where p is an arbitrary integer number; n is also integer, as above.

Obviously, for $p \leq n$ we have

$$\lim_{h \to 0} f_h^{(p)}(t) = f^{(p)}(t) = \frac{d^p f}{dt^p}, \tag{2.7}$$

because in such a case, as follows from (2.5), all the coefficients in the numerator after $\binom{p}{p}$ are equal to 0.

Let us consider negative values of p. For convenience, let us denote

$$\begin{bmatrix} p \\ r \end{bmatrix} = \frac{p(p+1)\ldots(p+r-1)}{r!}. \tag{2.8}$$

Then we have

$$\binom{-p}{r} = \frac{-p(-p-1)\ldots(-p-r+1)}{r!} = (-1)^r \begin{bmatrix} p \\ r \end{bmatrix} \tag{2.9}$$

and replacing p in (2.6) with $-p$ we can write

$$f_h^{(-p)}(t) = \frac{1}{h^p} \sum_{r=0}^{n} \begin{bmatrix} p \\ r \end{bmatrix} f(t - rh), \tag{2.10}$$

where p is a positive integer number.

If n is fixed, then $f_h^{(-p)}(t)$ tends to the uninteresting limit 0 as $h \to 0$. To arrive at a non-zero limit, we have to suppose that $n \to \infty$ as $h \to 0$. We can take $h = \frac{t-a}{n}$, where a is a real constant, and consider the limit value, either finite or infinite, of $f_h^{(-p)}(t)$, which we will denote as

$$\lim_{\substack{h \to 0 \\ nh = t-a}} f_h^{(-p)}(t) = {}_aD_t^{-p} f(t). \tag{2.11}$$

Here ${}_aD_t^{-p} f(t)$ denotes, in fact, a certain operation performed on the function $f(t)$; a and t are the *terminals* — the limits relating to this operation.

Let us consider several particular cases.

For $p = 1$ we have:

$$f_h^{(-1)}(t) = h \sum_{r=0}^{n} f(t - rh). \tag{2.12}$$

Taking into account that $t - nh = a$ and that the function $f(t)$ is assumed to be continuous, we conclude that

$$\lim_{\substack{h \to 0 \\ nh = t-a}} f_h^{(-1)}(t) = {}_aD_t^{-1} f(t) = \int_0^{t-a} f(t - z)dz = \int_a^t f(\tau)d\tau. \tag{2.13}$$

Let us take $p = 2$. In this case

$$\begin{bmatrix} 2 \\ r \end{bmatrix} = \frac{2 \cdot 3 \cdot \ldots \cdot (2 + r - 1)}{r!} = r + 1,$$

and we have:

$$f_h^{(-2)}(t) = h \sum_{r=0}^{n} (rh) f(t - rh). \tag{2.14}$$

Denoting $t + h = y$ we can write

$$f_h^{(-2)}(t) = h \sum_{r=1}^{n+1} (rh) f(t - rh), \tag{2.15}$$

and taking $h \to 0$ we obtain

$$\lim_{\substack{h \to 0 \\ nh = t-a}} f_h^{(-2)}(t) = {}_aD_t^{-2} f(t) = \int_0^{t-a} z f(t - z) dt = \int_a^t (t - \tau) f(\tau) d\tau, \tag{2.16}$$

because $y \to t$ as $h \to 0$.

The third particular case, namely $p = 3$, will show us the general expression for ${}_aD_t^{-p}$.

Taking into account that

$$\begin{bmatrix} 2 \\ r \end{bmatrix} = \frac{3 \cdot 4 \cdot \ldots \cdot (3 + r - 1)}{r!} = \frac{(r+1)(r+2)}{1 \cdot 2},$$

we have

$$f_h^{(-3)}(t) = \frac{h}{1 \cdot 2} \sum_{r=0}^{n} (r+1)(r+2) h^2 f(t - rh). \tag{2.17}$$

Denoting, as above, $t + h = y$, we write

$$f_h^{(-3)}(t) = \frac{h}{1 \cdot 2} \sum_{r=1}^{n+1} r(r+1) h^2 f(y - rh) \tag{2.18}$$

Expression (2.18) can be written as

$$f_h^{(-3)}(t) = \frac{h}{1 \cdot 2} \sum_{r=1}^{n+1} (rh)^2 f(y - rh) + \frac{h^2}{1 \cdot 2} \sum_{r=1}^{n+1} rh f(y - rh). \tag{2.19}$$

Taking now $h \to 0$, we obtain

$${}_aD_t^{-3} f(t) = \frac{1}{2!} \int_0^{t-a} z^2 f(t - z) dz = \int_a^t (t - \tau)^2 f(\tau) d\tau, \tag{2.20}$$

because $y \to t$ as $h \to 0$ and

$$\lim_{\substack{h\to 0\\ nh=t-a}} \frac{h^2}{1\cdot 2}\sum_{r=1}^{n+1} rhf(y-rh) = \lim_{\substack{h\to 0\\ nh=t-a}} h\int_a^t (t-\tau)f(\tau)d\tau = 0.$$

Relationships (2.13)–(2.20) suggest the following general expression:

$$ {}_aD_t^{-p}f(t) = \lim_{\substack{h\to 0\\ nh=t-a}} h^p \sum_{r=0}^{n} \left[{p \atop r} \right] f(t-rh) = \frac{1}{(p-1)!}\int_a^t (t-\tau)^{p-1}f(\tau)d\tau. \tag{2.21}$$

To prove the formula (2.21) by induction we have to show that if it holds for some p, then it holds also for $p+1$.

Let us introduce the function

$$f_1(t) = \int_a^t f(\tau)d\tau, \tag{2.22}$$

which has the obvious property $f_1(a) = 0$, and consider

$$\begin{aligned} {}_aD_t^{-p-1}f(t) &= \lim_{\substack{h\to 0\\ nh=t-a}} h^{p+1}\sum_{r=0}^{n}\left[{p+1 \atop r} \right] f(t-rh) \\ &= \lim_{\substack{h\to 0\\ nh=t-a}} h^{p}\sum_{r=0}^{n}\left[{p+1 \atop r} \right] f_1(t-rh) \\ &\quad - \lim_{\substack{h\to 0\\ nh=t-a}} h^{p}\sum_{r=0}^{n}\left[{p+1 \atop r} \right] f_1(t-(r+1)h) \end{aligned} \tag{2.23}$$

Using (2.8) it is easy to verify that

$$\left[{p+1 \atop r} \right] = \left[{p \atop r} \right] + \left[{p+1 \atop r-1} \right], \tag{2.24}$$

where we must put

$$\left[{p+1 \atop -1} \right] = 0.$$

Relationship (2.24) applied to the first sum in (2.23) and the replacement of r by $r-1$ in the second sum gives:

$$ {}_aD_t^{-p-1}f(t) = \lim_{\substack{h\to 0\\ nh=t-a}} h^p \sum_{r=0}^{n}\left[{p \atop r} \right] f_1(t-rh)$$

$$
\begin{aligned}
&+ \lim_{\substack{h\to 0\\ nh=t-a}} h^p \sum_{r=0}^{n} \begin{bmatrix} p+1 \\ r-1 \end{bmatrix} f_1(t-rh) \\
&- \lim_{\substack{h\to 0\\ nh=t-a}} h^p \sum_{r=1}^{n+1} \begin{bmatrix} p+1 \\ r-1 \end{bmatrix} f_1(t-rh) \\
= \;& {}_aD_t^{-p} f_1(t) - \lim_{\substack{h\to 0\\ nh=t-a}} h^p \begin{bmatrix} p+1 \\ n \end{bmatrix} f_1(t-(n+1)h) \\
= \;& {}_aD_t^{-p} f_1(t) - (t-a)^p \lim_{n\to\infty} \begin{bmatrix} p+1 \\ n \end{bmatrix} \frac{1}{n^p} f_1(a - \frac{t-a}{n}).
\end{aligned}
$$

It follows from the definition (2.22) of the function $f_1(t)$ that

$$
\lim_{n\to\infty} f_1(a - \frac{t-a}{n}) = 0.
$$

Taking into account the known limit (1.7)

$$
\lim_{n\to\infty} \begin{bmatrix} p+1 \\ n \end{bmatrix} \frac{1}{n^p} = \lim_{n\to\infty} \frac{(p+1)(p+2)\dots(p+n)}{n^p n!} = \frac{1}{\Gamma(p+1)},
$$

we obtain

$$
\begin{aligned}
{}_aD_t^{-p-1} f(t) &= {}_aD_t^{-p} f_1(t) = \frac{1}{(p-1)!} \int_a^t (t-\tau)^{p-1} f_1(\tau) d\tau \\
&= - \frac{(t-\tau)^p f_1(\tau)}{p!} \Bigg|_{\tau=a}^{\tau=t} + \frac{1}{p!} \int_a^t (t-\tau)^p f(\tau) d\tau \\
&= \frac{1}{p!} \int_a^t (t-\tau)^p f(\tau) d\tau, \qquad (2.25)
\end{aligned}
$$

which ends the proof of formula (2.21) by induction.

Now let us show that formula (2.21) is a representation of a p-fold integral.

Integrating the relationship

$$
\frac{d}{dt} \left({}_aD_t^{-p} f(t) \right) = \frac{1}{(p-2!)} \int_a^t (t-\tau)^{p-2} f(\tau) d\tau = {}_aD_t^{-p+1} f(t)
$$

from a to t we obtain:

$$
\begin{aligned}
{}_aD_t^{-p} f(t) &= \int_a^t \left({}_aD_t^{-p+1} f(t) \right) dt, \\
{}_aD_t^{-p+1} f(t) &= \int_a^t \left({}_aD_t^{-p+2} f(t) \right) dt, \text{etc.},
\end{aligned}
$$

and therefore

$$
\begin{aligned}
{}_aD_t^{-p} f(t) &= \int_a^t dt \int_a^t \left({}_aD_t^{-p+2} f(t) \right) dt \\
&= \int_a^t dt \int_a^t dt \int_a^t \left({}_aD_t^{-p+3} f(t) \right) dt \\
&= \underbrace{\int_a^t dt \int_a^t dt \ldots \int_a^t}_{p \text{ times}} f(t) dt. \qquad (2.26)
\end{aligned}
$$

We see that the derivative of an integer order n (2.4) and the p-fold integral (2.21) of the continuous function $f(t)$ are particular cases of the general expression

$$
{}_aD_t^p f(t) = \lim_{\substack{h \to 0 \\ nh = t-a}} h^{-p} \sum_{r=0}^{n} (-1)^r \binom{p}{r} f(t - rh), \qquad (2.27)
$$

which represents the derivative of order m if $p = m$ and the m-fold integral if $p = -m$.

This observation naturally leads to the idea of a generalization of the notions of differentiation and integration by allowing p in (2.27) to be an arbitrary real or even complex number. We will restrict our attention to real values of p.

2.2.2 Integrals of Arbitrary Order

Let us consider the case of $p < 0$. For convenience let us replace p by $-p$ in the expression (2.27). Then (2.27) takes the form

$$
{}_aD_t^{-p} f(t) = \lim_{\substack{h \to 0 \\ nh = t-a}} h^{p} \sum_{r=0}^{n} \left[\begin{matrix} p \\ r \end{matrix} \right] f(t - rh), \qquad (2.28)
$$

where, as above, the values of h and n relate as $nh = t - a$.

To prove the existence of the limit in (2.28) and to evaluate that limit we need the following theorem (A. V. Letnikov, [124]):

THEOREM 2.1 ∘ *Let us take a sequence β_k, $(k = 1, 2, \ldots)$ and suppose that*

$$\lim_{k\to\infty} \beta_k = 1, \tag{2.29}$$

$$\lim_{n\to\infty} \alpha_{n,k} = 0 \quad \textit{for all } k, \tag{2.30}$$

$$\lim_{n\to\infty} \sum_{k=1}^{n} \alpha_{n,k} = A \quad \textit{for all } k, \tag{2.31}$$

$$\sum_{k=1}^{n} |\alpha_{n,k}| < K \quad \textit{for all } n. \tag{2.32}$$

Then

$$\lim_{n\to\infty} \sum_{k=1}^{n} \alpha_{n,k}\beta_k = A. \qquad \bullet \tag{2.33}$$

Proof. The condition (2.29) allows us to put

$$\beta_k = 1 - \sigma_k, \qquad \text{where} \quad \lim_{k\to\infty} \sigma_k = 0. \tag{2.34}$$

It follows from the condition (2.30) that for every fixed r

$$\lim_{n\to\infty} \sum_{k=1}^{r-1} \alpha_{n,k}\beta_k = 0 \tag{2.35}$$

and

$$\lim_{n\to\infty} \sum_{k=1}^{r-1} \alpha_{n,k} = 0. \tag{2.36}$$

Using subsequently (2.35), (2.34), (2.31), and (2.36) we obtain

$$\begin{aligned}
\lim_{n\to\infty} \sum_{k=1}^{n} \alpha_{n,k}\beta_k &= \lim_{n\to\infty} \sum_{k=r}^{n} \alpha_{n,k}\beta_k \\
&= \lim_{n\to\infty} \sum_{k=r}^{n} \alpha_{n,k} - \lim_{n\to\infty} \sum_{k=r}^{n} \alpha_{n,k}\sigma_k \\
&= \lim_{n\to\infty} \sum_{k=1}^{n} \alpha_{n,k} - \lim_{n\to\infty} \sum_{k=r}^{n} \alpha_{n,k}\sigma_k \\
&= A - \lim_{n\to\infty} \sum_{k=r}^{n} \alpha_{n,k}\sigma_k.
\end{aligned}$$

Now, using (2.36) and (2.32), we can perform the following estimation:

$$\begin{aligned}\left|A - \lim_{n\to\infty}\sum_{k=1}^{n}\alpha_{n,k}\beta_k\right| &< \lim_{n\to\infty}\sum_{k=r}^{n}|\alpha_{n,k}|\cdot|\sigma_k| \\ &< \sigma^* \lim_{n\to\infty}\sum_{k=r}^{n}|\alpha_{n,k}| = \sigma^* \lim_{n\to\infty}\sum_{k=1}^{n}|\alpha_{n,k}| \\ &< \sigma^* K\end{aligned}$$

where $\sigma^* = \max_{k\geq r}|\sigma_k|$.

It follows from (2.34) that for each arbitrarily small $\epsilon > 0$ there exists r such that $\sigma^* < \epsilon/K$ and, therefore,

$$\left|A - \lim_{n\to\infty}\sum_{k=1}^{n}\alpha_{n,k}\beta_k\right| < \epsilon,$$

and the statement (2.33) of the theorem holds.

Theorem 2.1 has a simple consequence. Namely, if we take

$$\lim_{k\to\infty}\beta_k = B,$$

then

$$\lim_{n\to\infty}\sum_{k=1}^{n}\alpha_{n,k}\beta_k = AB. \tag{2.37}$$

Indeed, introducing the sequence

$$\tilde{\beta}_k = \frac{\beta_k}{B}, \qquad \lim_{k\to\infty}\tilde{\beta}_k = 1,$$

we can apply Theorem 2.1 to obtain

$$\lim_{n\to\infty}\sum_{k=1}^{n}\alpha_{n,k}\tilde{\beta}_k = \lim_{n\to\infty}\sum_{k=1}^{n}\alpha_{n,k}\frac{\beta_k}{B} = A,$$

from which the statement (2.37) follows.

To apply Theorem 2.1 for the evaluation of the limit (2.28), we write

$${}_aD_t^{-p}f(t) = \lim_{\substack{h\to 0\\ nh=t-a}} h^p \sum_{r=0}^{n}\left[\begin{matrix}p\\ r\end{matrix}\right] f(t-rh)$$

$$= \lim_{\substack{h\to 0\\ nh=t-a}} \sum_{r=0}^{n} \frac{1}{r^{p-1}} \begin{bmatrix} p \\ r \end{bmatrix} h(rh)^{p-1} f(t-rh)$$

$$= \frac{1}{\Gamma(p)} \lim_{\substack{h\to 0\\ nh=t-a}} \sum_{r=0}^{n} \frac{\Gamma(p)}{r^{p-1}} \begin{bmatrix} p \\ r \end{bmatrix} h(rh)^{p-1} f(t-rh)$$

$$= \frac{1}{\Gamma(p)} \lim_{n\to\infty} \sum_{r=0}^{n} \frac{\Gamma(p)}{r^{p-1}} \begin{bmatrix} p \\ r \end{bmatrix} \frac{t-a}{n} \left(r\frac{t-a}{n} \right)^{p-1} f(t - r\frac{t-a}{n})$$

and take

$$\beta_r = \frac{\Gamma(p)}{r^{p-1}} \begin{bmatrix} p \\ r \end{bmatrix},$$

$$\alpha_{n,r} = \frac{t-a}{n} \left(r\frac{t-a}{n} \right)^{p-1} f(t - r\frac{t-a}{n}).$$

Using (1.7) we have

$$\lim_{r\to\infty} \beta_r = \lim_{r\to\infty} \frac{\Gamma(p)}{r^{p-1}} \begin{bmatrix} p \\ r \end{bmatrix} = 1. \tag{2.38}$$

Obviously, if the function $f(t)$ is continuous in the closed interval $[a, t]$, then

$$\lim_{n\to\infty} \sum_{r=0}^{n} \alpha_{n,r} = \lim_{n\to\infty} \sum_{r=0}^{n} \frac{t-a}{n} \left(r\frac{t-a}{n} \right)^{p-1} f(t - r\frac{t-a}{n})$$

$$= \lim_{h\to 0} \sum_{r=0}^{n} h(rh)^{p-1} f(t-rh)$$

$$= \int_a^t (t-\tau)^{p-1} f(\tau) d\tau. \tag{2.39}$$

Taking into account (2.38) and (2.39) and applying Theorem 2.1 we conclude that

$${}_aD_t^{-p} f(t) = \lim_{\substack{h\to 0\\ nh=t-a}} h^p \sum_{r=0}^{n} \begin{bmatrix} p \\ r \end{bmatrix} f(t-rh) = \frac{1}{\Gamma(p)} \int_a^t (t-\tau)^{p-1} f(\tau) d\tau. \tag{2.40}$$

If the derivative $f'(t)$ is continuous in $[a, b]$, then integrating by parts we can write (2.40) in the form

$${}_aD_t^{-p} f(t) = \frac{f(a)(t-a)^p}{\Gamma(p+1)} + \frac{1}{\Gamma(p+1)} \int_a^t (t-\tau)^p f'(\tau) d\tau, \tag{2.41}$$

and if the function $f(t)$ has $m+1$ continuous derivatives, then

$$ {}_aD_t^{-p}f(t)=\sum_{k=0}^{m}\frac{f^{(k)}(a)(t-a)^{p+k}}{\Gamma(p+k+1)}+\frac{1}{\Gamma(p+k+1)}\int_a^t(t-\tau)^{p+m}f^{(m+1)}(\tau)d\tau. \tag{2.42} $$

The formula (2.42) immediately provides us with the asymptotics of ${}_aD_t^{-p}f(t)$ at $t=a$.

2.2.3 Derivatives of Arbitrary Order

Let us now consider the case of $p>0$. Our aim is, as above, to evaluate the limit

$$ {}_aD_t^{p}f(t)=\lim_{\substack{h\to 0\\ nh=t-a}} h^{-p}\sum_{r=0}^{n}(-1)^r\binom{p}{r}f(t-rh)=\lim_{\substack{h\to 0\\ nh=t-a}} f_h^{(p)}(t) \tag{2.43} $$

where

$$ f_h^{(p)}(t)=h^{-p}\sum_{r=0}^{n}(-1)^r\binom{p}{r}f(t-rh). \tag{2.44} $$

To evaluate the limit (2.43), let us first transform the expression for $f_h^{(p)}(t)$ in the following way.

Using the known property of the binomial coefficients

$$ \binom{p}{r}=\binom{p-1}{r}+\binom{p-1}{r-1} \tag{2.45} $$

we can write

$$ \begin{aligned} f_h^{(p)}(t) &= h^{-p}\sum_{r=0}^{n}(-1)^r\binom{p-1}{r}f(t-rh) \\ &\quad + h^{-p}\sum_{r=1}^{n}(-1)^r\binom{p-1}{r-1}f(t-rh) \\ &= h^{-p}\sum_{r=0}^{n}(-1)^r\binom{p-1}{r}f(t-rh) \\ &\quad + h^{-p}\sum_{r=0}^{n-1}(-1)^{r+1}\binom{p-1}{r}f(t-(r+1)h) \\ &= (-1)^n\binom{p-1}{n}h^{-p}f(a) \\ &\quad + h^{-p}\sum_{r=0}^{n-1}(-1)^r\binom{p-1}{r}\Delta f(t-rh), \end{aligned} \tag{2.46} $$

where we denote

$$\Delta f(t-rh) = f(t-rh) - f(t-(r+1)h).$$

Obviously, $\Delta f(t-rh)$ is a first-order backward difference of the function $f(\tau)$ at the point $\tau = t - rh$.

Applying the property (2.45) of the binomial coefficients repeatedly m times, we obtain starting from (2.46):

$$\begin{aligned}
f_h^{(p)}(t) &= (-1)^n \binom{p-1}{n} h^{-p} f(a) + (-1)^{n-1}\binom{p-2}{n-1} h^{-p}\Delta f(a+h) \\
&\quad + h^{-p}\sum_{r=0}^{n-2}(-1)^r \binom{p-2}{r}\Delta^2 f(t-rh) \\
&= (-1)^n \binom{p-1}{n} h^{-p} f(a) + (-1)^{n-1}\binom{p-2}{n-1} h^{-p}\Delta f(a+h) \\
&\quad + (-1)^{n-2}\binom{p-3}{n-3} h^{-p}\Delta^2 f(a+2h) \qquad (2.47) \\
&\quad + h^{-p}\sum_{r=0}^{n-3}(-1)^r \binom{p-3}{r}\Delta^3 f(t-rh) \\
&= \ldots \\
&= \sum_{k=0}^{m}(-1)^{n-k}\binom{p-k-1}{n-k} h^{-p}\Delta^k f(a+kh) \\
&\quad + h^{-p}\sum_{r=0}^{n-m-1}(-1)^r \binom{p-m-1}{r}\Delta^{m+1} f(t-rh). \qquad (2.48)
\end{aligned}$$

Let us evaluate the limit of the k-th term in the first sum in (2.48):

$$\begin{aligned}
&\lim_{\substack{h\to 0\\ nh=t-a}} (-1)^{n-k}\binom{p-k-1}{n-k} h^{-p}\Delta^k f(a+kh) \\
&= \lim_{\substack{h\to 0\\ nh=t-a}} (-1)^{n-k}\binom{p-k-1}{n-k}(n-k)^{p-k} \\
&\qquad \times \left(\frac{n}{n-k}\right)^{p-k}(nh)^{-p+k}\frac{\Delta^k f(a+kh)}{h^k} \\
&= (t-a)^{-p+k}\lim_{n\to\infty}(-1)^{n-k}\binom{p-k-1}{n-k}(n-k)^{p-k} \\
&\qquad \times \lim_{n\to\infty}\left(\frac{n}{n-k}\right)^{p-k} \times \lim_{h\to 0}\frac{\Delta^k f(a+kh)}{h^k}
\end{aligned}$$

$$= \frac{f^{(k)}(a)(t-a)^{-p+k}}{\Gamma(-p+k+1)}, \tag{2.49}$$

because using (1.7) gives

$$\lim_{n\to\infty} (-1)^{n-k} \binom{p-k-1}{n-k} (n-k)^{p-k}$$

$$= \lim_{n\to\infty} \frac{(-p+k+1)(-p+k+2)\dots(-p+n)}{(n-k)^{-p+k}(n-k)!} = \frac{1}{\Gamma(-p+k+1)}$$

and

$$\lim_{n\to\infty} \left(\frac{n}{n-k}\right)^{p-k} = 1,$$

$$\lim_{h\to 0} \frac{\Delta^k f(a+kh)}{h^k} = f^{(k)}(a).$$

Knowing the limit (2.49) we can easily write the limit of the first sum in (2.48).

To evaluate the limit of the second sum in (2.48) let us write it in the form

$$\frac{1}{\Gamma(-p+m+1)} \sum_{r=0}^{n-m-1} (-1)^r \Gamma(-p+m+1) \binom{p-m-1}{r} r^{-m+p}$$
$$\times h(rh)^{m-p} \frac{\Delta^{m+1} f(t-rh)}{h^{m+1}}. \tag{2.50}$$

To apply Theorem 2.1 we take

$$\beta_r = (-1)^r \Gamma(-p+m+1) \binom{p-m-1}{r} r^{-m+p},$$

$$\alpha_{n,r} = h(rh)^{m-p} \frac{\Delta^{m+1} f(t-rh)}{h^{m+1}}, \quad h = \frac{t-a}{n}.$$

Using (1.7) we verify that

$$\lim_{r\to\infty} \beta_r = \lim_{r\to\infty} (-1)^r \Gamma(-p+m+1) \binom{p-m-1}{r} r^{-m+p} = 1. \tag{2.51}$$

In addition, if $m-p>-1$, then

$$\lim_{n\to\infty} \sum_{r=0}^{n-m-1} \alpha_{n,r} = \lim_{\substack{h\to 0\\ nh=t-a}} \sum_{r=0}^{n-m-1} h(rh)^{m-p} \frac{\Delta^{m+1} f(t-rh)}{h^{m+1}}$$

$$= \int_a^t (t-\tau)^{m-p} f^{(m+1)}(\tau) d\tau. \tag{2.52}$$

Taking into account (2.51) and (2.52) and applying Theorem 2.1 we conclude that

$$\lim_{\substack{h\to 0\\ nh=t-a}} h^{-p} \sum_{r=0}^{n-m-1} (-1)^r \binom{p-m-1}{r} \Delta^{m+1} f(t-rh)$$
$$= \frac{1}{\Gamma(-p+m+1)} \int_a^t (t-\tau)^{m-p} f^{(m+1)}(\tau) d\tau. \quad (2.53)$$

Using (2.49) and (2.53) we finally obtain the limit (2.43):

$$\begin{aligned} {}_aD_t^p f(t) &= \lim_{\substack{h\to\infty\\ nh=t-a}} f_h^{(p)}(t) \\ &= \sum_{k=0}^{m} \frac{f^{(k)}(a)(t-a)^{-p+k}}{\Gamma(-p+k+1)} \\ &\quad + \frac{1}{\Gamma(-p+m+1)} \int_a^t (t-\tau)^{m-p} f^{(m+1)}(\tau) d\tau. \quad (2.54) \end{aligned}$$

The formula (2.54) has been obtained under the assumption that the derivatives $f^{(k)}(t)$, $(k = 1, 2, \ldots, m+1)$ are continuous in the closed interval $[a, t]$ and that m is an integer number satisfying the condition $m > p - 1$. The smallest possible value for m is determined by the inequality

$$m < p < m + 1.$$

2.2.4 Fractional Derivative of $(t-a)^\beta$

Let us evaluate the Grünwald–Letnikov fractional derivative ${}_aD_t^p f(t)$ of the power function

$$f(t) = (t-a)^\nu,$$

where ν is a real number.

We will start by considering negative values of p, which means that we will start with the evaluation of the fractional integral of order $-p$. Let us use the formula (2.40):

$${}_aD_t^p (t-a)^\nu = \frac{1}{\Gamma(-p)} \int_a^t (t-\tau)^{-p-1} (\tau-a)^\nu d\tau, \quad (2.55)$$

and suppose $\nu > -1$ for the convergence of the integral. Performing in (2.55) the substitution $\tau = a + \xi(t-a)$ and then using the definition of the beta function (1.20) we obtain:

$$
\begin{aligned}
{}_aD_t^p(t-a)^\nu &= \frac{1}{\Gamma(-p)}(t-a)^{\nu-p}\int_0^1 \xi^\nu(1-\xi)^{-p-1}d\xi \\
&= \frac{1}{\Gamma(-p)}B(-p,\nu+1)(t-a)^{\nu-p} \\
&= \frac{\Gamma(\nu+1)}{\Gamma(\nu-p+1)}(t-a)^{\nu-p}, \quad (p<0,\ \nu>-1). \qquad (2.56)
\end{aligned}
$$

Now let us consider the case $0 \le m \le p < m+1$. To apply the formula (2.54), we must require $\nu > m$ for the convergence of the integral in (2.54). Then we have:

$$
{}_aD_t^p(t-a)^\nu = \frac{1}{\Gamma(-p+m+1)}\int_a^t (t-\tau)^{m-p}\frac{d^{m+1}(\tau-a)^\nu}{d\tau^{m+1}}d\tau, \qquad (2.57)
$$

because all non-integral addends are equal to 0.

Taking into account that

$$
\frac{d^{m+1}(\tau-a)^\nu}{d\tau^{m+1}} = \nu(\nu-1)\ldots(\nu-m)(\tau-a)^{\nu-m-1} = \frac{\Gamma(\nu+1)}{\nu-m}(\tau-a)^{\nu-m-1}
$$

and performing the substitution $\tau = a + \xi(t-a)$ we obtain:

$$
\begin{aligned}
{}_aD_t^p(t-a)^\nu &= \frac{\Gamma(\nu+1)}{\Gamma(\nu-m)\Gamma(-p+m+1)}\int_a^t (t-\tau)^{m-p}(\tau-a)^{\nu-m-1}d\tau \\
&= \frac{\Gamma(\nu+1)B(-p+m+1,\nu-m)}{\Gamma(\nu-m)\Gamma(-p+m+1)}(t-a)^{\nu-p} \\
&= \frac{\Gamma(\nu+1)}{\Gamma(-p+\nu+1)}(t-a)^{\nu-p}. \qquad (2.58)
\end{aligned}
$$

Noting that the expression (2.58) is formally identical to the expression (2.56) we can conclude that the Grünwald–Letnikov fractional derivative of the power function $f(t) = (t-a)^\nu$ is given by the formula

$$
{}_aD_t^p(t-a)^\nu = \frac{\Gamma(\nu+1)}{\Gamma(-p+\nu+1)}(t-a)^{\nu-p}, \qquad (2.59)
$$

$$(p < 0, \quad \nu > -1) \quad \text{or} \quad (0 \leq m \leq p < m+1, \quad \nu > m).$$

We will return to formula (2.59) for the Grünwald–Letnikov fractional derivative of the power function later, when we consider some other approaches to fractional differentiation. The formula will be the same, but the conditions for its applicability will be different.

From the theoretical point of view, the class of functions for which the considered Grünwald–Letnikov definition of the fractional derivative is defined ($(m+1)$-times continuously differentiable functions) is very narrow. However, in most applied problems describing continuous physical, chemical and other processes we deal with such very smooth functions.

2.2.5 Composition with Integer-order Derivatives

Noting that we have only one restriction for m in the formula (2.54), namely the condition $m > p-1$, let us write s instead of m and rewrite (2.54) as

$$
\begin{aligned}
{}_aD_t^p f(t) = &\sum_{k=0}^{s} \frac{f^{(k)}(a)(t-a)^{-p+k}}{\Gamma(-p+k+1)} \\
&+ \frac{1}{\Gamma(-p+s+1)} \int_a^t (t-\tau)^{s-p} f^{(s+1)}(\tau)d\tau. \qquad (2.60)
\end{aligned}
$$

In what follows we assume that $m < p < m+1$.

Let us evaluate the derivative of integer order n of the fractional derivative of fractional order p in the form (2.60), where we take $s \geq m+n-1$. The result is:

$$
\begin{aligned}
\frac{d^n}{dt^n}\left({}_aD_t^p f(t)\right) = &\sum_{k=0}^{s} \frac{f^{(k)}(a)(t-a)^{-p-n+k}}{\Gamma(-p-n+k+1)} \\
&+ \frac{1}{\Gamma(-p-n+s+1)} \int_a^t (t-\tau)^{s-p-n} f^{(s+1)}(\tau)d\tau \qquad (2.61)\\
= &\, {}_aD_t^{p+n} f(t). \qquad (2.62)
\end{aligned}
$$

Since $s \geq m+n-1$ is arbitrary, let us take $s = m+n-1$. This gives:

$$\frac{d^n}{dt^n}\left({}_aD_t^p f(t)\right) = {}_aD_t^{p+n} f(t)$$

$$= \sum_{k=0}^{m+n-1} \frac{f^{(k)}(a)(t-a)^{-p-n+k}}{\Gamma(-p-n+k+1)}$$

$$+ \frac{1}{\Gamma(m-p)} \int_a^t (t-\tau)^{m-p-1} f^{(m+n)}(\tau) d\tau \quad (2.63)$$

Let us now consider the reverse order of operations and evaluate the fractional derivative of order p of an integer-order derivative $\frac{d^n f(t)}{dt^n}$. Using the formula (2.60) we obtain:

$${}_aD_t^p \left(\frac{d^n f(t)}{dt^n} \right) = \sum_{k=0}^{s} \frac{f^{(n+k)}(a)(t-a)^{-p+k}}{\Gamma(-p+k+1)}$$

$$+ \frac{1}{\Gamma(-p+s+1)} \int_a^t (t-\tau)^{s-p} f^{(n+s+1)}(\tau) d\tau. \quad (2.64)$$

Putting here $s = m-1$ we obtain:

$${}_aD_t^p \left(\frac{d^n f(t)}{dt^n} \right) = \sum_{k=0}^{m-1} \frac{f^{(n+k)}(a)(t-a)^{-p+k}}{\Gamma(-p+k+1)}$$

$$+ \frac{1}{\Gamma(m-p)} \int_a^t (t-\tau)^{m-p-1} f^{(m+n)}(\tau) d\tau, \quad (2.65)$$

and comparing (2.63) and (2.65) we arrive at the conclusion that

$$\frac{d^n}{dt^n} \left({}_aD_t^p f(t) \right) = {}_aD_t^p \left(\frac{d^n f(t)}{dt^n} \right) + \sum_{k=0}^{n-1} \frac{f^{(k)}(a)(t-a)^{-p-n+k}}{\Gamma(-p-n+k+1)}. \quad (2.66)$$

The relationship (2.66) says that the operations $\frac{d^n}{dt^n}$ and ${}_aD_t^p$ are commutative, i.e., that

$$\frac{d^n}{dt^n} \left({}_aD_t^p f(t) \right) = {}_aD_t^p \left(\frac{d^n f(t)}{dt^n} \right) = {}_aD_t^{p+n} f(t), \quad (2.67)$$

only if at the lower terminal $t = a$ of the fractional differentiation we have

$$f^{(k)}(a) = 0, \qquad (k = 0, 1, 2, \ldots, n-1). \quad (2.68)$$

2.2.6 Composition with Fractional Derivatives

Let us now consider the fractional derivative of order q of a fractional derivative of order p:

$$ {}_aD_t^q\Big({}_aD_t^p f(t)\Big). $$

Two cases will be considered separately: $p < 0$ and $p > 0$. The first case means that — depending on the sign of q — differentiation of order $q > 0$ or integration of order $-q > 0$ is applied to the fractional integral of order $-p > 0$. In the second case, the object of the outer operation is the fractional derivative of order $p > 0$.

In both cases we will obtain an analogue of the well-known property of integer-order differentiation:

$$ \frac{d^n}{dt^n}\left(\frac{d^m f(t)}{dt^m}\right) = \frac{d^m}{dt^m}\left(\frac{d^n f(t)}{dt^n}\right) = \frac{d^{m+n} f(t)}{dt^{m+n}}. $$

Case $p < 0$

Let us first take $q < 0$. Then we have:

$$ \begin{aligned} {}_aD_t^q\Big({}_aD_t^p f(t)\Big) &= \frac{1}{\Gamma(-q)} \int_a^t (t-\tau)^{-q-1}\Big({}_aD_\tau^p f(\tau)\Big)d\tau \\ &= \frac{1}{\Gamma(-q)\Gamma(-p)} \int_a^t (t-\tau)^{-q-1}d\tau \int_a^\tau (\tau-\xi)^{-q-1} f(\xi)d\xi \\ &= \frac{1}{\Gamma(-q)\Gamma(-p)} \int_a^t f(\xi)d\xi \int_\xi^t (t-\tau)^{-q-1}(\tau-\xi)^{-p-1}d\tau \\ &= \frac{1}{\Gamma(-p-q)} \int_a^t (t-\xi)^{-p-q-1} f(\xi)d\xi \\ &= {}_aD_t^{p+q} f(t), \end{aligned} \qquad (2.69) $$

where the integral

$$ \begin{aligned} \int_\xi^t (t-\tau)^{-q-1}(\tau-\xi)^{-p-1}d\tau &= (t-\xi)^{-p-q-1} \int_0^1 (1-z)^{-q-1} z^{-p-1}dz \\ &= \frac{\Gamma(-q)\Gamma(-p)}{\Gamma(-p-q)}(t-\xi)^{-p-q-1} \end{aligned} $$

is evaluated with the help of the substitution $\tau = \xi + z(t - \xi)$ and the definition of the beta function (1.20).

Let us now suppose that $0 < n < q < n + 1$. Noting that $q = (n+1) + (q - n - 1)$, where $q - n - 1 < 0$, and using the formulas (2.62) and (2.69) we obtain:

$$
\begin{aligned}
{}_aD_t^q\Big({}_aD_t^p f(t)\Big) &= \frac{d^{n+1}}{dt^{n+1}}\Big\{{}_aD_t^{q-n-1}\big({}_aD_t^p f(t)\big)\Big\} \\
&= \frac{d^{n+1}}{dt^{n+1}}\Big\{{}_aD_t^{p+q-n-1} f(t)\Big\} \\
&= {}_aD_t^{p+q} f(t). \qquad (2.70)
\end{aligned}
$$

Combining (2.69) and (2.70) we conclude that if $p < 0$, then for any real q

$$
{}_aD_t^q\Big({}_aD_t^p f(t)\Big) = {}_aD_t^{p+q} f(t).
$$

Case $p > 0$

Let us assume that $0 \leq m < p < m + 1$. Then, according to formula (2.54), we have

$$
\begin{aligned}
{}_aD_t^p f(t) &= \lim_{\substack{h \to \infty \\ nh = t-a}} f_h^{(p)}(t) \\
&= \sum_{k=0}^{m} \frac{f^{(k)}(a)(t-a)^{-p+k}}{\Gamma(-p+k+1)} \\
&\quad + \frac{1}{\Gamma(-p+m+1)} \int_a^t (t-\tau)^{m-p} f^{(m+1)}(\tau) d\tau. \qquad (2.71)
\end{aligned}
$$

Let us take $q < 0$ and evaluate

$$
{}_aD_t^q\Big({}_aD_t^p f(t)\Big).
$$

Examining the right-hand side of (2.71) we see that the functions $(t-a)^{-p+k}$ have non-integrable singularities for $k = 0, 1, \ldots, m-1$. Therefore, the derivative of real order q of ${}_aD_t^p f(t)$ exists only if

$$
f^{(k)}(a) = 0, \qquad (k = 0, 1, \ldots, m-1). \qquad (2.72)
$$

The integral in the right-hand side of (2.71) is equal to ${}_aD_t^{p-m-1}f(t)$ (the fractional integral of order $-p+m+1$ of the function $f(t)$). Therefore, under the conditions (2.72) the representation (2.71) of the p-th derivative of $f(t)$ takes the following form:

$$ {}_aD_t^p f(t) = \frac{f^{(m)}(a)(t-a)^{-p+m}}{\Gamma(-p+m+1)} + {}_aD_t^{p-m-1} f^{(m+1)}(t). \tag{2.73} $$

Now we can find the derivative of order $q < 0$ (in other words, the integral of order $-q > 0$) of the derivative of order p given by (2.73):

$$ {}_aD_t^q \Big({}_aD_t^p f(t) \Big) = \frac{f^{(m)}(a)(t-a)^{-p-q+m}}{\Gamma(-p-q+m+1)} + \frac{1}{\Gamma(-p-q+m+1)} \int\limits_a^t \frac{f^{(m+1)}(\tau)d\tau}{(t-\tau)^{p+q-m}}, \tag{2.74} $$

because

$$ \begin{aligned} {}_aD_t^q \Big({}_aD_t^{p-m-1} f^{(m+1)}(t) \Big) &= {}_aD_t^{p+q-m-1} f^{(m+1)}(t) \\ &= \frac{1}{\Gamma(-p-q+m+1)} \int\limits_a^t \frac{f^{(m+1)}(\tau)d\tau}{(t-\tau)^{p+q-m}}. \end{aligned} $$

Taking into account the conditions (2.72) and the formula (2.71) we arrive at

$$ {}_aD_t^q \Big({}_aD_t^p f(t) \Big) = {}_aD_t^{p+q} f(t). \tag{2.75} $$

Let us now take $0 \le n < q < n+1$. Assuming that $f(t)$ satisfies the conditions (2.72) and taking into account that $q-n-1 < 0$ and, therefore, the formula (2.75) can be used, we obtain:

$$ \begin{aligned} {}_aD_t^q \Big({}_aD_t^p f(t) \Big) &= \frac{d^{n+1}}{dt^{n+1}} \Big\{ {}_aD_t^{q-n-1} \Big({}_aD_t^p f(t) \Big) \Big\} \\ &= \frac{d^{n+1}}{dt^{n+1}} \Big\{ {}_aD_t^{p+1-n-1} f(t) \Big\} \\ &= {}_aD_t^{p+q} f(t), \end{aligned} \tag{2.76} $$

which is the same as (2.75).

Therefore, we can conclude that if $p < 0$, then the relationship (2.75) holds for arbitrary real q; if $0 \le m < p < m+1$, then the relationship

(2.75) holds also for arbitrary real q, if the function $f(t)$ satisfies the conditions (2.72).

Moreover, if $0 \le m < p < m+1$ and $0 \le n < q < n+1$ and the function $f(t)$ satisfies the conditions

$$f^{(k)}(a) = 0, \quad (k = 0, 1, \ldots, r-1), \tag{2.77}$$

where $r = \max(n, m)$, then the operators of fractional differentiation ${}_aD_t^p$ and ${}_aD_t^q$ commute:

$${}_aD_t^q\Big({}_aD_t^p f(t)\Big) = {}_aD_t^p\Big({}_aD_t^q f(t)\Big) = {}_aD_t^{p+q} f(t). \tag{2.78}$$

2.3 Riemann–Liouville Fractional Derivatives

Manipulation with the Grünwald–Letnikov fractional derivatives defined as a limit of a fractional-order backward difference is not convenient. The obtained expression (2.54) looks better because of the presense of the integral in it; but what about the non-integral terms? The answer is simple and elegant: to consider the expression (2.54) as a particular case of the integro-differential expression

$${}_a\mathbf{D}_t^p f(t) = \left(\frac{d}{dt}\right)^{m+1} \int_a^t (t-\tau)^{m-p} f(\tau)d\tau, \qquad (m \le p < m+1). \tag{2.79}$$

The expression (2.79) it is the most widely known definition of the fractional derivative; it is usually called the Riemann–Liouville definition.

Obviously, the expression (2.54), which has been obtained for the Grünwald–Letnikov fractional derivative under the assumption that the function $f(t)$ must be $m+1$ times continuously differentiable, can be obtained from (2.79) *under the same assumption* by performing repeatedly integration by parts and differentiation. This gives

$$\begin{aligned} {}_a\mathbf{D}_t^p f(t) &= \left(\frac{d}{dt}\right)^{m+1} \int_a^t (t-\tau)^{m-p} f(\tau)d\tau \\ &= \sum_{k=0}^{m} \frac{f^{(k)}(a)(t-a)^{-p+k}}{\Gamma(-p+k+1)} \end{aligned}$$

$$+ \frac{1}{\Gamma(-p+m+1)} \int_a^t (t-\tau)^{m-p} f^{(m+1)}(\tau) d\tau$$
$$= {}_aD_t^p f(t), \qquad (m \le p < m+1). \tag{2.80}$$

Therefore, if we consider a class of functions $f(t)$ having $m+1$ continuous derivatives for $t \ge 0$, then the Grünwald–Letnikov definition (2.43) (or, what is in this case the same, its integral form (2.54)) is equivalent to the Riemann–Liouville definition (2.79).

From the pure mathematical point of view such a class of functions is narrow; however, this class of functions is very important for applications, because the character of the majority of dynamical processes is smooth enough and does not allow discontinuities. Understanding this fact is important for the proper use of the methods of the fractional calculus in applications, especially because of the fact that the Riemann–Liouville definition (2.79) provides an excellent opportunity to weaken the conditions on the function $f(t)$. Namely, it is enough to require the integrability of $f(t)$; then the integral (2.79) exists for $t > a$ and can be differentiated $m+1$ times. The weak conditions on the function $f(t)$ in (2.79) are necessary, for example, for obtaining the solution of the Abel integral equation.

Let us look at how the Riemann–Liouville definition (2.79) appears as the result of the unification of the notions of integer-order integration and differentiation.

2.3.1 Unification of Integer-order Derivatives and Integrals

Let us suppose that the function $f(\tau)$ is continuous and integrable in every finite interval (a, t); the function $f(t)$ may have an integrable singularity of order $r < 1$ at the point $\tau = a$:

$$\lim_{\tau \to a} (\tau - a)^r f(t) = \text{const} \, (\neq 0).$$

Then the integral

$$f^{(-1)}(t) = \int_a^t f(\tau) d\tau \tag{2.81}$$

exists and has a finite value, namely equal to 0, as $t \to a$. Indeed, performing the substitution $\tau = a + y(t-a)$ and then denoting $\epsilon = t - a$, we obtain

$$\begin{aligned}\lim_{t\to a} f^{(-1)}(t) &= \lim_{t\to a} \int_a^t f(\tau)d\tau \\ &= \lim_{t\to a} (t-a) \int_0^1 f(a+y(t-a))dy \\ &= \lim_{\epsilon\to 0} \epsilon^{1-r} \int_0^1 (\epsilon y)^r f(a+y\epsilon) y^{-r} dy = 0, \qquad (2.82)\end{aligned}$$

because $r < 1$. Therefore, we can consider the two-fold integral

$$\begin{aligned}f^{(-2)}(t) &= \int_a^t d\tau_1 \int_a^{\tau_1} f(\tau)d\tau = \int_a^t f(\tau)d\tau \int_\tau^t d\tau_1 \\ &= \int_a^t (t-\tau) f(\tau)d\tau. \qquad (2.83)\end{aligned}$$

Integration of (2.83) gives the three-fold integral of $f(\tau)$:

$$\begin{aligned}f^{(-3)}(t) &= \int_a^t d\tau_1 \int_a^{\tau_1} d\tau_2 \int_a^{\tau_2} f(\tau_3)d\tau_3 \\ &= \int_a^t d\tau_1 \int_a^{\tau_1} (\tau_1 - \tau) f(\tau)d\tau \\ &= \frac{1}{2} \int_a^t (t-\tau)^2 f(\tau)d\tau, \qquad (2.84)\end{aligned}$$

and by induction in the general case we have the Cauchy formula

$$f^{(-n)}(t) = \frac{1}{\Gamma(n)} \int_a^t (t-\tau)^{n-1} f(\tau)d\tau. \qquad (2.85)$$

Let us now suppose that $n \geq 1$ is fixed and take integer $k \geq 0$. Obviously, we will obtain

$$f^{(-k-n)}(t) = \frac{1}{\Gamma(n)} D^{-k} \int_a^t (t-\tau)^{n-1} f(\tau)d\tau, \qquad (2.86)$$

where the symbol D^{-k} $(k \geq 0)$ denotes k iterated integrations.

On the other hand, for a fixed $n \geq 1$ and integer $k \geq n$ the $(k-n)$-th derivative of the function $f(t)$ can be written as

$$f^{(k-n)}(t) = \frac{1}{\Gamma(n)} D^k \int_a^t (t-\tau)^{n-1} f(\tau) d\tau, \tag{2.87}$$

where the symbol D^k $(k \geq 0)$ denotes k iterated differentiations.

We see that the formulas (2.86) and (2.87) can be considered as particular cases of one of them, namely (2.87), in which n $(n \geq 1)$ is fixed and the symbol D^k means k integrations if $k \leq 0$ and k differentiations if $k > 0$. If $k = n-1,\ n-2,\ \ldots$, then the formula (2.87) gives iterated integrals of $f(t)$; for $k = n$ it gives the function $f(t)$; for $k = n+1,\ n+2,\ n+3,\ \ldots$ it gives derivatives of order $k - n = 1,\ 2,\ 3,\ \ldots$ of the function $f(t)$.

2.3.2 Integrals of Arbitrary Order

To extend the notion of n-fold integration to non-integer values of n, we can start with the Cauchy formula (2.85) and replace the integer n in it by a real $p > 0$:

$$ {}_a\mathbf{D}_t^{-p} f(t) = \frac{1}{\Gamma(p)} \int_a^t (t-\tau)^{p-1} f(\tau) d\tau. \tag{2.88}$$

In (2.85) the integer n must satisfy the condition $n \geq 1$; the corresponding condition for p is weaker: for the existence of the integral (2.88) we must have $p > 0$.

Moreover, under certain reasonable assumptions

$$\lim_{p \to 0} {}_a\mathbf{D}_t^{-p} f(t) = f(t), \tag{2.89}$$

so we can put

$$ {}_a\mathbf{D}_t^{0} f(t) = f(t). \tag{2.90}$$

The proof of the relationship (2.89) is very simple if $f(t)$ has continuous derivatives for $t \geq 0$. In such a case, integration by parts and the use of (1.3) gives

$$ {}_a\mathbf{D}_t^{-p} f(t) = \frac{(t-a)^p f(a)}{\Gamma(p+1)} + \frac{1}{\Gamma(p+1)} \int_a^t (t-\tau)^p f'(\tau) d\tau,$$

and we obtain

$$\lim_{p\to 0} {}_a\mathbf{D}_t^{-p} f(t) = f(a) + \int_a^t f'(\tau)d\tau = f(a) + \Big(f(t) - f(a)\Big) = f(t).$$

If $f(t)$ is only continuous for $t \geq a$, then the proof of (2.89) is somewhat longer. In such a case, let us write ${}_a\mathbf{D}_t^{-p} f(t)$ in the form:

$$\begin{aligned} {}_a\mathbf{D}_t^{-p} f(t) &= \frac{1}{\Gamma(p)} \int_a^t (t-\tau)^{p-1} (f(\tau) - f(t))\, d\tau + \frac{f(t)}{\Gamma(p)} \int_a^t (t-\tau)^{p-1} d\tau \\ &= \frac{1}{\Gamma(p)} \int_a^{t-\delta} (t-\tau)^{p-1} (f(\tau) - f(t))\, d\tau \qquad (2.91) \\ &\quad + \frac{1}{\Gamma(p)} \int_{t-\delta}^t (t-\tau)^{p-1} (f(\tau) - f(t))\, d\tau \qquad (2.92) \\ &\quad + \frac{f(t)(t-a)^p}{\Gamma(p+1)}. \qquad (2.93) \end{aligned}$$

Let us consider the integral (2.92). Since $f(t)$ is continuous, for every $\delta > 0$ there exists $\epsilon > 0$ such that

$$|f(\tau) - f(t)| < \epsilon.$$

Then we have the following estimate of the integral (2.92):

$$|I_2| < \frac{\epsilon}{\Gamma(p)} \int_{t-\delta}^t (t-\tau)^{p-1} d\tau < \frac{\epsilon \delta^p}{\Gamma(p+1)}, \qquad (2.94)$$

and taking into account that $\epsilon \to 0$ as $\delta \to 0$ we obtain that for all $p \geq 0$

$$\lim_{\delta \to 0} |I_2| = 0. \qquad (2.95)$$

Let us now take an arbitrary $\epsilon > 0$ and choose δ such that

$$|I_2| < \epsilon \qquad (2.96)$$

for all $p \geq 0$. For this fixed δ we obtain the following estimate of the integral (2.91):

$$|I_1| \leq \frac{M}{\Gamma(p)} \int_a^{t-\delta} (t-\tau)^{p-1} d\tau \leq \frac{M}{\Gamma(p+1)} \Big(\delta^p - (t-a)^p\Big), \qquad (2.97)$$

from which it follows that for fixed $\delta > 0$

$$\lim_{p\to 0} |I_1| = 0. \tag{2.98}$$

Considering

$$\left| {}_a\mathbf{D}_t^{-p} f(t) - f(t) \right| \le |I_1| + |I_2| + |f(t)| \cdot \left| \frac{(t-a)^p}{\Gamma(p+1)} - 1 \right|$$

and taking into account the limits (2.95) and (2.95) and the estimate (2.96) we obtain

$$\limsup_{p\to 0} \left| {}_a\mathbf{D}_t^{-p} f(t) - f(t) \right| \le \epsilon,$$

where ϵ can be chosen as small as we wish. Therefore,

$$\limsup_{p\to 0} \left| {}_a\mathbf{D}_t^{-p} f(t) - f(t) \right| = 0,$$

and (2.89) holds if $f(t)$ is continuous for $t \ge a$.

If $f(t)$ is continuous for $t \ge a$, then integration of arbitrary real order defined by (2.88) has the following important property:

$${}_a\mathbf{D}_t^{-p} \left({}_a\mathbf{D}_t^{-q} f(t) \right) = {}_a\mathbf{D}_t^{-p-q} f(t). \tag{2.99}$$

Indeed, we have

$$\begin{aligned}
{}_a\mathbf{D}_t^{-p} \left({}_a\mathbf{D}_t^{-q} f(t) \right) &= \frac{1}{\Gamma(q)} \int_a^t (t-\tau)^{q-1} \, {}_a\mathbf{D}_\tau^{-p} f(\tau) d\tau \\
&= \frac{1}{\Gamma(p)\Gamma(q)} \int_a^t (t-\tau)^{q-1} d\tau \int_a^\tau (\tau-\xi)^{p-1} f(\xi) d\xi \\
&= \frac{1}{\Gamma(p)\Gamma(q)} \int_a^t f(\xi) d\xi \int_\xi^t (t-\tau)^{q-1} (\tau-\xi)^{p-1} d\tau \\
&= \frac{1}{\Gamma(p+q)} \int_a^t (t-\xi)^{p+q-1} f(\xi) d\xi \\
&= {}_a\mathbf{D}_t^{-p-q} f(t).
\end{aligned}$$

(For the evaluation of the integral from ξ to t we used the substitution $\tau = \xi + \zeta(t - \xi)$ allowing us to express it in terms of the beta function (1.20).)

Obviously, we can interchange p and q, so we have

$$ {}_a\mathbf{D}_t^{-p}\Big({}_a\mathbf{D}_t^{-q} f(t)\Big) = {}_a\mathbf{D}_t^{-q}\Big({}_a\mathbf{D}_t^{-p} f(t)\Big) = {}_a\mathbf{D}_t^{-p-q} f(t). \quad (2.100) $$

One may note that the rule (2.100) is similar to the well-known property of integer-order derivatives:

$$ \frac{d^m}{dt^m}\left(\frac{d^n f(t)}{dt^n}\right) = \frac{d^n}{dt^n}\left(\frac{d^m f(t)}{dt^m}\right) = \frac{d^{m+n} f(t)}{dt^{m+n}}. \quad (2.101) $$

2.3.3 Derivatives of Arbitrary Order

The representation (2.87) for the derivative of an integer order $k - n$ provides an opportunity for extending the notion of differentiation to non-integer order. Namely, we can leave integer k and replace integer n with a real α so that $k - \alpha > 0$. This gives

$$ {}_a\mathbf{D}_t^{k-\alpha} f(t) = \frac{1}{\Gamma(\alpha)} \frac{d^k}{dt^k} \int_a^t (t-\tau)^{\alpha-1} f(\tau)d\tau, \quad (0 < \alpha \leq 1), \quad (2.102) $$

where the only substantial restriction for α is $\alpha > 0$, which is necessary for the convergence of the integral in (2.102). This restriction, however, can be — without loss of generality — replaced with the narrower condition $0 < \alpha \leq 1$; this can be easily shown with the help of the property (2.100) of the integrals of arbitrary real order and the definition (2.102).

Denoting $p = k - \alpha$ we can write (2.102) as

$$ {}_a\mathbf{D}_t^{p} f(t) = \frac{1}{\Gamma(k-p)} \frac{d^k}{dt^k} \int_a^t (t-\tau)^{k-p-1} f(\tau)d\tau, \quad (k-1 \leq p < k) \quad (2.103) $$

or

$$ {}_a\mathbf{D}_t^{p} f(t) = \frac{d^k}{dt^k}\Big({}_a\mathbf{D}_t^{-(k-p)} f(t)\Big), \quad (k-1 \leq p < k). \quad (2.104) $$

If $p = k - 1$, then we obtain a conventional integer-order derivative of order $k - 1$:

$$ \begin{aligned} {}_a\mathbf{D}_t^{k-1} f(t) &= \frac{d^k}{dt^k}\Big({}_a\mathbf{D}_t^{-(k-(k-1))} f(t)\Big) \\ &= \frac{d^k}{dt^k}\Big({}_a\mathbf{D}_t^{-1} f(t)\Big) = f^{(k-1)}(t). \end{aligned} $$

Moreover, using (2.90) we see that for $p = k \geq 1$ and $t > a$

$$ {}_a\mathbf{D}_t^p f(t) = \frac{d^k}{dt^k}\Big({}_a\mathbf{D}_t^0 f(t)\Big) = \frac{d^k f(t)}{dt^k} = f^{(k)}(t), \qquad (2.105) $$

which means that for $t > a$ the Riemann–Liouville fractional derivative (2.103) of order $p = k > 1$ coincides with the conventional derivative of order k.

Let us now consider some properties of the Riemann–Liouville fractional derivatives.

The first — and maybe the most important — property of the Riemann–Liouville fractional derivative is that for $p > 0$ and $t > a$

$$ {}_a\mathbf{D}_t^p\Big({}_a\mathbf{D}_t^{-p} f(t)\Big) = f(t), \qquad (2.106) $$

which means that the Riemann–Liouville fractional differentiation operator is a left inverse to the Riemann–Liouville fractional integration operator of the same order p.

To prove the property (2.106), let us consider the case of integer $p = n \geq 1$:

$$ \begin{aligned} {}_a\mathbf{D}_t^n\Big({}_a\mathbf{D}_t^{-n} f(t)\Big) &= \frac{d^n}{dt^n}\int_a^t (t-\tau)^{n-1} f(\tau)d\tau \\ &= \frac{d}{dt}\int_a^t f(\tau)d\tau = f(t). \end{aligned} $$

Taking now $k - 1 \leq p < k$ and using the composition rule (2.100) for the Riemann–Liouville fractional integrals, we can write

$$ {}_a\mathbf{D}_t^{-k} f(t) = {}_a\mathbf{D}_t^{-(k-p)}\Big({}_a\mathbf{D}_t^{-p} f(t)\Big), \qquad (2.107) $$

and, therefore,

$$ \begin{aligned} {}_a\mathbf{D}_t^p\Big({}_a\mathbf{D}_t^{-p} f(t)\Big) &= \frac{d^k}{dt^k}\Big\{ {}_a\mathbf{D}_t^{-(k-p)}\Big({}_a\mathbf{D}_t^{-p} f(t)\Big)\Big\} \\ &= \frac{d^k}{dt^k}\Big\{ {}_a\mathbf{D}_t^{-p} f(t)\Big\} = f(t), \end{aligned} $$

which ends the proof of the property (2.106).

As with conventional integer-order differentiation and integration, fractional differentiation and integration do not commute.

If the fractional derivative ${}_a\mathbf{D}_t^p f(t)$, $(k-1 \le p < k)$, of a function $f(t)$ is integrable, then

$$ {}_a\mathbf{D}_t^{-p}\Big({}_a\mathbf{D}_t^p f(t)\Big) = f(t) - \sum_{j=1}^{k}\Big[{}_a\mathbf{D}_t^{p-j} f(t)\Big]_{t=a} \frac{(t-a)^{p-j}}{\Gamma(p-j+1)}. \tag{2.108} $$

Indeed, on the one hand we have

$$ \begin{aligned} {}_a\mathbf{D}_t^{-p}\Big({}_a\mathbf{D}_t^p f(t)\Big) &= \frac{1}{\Gamma(p)}\int_a^t (t-\tau)^{p-1}\, {}_a\mathbf{D}_\tau^p f(\tau)d\tau \\ &= \frac{d}{dt}\left\{\frac{1}{\Gamma(p+1)}\int_a^t (t-\tau)^{p}\, {}_a\mathbf{D}_\tau^p f(\tau)d\tau\right\}. \end{aligned} \tag{2.109} $$

On the other hand, repeatedly integrating by parts and then using (2.100) we obtain

$$ \begin{aligned} &\frac{1}{\Gamma(p+1)}\int_a^t (t-\tau)^{p}\, {}_a\mathbf{D}_\tau^p f(\tau)d\tau \\ &= \frac{1}{\Gamma(p+1)}\int_a^t (t-\tau)^{p}\frac{d^k}{d\tau^k}\left\{{}_a\mathbf{D}_\tau^{-(k-p)} f(\tau)\right\}d\tau \\ &= \frac{1}{\Gamma(p-k+1)}\int_a^t (t-\tau)^{p-k}\left\{{}_a\mathbf{D}_\tau^{-(k-p)} f(\tau)\right\}d\tau \\ &\quad -\sum_{j=1}^{k}\left[\frac{d^{k-j}}{dt^{k-j}}\Big({}_a\mathbf{D}_t^{-(k-p)} f(t)\Big)\right]_{t=a}\frac{(t-a)^{p-j+1}}{\Gamma(2+p-j)} \\ &= \frac{1}{\Gamma(p-k+1)}\int_a^t (t-\tau)^{p-k}\left\{{}_a\mathbf{D}_\tau^{-(k-p)} f(\tau)\right\}d\tau \\ &\quad -\sum_{j=1}^{k}\Big[{}_a\mathbf{D}_t^{p-j} f(t)\Big]_{t=a}\frac{(t-a)^{p-j+1}}{\Gamma(2+p-j)} \qquad (2.110) \\ &= {}_a\mathbf{D}_t^{-(p-k+1)}\Big({}_a\mathbf{D}_t^{-(k-p)} f(t)\Big) \\ &\quad -\sum_{j=1}^{k}\Big[{}_a\mathbf{D}_t^{p-j} f(t)\Big]_{t=a}\frac{(t-a)^{p-j+1}}{\Gamma(2+p-j)} \qquad (2.111) \end{aligned} $$

$$= {}_a\mathbf{D}_t^{-1} f(t) - \sum_{j=1}^{k} \left[{}_a\mathbf{D}_t^{p-j} f(t) \right]_{t=a} \frac{(t-a)^{p-j+1}}{\Gamma(2+p-j)}. \qquad (2.112)$$

The existence of all terms in (2.110) follows from the integrability of ${}_a\mathbf{D}_t^p f(t)$, because due to this condition the fractional derivatives ${}_a\mathbf{D}_t^{p-j} f(t)$, $(j = 1, 2, \ldots, k)$, are all bounded at $t = a$.

Combining (2.109) and (2.112) ends the proof of the relationship (2.108).

An important particular case must be mentioned. If $0 < p < 1$, then

$${}_a\mathbf{D}_t^{-p} \left({}_a\mathbf{D}_t^p f(t) \right) = f(t) - \left[{}_a\mathbf{D}_t^{p-1} f(t) \right]_{t=a} \frac{(t-a)^{p-1}}{\Gamma(p)}. \qquad (2.113)$$

The property (2.106) is a particular case of a more general property

$${}_a\mathbf{D}_t^p \left({}_a\mathbf{D}_t^{-q} f(t) \right) = {}_a\mathbf{D}_t^{p-q} f(t), \qquad (2.114)$$

where we assume that $f(t)$ is continuous and, if $p \geq q \geq 0$, that the derivative ${}_a\mathbf{D}_t^{p-q} f(t)$ exists.

Two cases must be considered: $q \geq p \geq 0$ and $p > q \geq 0$.

If $q \geq p \geq 0$, then using the properties (2.100) and (2.106) we obtain

$$\begin{aligned} {}_a\mathbf{D}_t^p \left({}_a\mathbf{D}_t^{-q} f(t) \right) &= {}_a\mathbf{D}_t^p \left({}_a\mathbf{D}_t^{-p} \, {}_a\mathbf{D}_t^{-(q-p)} \right) \\ &= {}_a\mathbf{D}_t^{-(q-p)} = {}_a\mathbf{D}_t^{p-q} f(t). \end{aligned}$$

Now let us consider the case $p > q \geq 0$. Let us denote by m and n integers such that $0 \leq m - 1 \leq p < m$ and $0 \leq n \leq p - q < n$. Obviously, $n \leq m$. Then, using the definition (2.103) and the property (2.100) we obtain

$$\begin{aligned} {}_a\mathbf{D}_t^p \left({}_a\mathbf{D}_t^{-q} f(t) \right) &= \frac{d^m}{dt^m} \left\{ {}_a\mathbf{D}_t^{-(m-p)} \left({}_a\mathbf{D}_t^{-q} f(t) \right) \right\} \\ &= \frac{d^m}{dt^m} \left\{ {}_a\mathbf{D}_t^{p-q-m} f(t) \right\} \\ &= \frac{d^n}{dt^n} \left\{ {}_a\mathbf{D}_t^{p-q-n} f(t) \right\} = {}_a\mathbf{D}_t^{p-q} f(t). \end{aligned}$$

The above mentioned property (2.108) is a particular case of the more general property

$${}_a\mathbf{D}_t^{-p} \left({}_a\mathbf{D}_t^q f(t) \right) = {}_a\mathbf{D}_t^{q-p} f(t) - \sum_{j=1}^{k} \left[{}_a\mathbf{D}_t^{q-j} f(t) \right]_{t=a} \frac{(t-a)^{p-j}}{\Gamma(1+p-j)}, \qquad (2.115)$$

$$(0 \le k-1 \le q < k).$$

To prove the formula (2.115) we first use property (2.100) (if $q \le p$) or property (2.114) (if $q \ge p$) and then property (2.108). This gives:

$$\begin{aligned}
{}_a\mathbf{D}_t^{-p}\Big({}_a\mathbf{D}_t^{q} f(t)\Big) &= {}_a\mathbf{D}_t^{q-p}\Big\{{}_a\mathbf{D}_t^{-q}\Big({}_a\mathbf{D}_t^{q} f(t)\Big)\Big\} \\
&= {}_a\mathbf{D}_t^{q-p}\left\{ f(t) - \sum_{j=1}^{k}\Big[{}_a\mathbf{D}_t^{q-j} f(t)\Big]_{t=a}\frac{(t-a)^{q-j}}{\Gamma(p-j+1)}\right\} \\
&= {}_a\mathbf{D}_t^{q-p} f(t) - \sum_{j=1}^{k}\Big[{}_a\mathbf{D}_t^{q-j} f(t)\Big]_{t=a}\frac{(t-a)^{p-j}}{\Gamma(1+p-j)},
\end{aligned}$$

where we used the known derivative of the power function (2.117):

$${}_a\mathbf{D}_t^{q-p}\left\{\frac{(t-a)^{q-j}}{\Gamma(1+q-j)}\right\} = \frac{(t-a)^{p-j}}{\Gamma(1+p-j)}.$$

2.3.4 Fractional Derivative of $(t-a)^{\beta}$

Let us now evaluate the Riemann–Liouville fractional derivative ${}_a\mathbf{D}_t^{p} f(t)$ of the power function

$$f(t) = (t-a)^{\nu},$$

where ν is a real number.

For this purpose let us assume that $n-1 \le p < n$ and recall that by the definition of the Riemann–Liouville derivative

$${}_a\mathbf{D}_t^{p} f(t) = \frac{d^n}{dt^n}\Big({}_a\mathbf{D}_t^{-(n-p)} f(t)\Big), \qquad (n-1 \le p < n). \tag{2.116}$$

Substituting into the formula (2.116) the fractional integral of order $\alpha = n-p$ of this function, which we have evaluated earlier (see formula (2.56), p. 56), i.e.

$${}_a\mathbf{D}_t^{-\alpha}\Big((t-a)^{\nu}\Big) = \frac{\Gamma(1+\nu)}{\Gamma(1+\nu+\alpha)}(t-a)^{\nu+\alpha},$$

we obtain:

$${}_a\mathbf{D}_t^{p}\Big((t-a)^{\nu}\Big) = \frac{\Gamma(1+\nu)}{\Gamma(1+\nu-p)}(t-a)^{\nu-p}, \tag{2.117}$$

and the only restriction for $f(t) = (t-a)^{\nu}$ is its integrability, namely $\nu > -1$.

2.3.5 Composition with Integer-order Derivatives

In many applied problems the composition of the Riemann–Liouville fractional derivative with integer-order derivatives appears.

Let us consider the n-th derivative of the Riemann–Liouville fractional derivative of real order p.

Using the definition (2.102) of the Riemann–Liouville derivative we obtain:

$$\frac{d^n}{dt^n}\Big({}_a\mathbf{D}_t^{k-\alpha} f(t) \Big) = \frac{1}{\Gamma(\alpha)} \frac{d^{n+k}}{dt^{n+k}} \int_a^t (t-\tau)^{\alpha-1} f(\tau) d\tau = {}_a\mathbf{D}_t^{n+k-\alpha} f(t), \tag{2.118}$$

$$(0 < \alpha \leq 1),$$

and denoting $p = k - \alpha$ we have

$$\frac{d^n}{dt^n}\Big({}_a\mathbf{D}_t^{p} f(t) \Big) = {}_a\mathbf{D}_t^{n+p} f(t). \tag{2.119}$$

To consider the reversed order of operations, we must take into account that

$$\begin{aligned} {}_a\mathbf{D}_t^{-n} f^{(n)}(t) &= \frac{1}{(n-1)!} \int_a^t (t-\tau)^{n-1} f^{(n)}(\tau) d\tau \\ &= f(t) - \sum_{j=0}^{n-1} \frac{f^{(j)}(a)(t-a)^j}{\Gamma(j+1)} \end{aligned} \tag{2.120}$$

and that

$$ {}_a\mathbf{D}_t^{p} g(t) = {}_a\mathbf{D}_t^{p+n}\Big({}_a\mathbf{D}_t^{-n} g(t) \Big). \tag{2.121}$$

Using (2.120), (2.121) and (2.117) we obtain:

$$\begin{aligned} {}_a\mathbf{D}_t^{p}\Big(\frac{d^n f(t)}{dt^n}\Big) &= {}_a\mathbf{D}_t^{p+n}\Big({}_a\mathbf{D}_t^{-n} f^{(n)}(t) \Big) \\ &= {}_a\mathbf{D}_t^{p+n}\Big(f(t) - \sum_{j=0}^{n-1} \frac{f^{(j)}(a)(t-a)^j}{\Gamma(j+1)} \Big) \\ &= {}_a\mathbf{D}_t^{p+n} f(t) - \sum_{j=0}^{n-1} \frac{f^{(j)}(a)(t-a)^{j-p-n}}{\Gamma(1+j-p-n)}, \end{aligned} \tag{2.122}$$

which is the same as the relationship (2.66).

Therefore, as in the case of the Grünwald–Letnikov derivatives, we see that the Riemann–Liouville fractional derivative operator ${}_a\mathbf{D}_t^p$ commutes with $\frac{d^n}{dt^n}$, i.e., that

$$\frac{d^n}{dt^n}\left({}_a\mathbf{D}_t^p f(t)\right) = {}_a\mathbf{D}_t^p\left(\frac{d^n f(t)}{dt^n}\right) = {}_a\mathbf{D}_t^{p+n} f(t), \tag{2.123}$$

only if at the lower terminal $t = a$ of the fractional differentiation the function $f(t)$ satisfies the conditions

$$f^{(k)}(a) = 0, \qquad (k = 0,\, 1,\, 2,\, \ldots,\, n-1). \tag{2.124}$$

2.3.6 Composition with Fractional Derivatives

Let us now turn our attention to the composition of two fractional Riemann–Liouville derivative operators: ${}_a\mathbf{D}_t^p$, $(m-1 \le p < m)$, and ${}_a\mathbf{D}_t^q$, $(n-1 \le q < n)$.

Using subsequently the definition of the Riemann–Liouville fractional derivative (2.104), the formula (2.108) and the composition with integer-order derivatives (2.119) we obtain:

$$\begin{aligned}
{}_a\mathbf{D}_t^p\Big({}_a\mathbf{D}_t^q f(t)\Big) &= \frac{d^m}{dt^m}\Big\{{}_a\mathbf{D}_t^{-(m-p)}\Big({}_a\mathbf{D}_t^q f(t)\Big)\Big\} \\
&= \frac{d^m}{dt^m}\Big\{{}_a\mathbf{D}_t^{p+q-m} f(t) \\
&\qquad - \sum_{j=1}^{n}\Big[{}_a\mathbf{D}_t^{q-j} f(t)\Big]_{t=a}\frac{(t-a)^{m-p-j}}{\Gamma(1+m-p-j)}\Big\} \\
&= {}_a\mathbf{D}_t^{p+q} f(t) - \sum_{j=1}^{n}\Big[{}_a\mathbf{D}_t^{q-j} f(t)\Big]_{t=a}\frac{(t-a)^{-p-j}}{\Gamma(1-p-j)}.
\end{aligned} \tag{2.125}$$

Interchanging p and q (and therefore m and n), we can write:

$${}_a\mathbf{D}_t^q\Big({}_a\mathbf{D}_t^p f(t)\Big) = {}_a\mathbf{D}_t^{p+q} f(t) - \sum_{j=1}^{m}\Big[{}_a\mathbf{D}_t^{p-j} f(t)\Big]_{t=a}\frac{(t-a)^{-q-j}}{\Gamma(1-q-j)}. \tag{2.126}$$

The comparison of the relationships (2.125) and (2.126) says that in the general case the Riemann–Liouville fractional derivative operators ${}_a\mathbf{D}_t^p$ and ${}_a\mathbf{D}_t^p$ do not commute, with only one exception (besides the trivial case $p = q$): namely, for $p \neq q$ we have

$${}_a\mathbf{D}_t^p\Big({}_a\mathbf{D}_t^q f(t)\Big) = {}_a\mathbf{D}_t^q\Big({}_a\mathbf{D}_t^p f(t)\Big) = {}_a\mathbf{D}_t^{p+q} f(t), \tag{2.127}$$

only if both sums in the right-hand sides of (2.125) and (2.126) vanish. For this we have to require the simultaneous fulfillment of the conditions

$$\left[{}_a\mathbf{D}_t^{p-j} f(t)\right]_{t=a} = 0, \quad (j = 1, 2, \ldots, m), \tag{2.128}$$

and the conditions

$$\left[{}_a\mathbf{D}_t^{q-j} f(t)\right]_{t=a} = 0, \quad (j = 1, 2, \ldots, n). \tag{2.129}$$

As will be shown below in Section 2.3.7, if $f(t)$ has a sufficient number of continous derivatives, then the conditions (2.128) are equivalent to

$$f^{(j)}(a) = 0, \quad (j = 0, 1, 2, \ldots, m-1) \tag{2.130}$$

and the conditions (2.129) are equivalent to

$$f^{(j)}(a) = 0, \quad (j = 0, 1, 2, \ldots, n-1), \tag{2.131}$$

and the relationship (2.127) holds (i.e. the p-th and q-th derivatives commute) if

$$f^{(j)}(a) = 0, \quad (j = 0, 1, 2, \ldots, r-1), \tag{2.132}$$

where $r = \max(n, m)$.

2.3.7 Link to the Grünwald–Letnikov Approach

As we mentioned above, see p. 63, there exists a link between the Riemann–Liouville and the Grünwald–Letnikov approaches to differentiation of arbitrary real order. The exact conditions of the equivalence of these two approaches are the following.

Let us suppose that the function $f(t)$ is $(n-1)$-times continuously differentiable in the interval $[a, T]$ and that $f^{(n)}(t)$ is integrable in $[a, T]$. Then for every p $(0 < p < n)$ the Riemann–Liouville derivative ${}_a\mathbf{D}_t^p f(t)$ exists and coincides with the Grünwald–Letnikov derivative ${}_aD_t^p f(t)$, and if $0 \leq m-1 \leq p < m \leq n$, then for $a < t < T$ the following holds:

$${}_a\mathbf{D}_t^p f(t) = {}_aD_t^p f(t) = \sum_{j=0}^{m-1} \frac{f^{(j)}(a)(t-a)^{j-p}}{\Gamma(1+j-p)} + \frac{1}{\Gamma(m-p)} \int_a^t \frac{f^{(m)}(\tau)d\tau}{(t-\tau)^{p-m+1}}. \tag{2.133}$$

Indeed, on the one hand the right-hand side of formula (2.133) is equal to the Grünwald–Letnikov derivative ${}_aD_t^p f(t)$. On the other hand, it can be written as

$$\frac{d^m}{dt^m}\left\{\sum_{j=0}^{m-1} \frac{f^{(j)}(a)(t-a)^{m+j-p}}{\Gamma(1+m+j-p)} + \frac{1}{\Gamma(2m-p)} \int_a^t (t-\tau)^{2m-p-1} f^{(m)}(\tau)d\tau\right\},$$

which after m integrations by parts takes the form of the Riemann–Liouville derivative ${}_a\mathbf{D}_t^p f(t)$

$$\frac{d^m}{dt^m}\left\{\frac{1}{\Gamma(m-p)}\int_a^t (t-\tau)^{m-p-1} f(\tau)d\tau\right\} = \frac{d^m}{dt^m}\left\{{}_a\mathbf{D}_t^{-(m-p)} f(t)\right\}$$
$$= {}_a\mathbf{D}_t^p f(t).$$

The following particular case of the relationship (2.133) is important from the viewpoint of numerous applied problems.

If $f(t)$ is continuous and $f'(t)$ is integrable in the interval $[a, T]$, then for every p $(0 < p < 1)$ both Riemann–Liouville and Grünwald–Letnikov derivatives exist and can be written in the form

$${}_a\mathbf{D}_t^p f(t) = {}_aD_t^p f(t) = \frac{f(a)(t-a)^{-p}}{\Gamma(1-p)} + \frac{1}{\Gamma(1-p)}\int_a^t (t-\tau)^{-p} f'(\tau)d\tau. \tag{2.134}$$

Obviously, the derivative given by the expression (2.134) is integrable.

Another important property following from (2.133) is that the existence of the derivative of order $p > 0$ implies the existence of the derivative of order q for all q such that $0 < q < p$.

More precisely, if for a given continuous function $f(t)$ having integrable derivative the Riemann–Liouville (Grünwald–Letnikov) derivative ${}_a\mathbf{D}_t^p f(t)$ exists and is integrable, then for every q such that $(0 < q < p)$ the derivative ${}_a\mathbf{D}_t^q f(t)$ also exists and is integrable.

Indeed, if we denote $g(t) = {}_a\mathbf{D}_t^{-(1-p)} f(t)$, then we can write

$${}_a\mathbf{D}_t^p f(t) = \frac{d}{dt}\left({}_a\mathbf{D}_t^{-(1-p)} f(t)\right) = g'(t).$$

Noting that $g'(t)$ is integrable and taking into account the formula (2.134) and the inequality $0 < 1 + q - p < 1$ we conclude that the derivative ${}_a\mathbf{D}_t^{1+q-p} g(t)$ exists and is integrable. Then, using the property (2.114), we obtain:

$${}_a\mathbf{D}_t^{1+q-p} g(t) = {}_a\mathbf{D}_t^{1+q-p}\left({}_a\mathbf{D}_t^{-(1-p)} f(t)\right) = {}_a\mathbf{D}_t^q f(t).$$

The relationship (2.133) between the Grünwald–Letnikov and the Riemann–Liouville definitions also has another consequence which is very important for the formulation of applied problems, manipulation with fractional derivatives and the formulation of physically meaningful initial-value problems for fractional-order differential equations.

Under the same assumptions on the function $f(t)$ ($f(t)$ is $(m-1)$-times continuously differentiable and its m-th derivative is integrable in $[a, T]$) and on p $(m-1 \le p < m)$ the condition

$$\left[{}_a\mathbf{D}_t^p f(t)\right]_{t=a} = 0 \qquad (2.135)$$

is equivalent to the conditions

$$f^{(j)}(a) = 0, \quad (j = 0, 1, 2, \ldots, m-1). \qquad (2.136)$$

Indeed, if the conditions (2.136) are fulfilled, then putting $t \to a$ in (2.133) we immediately obtain (2.135).

On the other hand, if the condition (2.135) is fulfilled, then multiplying both sides of (2.133) subsequently by $(t-a)^{p-j}$, $(j = m-1, m-2, m-3, \ldots, 2, 1, 0)$ and taking the limits as $t \to a$ we obtain $f^{(m-1)}(a) = 0$, $f^{(m-2)}(a) = 0, \ldots, f''(a) = 0$, $f'(a) = 0$, $f(a) = 0$ — i.e., the conditions (2.136).

Therefore, (2.135) holds if and only if (2.136) holds.

From the equivalence of the conditions (2.135) and (2.136) it immediately follows that if for some $p > 0$ the p-th derivative of $f(t)$ is equal to zero at the terminal $t = a$, then all derivatives of order q $(0 < q < p)$ are also equal to zero at $t = a$:

$$\left[{}_a\mathbf{D}_t^q f(t)\right]_{t=a} = 0.$$

2.4 Some Other Approaches

Among other approaches to the generalization of the notion of differentiation and integration we decided to pay attention to the approach suggested by M. Caputo and to the approach based on generalized functions (distributions), because of its possible usefulness for the formulation and solution of applied problems and their transparency.

The approach developed by M. Caputo allows the formulation of initial conditions for initial-value problems for fractional-order differential equations in a form involving only the limit values of integer-order derivatives at the lower terminal (initial time) $t = a$, such as $y'(a)$, $y''(a)$ etc.

The generalized functions approach allows consideration and utilization of the Dirac delta function $\delta(t)$ and the Heaviside (unit-step) function $H(t)$; both functions are frequently used as models (or parts of models) for test signals and loading.

2.4.1 Caputo's Fractional Derivative

The definition (2.103) of the fractional differentiation of the Riemann–Liouville type played an important role in the development of the theory of fractional derivatives and integrals and for its applications in pure mathematics (solution of integer-order differential equations, definitions of new function classes, summation of series, etc.).

However, the demands of modern technology require a certain revision of the well-established pure mathematical approach. There have appeared a number of works, especially in the theory of viscoelasticity and in hereditary solid mechanics, where fractional derivatives are used for a better description of material properties. Mathematical modelling based on enhanced rheological models naturally leads to differential equations of fractional order — and to the necessity of the formulation of initial conditions to such equations.

Applied problems require definitions of fractional derivatives allowing the utilization of physically interpretable initial conditions, which contain $f(a)$, $f'(a)$, etc.

Unfortunately, the Riemann–Liouville approach leads to initial conditions containing the limit values of the Riemann–Liouville fractional derivatives at the lower terminal $t = a$, for example

$$\begin{aligned} \lim_{t \to a} {}_a\mathbf{D}_t^{\alpha-1} f(t) &= b_1, \\ \lim_{t \to a} {}_a\mathbf{D}_t^{\alpha-2} f(t) &= b_2, \\ &\ldots, \\ \lim_{t \to a} {}_a\mathbf{D}_t^{\alpha-n} f(t) &= b_n, \end{aligned} \tag{2.137}$$

where b_k, $k = 1, 2, \ldots, n$ are given constants.

In spite of the fact that initial value problems with such initial conditions can be successfully solved mathematically (see, for example, solutions given in [232] and in this book), their solutions are practically useless, because there is no known physical interpretation for such types of initial conditions.

Here we observe a conflict between the well-established and polished mathematical theory and practical needs.

A certain solution to this conflict was proposed by M. Caputo first in his paper [23] and two years later in his book [24], and recently (in Banach spaces) by El-Sayed [55, 56]). Caputo's definition can be written as

$$ {}_{a}^{C}D_{t}^{\alpha} f(t) = \frac{1}{\Gamma(\alpha - n)} \int_{a}^{t} \frac{f^{(n)}(\tau) d\tau}{(t-\tau)^{\alpha+1-n}}, \quad (n-1 < \alpha < n). \tag{2.138} $$

Under natural conditions on the function $f(t)$, for $\alpha \to n$ the Caputo derivative becomes a conventional n-th derivative of the function $f(t)$. Indeed, let us assume that $0 \leq n-1 < \alpha < n$ and that the function $f(t)$ has $n+1$ continuous bounded derivatives in $[a, T]$ for every $T > a$. Then

$$ \begin{aligned} \lim_{\alpha \to n} {}_{a}^{C}D_{t}^{\alpha} f(t) &= \lim_{\alpha \to n} \left(\frac{f^{(n)}(a)(t-a)^{n-\alpha}}{\Gamma(n-\alpha+1)} \right. \\ &\qquad \left. + \frac{1}{\Gamma(n-\alpha+1)} \int_{a}^{t} (t-\tau)^{n-\alpha} f^{(n+1)}(\tau) d\tau \right) \\ &= f^{n}(a) + \int_{a}^{t} f^{(n+1)}(\tau) d\tau = f^{(n)}(t), \quad n = 1, 2, \ldots \end{aligned} $$

This says that, similarly to the Grünwald–Letnikov and the Riemann–Liouville approaches, the Caputo approach also provides an interpolation between integer-order derivatives.

The main advantage of Caputo's approach is that the initial conditions for fractional differential equations with Caputo derivatives take on the same form as for integer-order differential equations, i.e. contain the limit values of integer-order derivatives of unknown functions at the lower terminal $t = a$.

To underline the difference in the form of the initial conditions which must accompany fractional differential equations in terms of the Riemann–Liouville and the Caputo derivatives, let us recall the corresponding Laplace transform formulas for the case $a = 0$.

The formula for the Laplace transform of the Riemann–Liouville fractional derivative is

$$ \int_{0}^{\infty} e^{-pt} \{ {}_{0}\mathbf{D}_{t}^{\alpha} f(t) \} \, dt = p^{\alpha} F(p) - \sum_{k=0}^{n-1} p^{k} \; {}_{0}\mathbf{D}_{t}^{\alpha-k-1} f(t) \Big|_{t=0}, \tag{2.139} $$

$$ (n-1 \leq \alpha < n), $$

whereas Caputo's formula, first obtained in [23], for the Laplace transform of the Caputo derivative is (see Section 2.8.3)

$$\int_0^{\infty} e^{-pt} \left\{ {}_0^C D_t^{\alpha} f(t) \right\} dt = p^{\alpha} F(p) - \sum_{k=0}^{n-1} p^{\alpha-k-1} f^{(k)}(0), \qquad (2.140)$$

$$(n-1 < \alpha \leq n).$$

We see that the Laplace transform of the Riemann–Liouville fractional derivative allows utilization of initial conditions of the type (2.137), which can cause problems with their physical interpretation. On the contrary, the Laplace transform of the Caputo derivative allows utilization of initial values of classical integer-order derivatives with known physical interpretations.

The Laplace transform method is frequently used for solving applied problems. To choose the appropriate Laplace transform formula, it is very important to understand which type of definition of fractional derivative (in other words, which type of initial conditions) must be used.

Another difference between the Riemann–Liouville definition (2.103) and the Caputo definition (2.138) is that the Caputo derivative of a constant is 0, whereas in the cases of a finite value of the lower terminal a the Riemann–Liouville fractional derivative of a constant C is not equal to 0, but

$${}_0\mathbf{D}_t^{\alpha} C = \frac{C t^{-\alpha}}{\Gamma(1-\alpha)}. \qquad (2.141)$$

This fact led, for example, Ochmann and Makarov [174] to using the Riemann–Liouville definition with $a = -\infty$, because, on the one hand, from the physical point of view they need the fractional derivative of a constant equal to zero and on the other hand formula (2.141) gives 0 if $a \to -\infty$. The physical meaning of this step is that the starting time of the physical process is set to $-\infty$. In such a case transient effects cannot be studied. However, taking $a = -\infty$ is the necessary abstraction for the consideration of the steady-state processes, for example for studying the response of the fractional-order dynamic system to the periodic input signal, wave propagation in viscoelastic materials, etc.

Putting $a = -\infty$ in both definitions and requiring reasonable behaviour of $f(t)$ and its derivatives for $t \to -\infty$, we arrive at the same formula

$${}_{-\infty}\mathbf{D}_t^{\alpha} f(t) = {}_{-\infty}^{\;\;C} D_t^{\alpha} f(t) = \frac{1}{\Gamma(n-\alpha)} \int_{-\infty}^{t} \frac{f^{(n)}(\tau) d\tau}{(t-\tau)^{\alpha+1-n}}, \qquad (2.142)$$

$$(n-1 < \alpha < n),$$

which shows that for the study of steady-state dynamical processes the Riemann–Liouville definition and the Caputo definition must give the same results.

There is also another difference between the Riemann–Liouville and the Caputo approaches, which we would like to mention here and which seems to be important for applications. Namely, for the Caputo derivative we have

$$ {}_a^C D_t^\alpha \left({}_a^C D_t^m f(t) \right) = {}_a^C D_t^{\alpha+m} f(t), \qquad (m = 0, 1, 2, \ldots; \quad n-1 < \alpha < n) \tag{2.143}$$

while for the Riemann–Liouville derivative

$$ {}_a\mathbf{D}_t^m \left({}_a\mathbf{D}_t^\alpha f(t) \right) = {}_a\mathbf{D}_t^{\alpha+m} f(t), \qquad (m = 0, 1, 2, \ldots; \quad n-1 < \alpha < n) \tag{2.144}$$

The interchange of the differentiation operators in formulas (2.143) and (2.144) is allowed under different conditions:

$$ {}_a^C D_t^\alpha \left({}_a^C D_t^m f(t) \right) = {}_a^C D_t^m \left({}_a^C D_t^\alpha f(t) \right) = {}_a^C D_t^{\alpha+m} f(t), \tag{2.145}$$

$$f^{(s)}(0) = 0, \quad s = n, n+1, \ldots, m$$

$$(m = 0, 1, 2, \ldots; n-1 < \alpha < n)$$

$$ {}_a\mathbf{D}_t^m \left({}_a\mathbf{D}_t^\alpha f(t) \right) = {}_a\mathbf{D}_t^\alpha \left({}_a\mathbf{D}_t^m f(t) \right) = {}_a\mathbf{D}_t^{\alpha+m} f(t), \tag{2.146}$$

$$f^{(s)}(0) = 0, \quad s = 0, 1, 2, \ldots, m$$

$$(m = 0, 1, 2, \ldots; n-1 < \alpha < n).$$

We see that contrary to the Riemann–Liouville approach, in the case of the Caputo derivative there are no restrictions on the values $f^{(s)}(0)$, $(s = 0, 1, \ldots, n-1)$.

2.4.2 Generalized Functions Approach

This approach is based on the observation that the Cauchy formula (2.85), see page 64,

$$f^{(-n)}(t) = \frac{1}{\Gamma(n)} \int_a^t (t-\tau)^{n-1} f(\tau) d\tau,$$

which allows replacement of the n-fold integral of the function $f(t)$ with a single integration, can be written as a convolution of the function $f(t)$ and the power function t^{n-1}:

$$f^{(-n)}(t) = f(t) * \frac{t^{n-1}}{\Gamma(n)}, \tag{2.147}$$

where both functions, $f(t)$ and t^{n-1}, are replaced with zero for $t < a$ and $t < 0$ correspondingly; the asterisk means the convolution:

$$f(t) * g(t) = \int_{-\infty}^{\infty} f(\tau) g(t - \tau) d\tau.$$

Let us consider the function $\Phi_p(t)$ defined by [76]

$$\Phi_p(t) = \begin{cases} \frac{t^{p-1}}{\Gamma(\gamma)}, & t > 0 \\ 0, & t \leq 0. \end{cases} \tag{2.148}$$

Using the function $\Phi_p(t)$ the formula (2.147) can be considered as a particular case of the more general convolution of the function $f(t)$ and the function $\Phi_p(t)$:

$$f^{(-p)}(t) = f(t) * \Phi_p(t). \tag{2.149}$$

To handle both positive and negative values of p in the same way, it is convenient to consider the function $\Phi_p(t)$ as a generalized function. Its properties are known [76]; for our purposes it is essential that

$$\lim_{p \to -k} \Phi_p(t) = \Phi_{-k}(t) \delta^{(k)}(t), \qquad (k = 0, 1, 2, \ldots), \tag{2.150}$$

where $\delta(t)$ is the Dirac delta function [76]. The Dirac delta function is often used in applied problems for the description of impulse loading (impulse forces). The convolution of the k-th derivative of the delta function and $f(t)$ is given by

$$\int_{-\infty}^{\infty} f(\tau) \delta^{(k)}(t - \tau) d\tau = f^{(k)}(t). \tag{2.151}$$

Obviously, if p is a positive integer ($p = n$), then the formula (2.149) reduces to (2.147). On the other hand, it follows from the relationship

(2.150) and the properties of the delta function that for negative integer values of p ($p = -n$, $n > 0$)

$$f^{(0)}(t) = f(t) * \Phi_0(t) = f(t) * \delta(t) = f(t),$$

$$f^{(1)}(t) = f(t) * \Phi_{-1}(t) = f(t) * \delta'(t) = f'(t),$$

$$\ldots \qquad \ldots \qquad \ldots$$

$$f^{(k)}(t) = f(t) * \Phi_{-k}(t) = f(t) * \delta^{(k)}(t) = f^{(k)}(t).$$

Therefore, both integer-order integrals and derivatives of a generalized function $f(t)$ can be obtained as particular cases of the convolution (2.149), which is also meaningful for non-integer values of p. This means that the formula (2.149) provides a unification of n-fold integrals and n-th order derivatives of a generalized function and an extention of these notions to real order p and that we can define the derivative of real order p of a generalized function $f(t)$, which is equal to zero for $t < a$, as

$${}_a\tilde{D}_t^p f(t) = f(t) * \Phi_p(t). \tag{2.152}$$

Another property of the function $\Phi_p(t)$, which leads to important consequences, is

$$\Phi_p(t-a) * \Phi_q(t) = \Phi_{p+q}(t-a). \tag{2.153}$$

To prove (2.153), let us first suppose that $p > 0$ and $q > 0$. Then using the substitution $\tau = a + \zeta(t-a)$ and the definition of the beta function (1.20) we obtain

$$\begin{aligned}
\Phi_p(t-a) * \Phi_q(t) &= \int_a^t \frac{(\tau-a)^{p-1}}{\Gamma(p)} \frac{(t-\tau)^{q-1}}{\Gamma(q)} d\tau \\
&= \frac{1}{\Gamma(p)\Gamma(q)} \int_0^t (\tau-a)^{p-1}(t-\tau)^{q-1} d\tau \\
&= \frac{(t-a)^{p+q-1}}{\Gamma(p)\Gamma(q)} \int_0^1 \zeta^{p-1}(1-\zeta)^{q-1} d\zeta \\
&= \frac{(t-a)^{p+q-1}}{\Gamma(p+q)},
\end{aligned} \tag{2.154}$$

and analytic continuation with respect to p and q gives (2.153).

It follows from (2.153) that if the function $f(t)$ is zero for $t < a$, then

$$\Big(f(t) * \Phi_p(t)\Big) * \Phi_q(t) = f(t) * \Big(\Phi_p(t) * \Phi_q(t)\Big) = f(t) * \Phi_{p+q}(t), \quad (2.155)$$

from which immediately follows the composition law

$${}_a\tilde{D}_t^p\Big({}_a\tilde{D}_t^q f(t)\Big) = {}_a\tilde{D}_t^q\Big({}_a\tilde{D}_t^p f(t)\Big) = {}_a\tilde{D}_t^{p+q} f(t). \quad (2.156)$$

for all p and q. The simplicity of the composition law (2.156) is, of course, a great advantage of the use of generalized functions.

From formula (2.153) we directly obtain the derivative of real order p of the generalized function

$$\Phi_{q+1}(t) = \frac{t_+^q}{\Gamma(q+1)} = \begin{cases} \frac{t^q}{\Gamma(q+1)}, & (t > 0) \\ 0, & (t \leq 0) \end{cases}$$

in the form

$${}_a\tilde{D}_t^p\Big(\frac{(t-a)^q}{\Gamma(q+1)}\Big) = \frac{(t-a)^{p-q}}{\Gamma(1+q-p)}, \quad (t > a). \quad (2.157)$$

In the particular case $q = 0$ we obtain the fractional derivative of the Heaviside unit-step function $H(t)$:

$${}_a\tilde{D}_t^p H(t-a) = \frac{(t-a)^{-p}}{\Gamma(1-p)}, \quad (t > a), \quad (2.158)$$

and, in general, for all $b < a$

$${}_b\tilde{D}_t^p H(t-a) = \begin{cases} \frac{(t-a)^{-p}}{\Gamma(1-p)} & (t > a) \\ 0, & (b \leq t \leq a). \end{cases} \quad (2.159)$$

Putting $q = -n-1$ $(n \geq 0)$ in (2.157), we obtain the fractional derivative of order p of the n-th derivative of the Dirac delta function:

$${}_a\tilde{D}_t^p \delta^{(n)}(t-a) = \frac{(t-a)^{-n-p-1}}{\Gamma(-n-p)}, \quad (t > a), \quad (2.160)$$

and, in general, for $b < a$ we have

$${}_b\tilde{D}_t^p \delta^{(n)}(t-a) = \begin{cases} \frac{(t-a)^{-n-p-1}}{\Gamma(-n-p)}, & (t > a) \\ 0, & (b \leq t \leq a). \end{cases} \quad (2.161)$$

Finally, if $q-p+1=-n$ $(n \geq 0)$ then from (2.157) it follows that

$$ {}_a\tilde{D}_t^p \left(\frac{(t-a)^{p-n-1}}{\Gamma(p-n)} \right) = \delta^{(n)}(t-a), \quad (t>a). \tag{2.162} $$

Relationships (2.158), (2.160) and (2.162) represent an interesting and useful link between the power function, the Heaviside unit-step function and the Dirac delta function.

The generalized function approach allows the establishment of an interesting link between the Riemann–Liouville and the Caputo approaches and their relationship to conventional and generalized integer-order derivatives.

Using the function $\Phi_p(t)$, the Riemann–Liouville definition (2.103) can be written as

$$ {}_a\mathbf{D}_t^p f(t) = \frac{d^n}{dt^n}\Big(f(t) * \Phi_{n-p}(t) \Big), \tag{2.163} $$

the Caputo definition can be written as

$$ {}_a^C D_t^p f(t) = \left(\frac{d^n f(t)}{dt^n} * \Phi_{n-p}(t) \right) \tag{2.164} $$

and the relationship (2.133) takes the form

$$ {}_a\mathbf{D}_t^p f(t) = {}_a^C D_t^p f(t) + \sum_{k=0}^{n-1} \Phi_{k-p+1}(t-a) f^{(k)}(a). \tag{2.165} $$

Taking $p \to n$, where n is a positive integer number and using (2.150), we obtain from (2.165) the following relationship:

$$ {}_a^L D_t^n f(t) = {}_a^C D_t^n f(t) + \sum_{k=0}^{n-1} \delta^{(n-k-1)}(t-a) f^{(k)}(a). \tag{2.166} $$

Comparing relationship (2.166) with the well known relationship between the classical derivative $f_C^{(n)}(t)$ and the generalized derivative $\tilde{f}^{(n)}(t)$

$$ \tilde{f}^{(n)}(t) = f_C^{(n)}(t) + \sum_{k=0}^{n-1} \delta^{(n-k-1)}(t-a) f^{(k)}(a), \tag{2.167} $$

where $\tilde{f}(t) = f(t)$ for $t \geq a$ and $\tilde{f}(t) \equiv 0$ for $t < a$, we conclude that *the Riemann–Liouville definition (2.79) serves as a generalization of the notion of the generalized (in the sense of generalized functions) derivative,* while *the Caputo derivative (2.138) is a generalization of differentiation in the classical sense.*

Similar results can be found in D. Matignon's work [143], where a relationship between the fractional derivative in the sense of distributions and the "smooth fractional derivative" (which coincides with Caputo's derivative) has been given, and in F. Mainardi's paper [135], where the relationship between the Riemann–Liouville and the Caputo definitions of fractional differentiation is also discussed.

2.5 Sequential Fractional Derivatives

The main idea of differentiation and integration of arbitrary order is the generalization of iterated integration and differentiation.

In all these approaches the general aim is the same: to "replace" the integer-valued parameter n of an operation denoted, for example, by the symbols

$$\frac{d^n}{dt^n}$$

with a non-integer parameter p. Other details vary (function classes, methods of "replacement" of n with p, some properties for non-integer values of p), but it is obvious that all efforts are made for the direct intermediate replacement of an integer n with a non-integer p.

However, there is also another way which is less well known but can be of great importance for many applications. This approach is based on the observation that, in fact, n-th order differentiation is simply a series of first-order differentiations:

$$\frac{d^n f(t)}{dt^n} = \underbrace{\frac{d}{dt}\frac{d}{dt}\cdots\frac{d}{dt}}_{n} f(t). \tag{2.168}$$

If there is a suitable method for "replacing" the derivative of first order $\frac{d}{dt}$ with the derivative of non-integer order D^α, where $0 \leq \alpha \leq 1$, then it is possible to consider the following analogue of (2.168):

$$D^{n\alpha} f(t) = \underbrace{D^\alpha D^\alpha \ldots D^\alpha}_{n} f(t). \tag{2.169}$$

K. S. Miller and B. Ross called the generalized differentiation defined by (2.169), where D^{α} is the Riemann–Liouville fractional derivative, *sequential differentiation* and considered differential equations with sequential fractional derivatives of type (2.169) in their book [153, Chapter VI, section 4].

Other mutations of sequential fractional derivatives can be obtained by interpreting D^{α} as the Grünwald–Letnikov derivative, the Caputo derivative or any other type of fractional derivative not considered here.

Instead of (2.169) it is possible to replace each first-order derivative in (2.168) by fractional derivatives of orders which are not necessarily equal, and to consider the more general expression:

$$\mathcal{D}^{\alpha} f(t) = D^{\alpha_1} D^{\alpha_2} \ldots D^{\alpha_n} f(t), \tag{2.170}$$

$$\alpha = \alpha_1 + \alpha_2 + \ldots + \alpha_n$$

which we will also call the *sequential fractional derivative.* Depending on the problem, the symbol $\mathcal{D}^{\alpha}$ in (2.170) can mean the Riemann–Liouville, the Grünwald–Letnikov, the Caputo or any other mutation of the operator of generalized differentiation. Moreover, from this point of view, the Riemann–Liouville fractional derivative and the Caputo fractional derivative are also just particular cases of the sequential derivative (2.170).

Indeed, the Riemann–Liouville fractional derivative can be written as

$${}_a\mathbf{D}_t^p f(t) = \underbrace{\frac{d}{dt}\frac{d}{dt}\cdots\frac{d}{dt}}_{n} {}_aD_t^{-(n-p)} f(t), \qquad (n-1 \leq p < n), \tag{2.171}$$

while the Caputo fractional differential operator can be written as

$${}_a^C D_t^p f(t) = {}_a\mathbf{D}_t^{-(n-p)} \underbrace{\frac{d}{dt}\frac{d}{dt}\cdots\frac{d}{dt}}_{n} f(t), \qquad (n-1 < p \leq n). \tag{2.172}$$

The properties of the Riemann–Liouville derivatives and the Caputo derivatives of the same cumulative order p are different due to the different *sequence* of differential operators $\frac{d}{dt}$ and ${}_a\mathbf{D}_t^{-(n-p)}$.

In the case of the Grünwald–Letnikov approach (p. 59) and the Riemann–Liouville approach (p. 68) we saw that for the fractional integrals it always holds that

$$D^p D^q f(t) = D^q D^p f(t) = D^{p+q} f(t), \qquad (p < 0, \quad q < 0). \tag{2.173}$$

Because of this, we do not see a reason for considering sequential *integral* operators.

However, in the general case, the property (2.173) does not hold for $p > 0$ and/or $q > 0$ (this explains the difference between the Riemann–Liouville and the Caputo fractional derivatives). Therefore, only consideration of sequential fractional *derivative* operators can be of interest and can give new results.

On the other hand, sequential fractional derivatives can appear in a natural way in the formulation of various applied problems in physics and applied science. Indeed, differential equations modelling processes or objects arise usually as a result of a substitution of one relationship involving derivatives into another one. If the derivatives in both relationships are fractional derivatives, then the resulting expression (equation) will contain — in the general case — sequential fractional derivatives.

It is worth mentioning that the sequential fractional integro-differential operators of the form (2.170), with $\alpha_1 < 0, \alpha_2 > 0, \dots, \alpha_n > 0$ were first considered and used for various purposes by M. M. Dzhrbashyan and A. B. Nersesyan at least since 1958 [46, 47, 49, 45, 50]. However, in this book we call sequential fractional derivatives also *Miller–Ross fractional derivatives*, because they clearly outlined the difference between the (single) Riemann–Liouville differentiation and sequential fractional differentiation [153, Chapter VI].

2.6 Left and Right Fractional Derivatives

Until now, we considered the fractional derivatives ${}_aD_t^p f(t)$ with fixed lower terminal a and moving upper terminal t. Moreover, we supposed that $a < t$. However, it is also possible to consider fractional derivatives with moving lower terminal t and fixed upper terminal b.

Let us suppose that the function $f(t)$ is defined in the interval $[a, b]$, where a and b can even be infinite.

The fractional derivative with the lower terminal at the left end of the interval $[a, b]$, ${}_aD_t^p f(t)$, is called the *left fractional derivative*. The

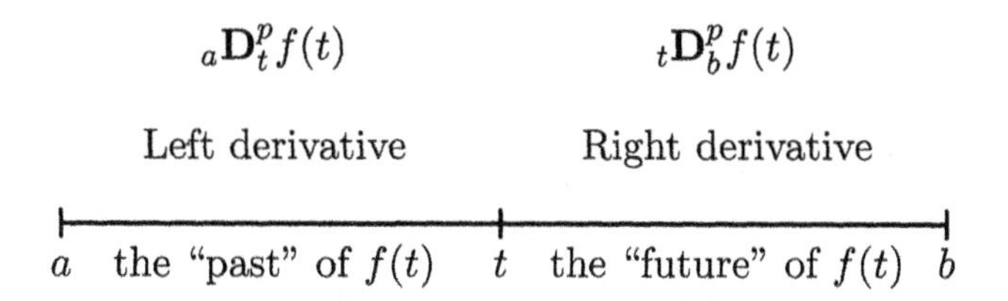

Figure 2.1: *The left and right derivatives as operations on the "past" and the "future" of* $f(t)$.

fractional derivative with the upper terminal at the right end of the interval $[a, b]$ is called the *right fractional derivative.* Obviously, the notions of left and right fractional derivatives can be introduced for any mutation of fractional differentiation — Riemann–Liouville, Grünwald–Letnikov, Caputo and others, which are not considered in this book.

For example, if $k-1 \leq p < k$, then the left Riemann–Liouville fractional derivative is, as we know, defined by

$$ {}_a\mathbf{D}_t^p f(t) = \frac{1}{\Gamma(k-p)} \left(\frac{d}{dt}\right)^k \int_a^t (t-\tau)^{k-p-1} f(\tau)d\tau. \tag{2.174} $$

The corresponding right Riemann–Liouville derivative is defined by [232, §2.3]

$$ {}_t\mathbf{D}_b^p f(t) = \frac{1}{\Gamma(k-p)} \left(-\frac{d}{dt}\right)^k \int_t^b (\tau-t)^{k-p-1} f(\tau)d\tau. \tag{2.175} $$

The right Caputo and Grünwald–Letnikov derivatives can be defined in a similar manner.

The notions of left and right fractional derivatives can be considered from the physical and the mathematical viewpoints.

Sometimes the following physical interpretation of the left and right derivative can be helpful.

Let us suppose that t is time and the function $f(t)$ describes a certain dynamical process developing in time. If we take $\tau < t$, where t is the present moment, then the state $f(\tau)$ of the process f belongs to the past of this process; if we take $\tau > t$, then $f(\tau)$ belongs to the future of the process f.

From such a point of view, the left derivative (2.174) is an operation performed on the past states of the process f and the right derivative is an operation performed on the future states of the process f.

The physical causality principle means that the present state of the process started at the instant $\tau = a$, i.e. the current value of $f(t)$, depends on all its previous (past) states $f(\tau)$ $(a \leq \tau < t)$. Since we are not aware of the dependence of the present state of any process on the results of its development in the future, only left derivatives are considered in this book. Perhaps once the right derivatives will also get a certain physical interpretation in terms of dynamical processes.

On the other hand, from the viewpoint of mathematics the right derivatives remind us of the operators conjugate to the operators of left differentiation. This means that the complete theory of fractional differential equations, especially the theory of boundary value problems for fractional differential equations, can be developed only with the use of both left and right derivatives.

At present, the above interpretation of fractional derivatives and integrals, related to dynamical processes, seems to be the most transparent and usable. There was an attempt undertaken by R. R. Nigmatullin [165] to derive a relationship between a static fractal structure and fractional integration, but it follows from R. S. Rutman's [231] critics that a suitable practically useful relationship between static fractals and fractional integration or differentiation still has not been established.

2.7 Properties of Fractional Derivatives

Let us turn our attention to the properties of fractional-order integration and differentiation, which are most frequently used in applications.

2.7.1 Linearity

Similarly to integer-order differentiation, fractional differentiation is a linear operation:

$$D^p\Big(\lambda f(t) + \mu g(t)\Big) = \lambda\, D^p f(t) + \mu\, D^p g(t), \tag{2.176}$$

where D^p denotes any mutation of the fractional differentiation considered in this book.

The linearity of fractional differentiation follows directly from the corresponding definition. For example, for the Grünwald–Letnikov fractional derivatives defined by (2.43) we have:

$$\begin{aligned}
{}_aD_t^p\Big(\lambda f(t) + \mu g(t)\Big) &= \lim_{\substack{h\to 0\\ nh=t-a}} h^{-p}\sum_{r=0}^{n}(-1)^r\binom{p}{r}\Big(\lambda f(t-rh) + \mu g(t-rh)\Big)\\
&= \lambda \lim_{\substack{h\to 0\\ nh=t-a}} h^{-p}\sum_{r=0}^{n}(-1)^r\binom{p}{r} f(t-rh)\\
&\quad + \mu \lim_{\substack{h\to 0\\ nh=t-a}} h^{-p}\sum_{r=0}^{n}(-1)^r\binom{p}{r} g(t-rh)\\
&= \lambda\, {}_aD_t^p f(t) + \mu\, {}_aD_t^p g(t).
\end{aligned}$$

Similarly, for Riemann–Liouville fractional derivatives of order p ($k-1 \leq p < k$) defined by (2.103) we have

$$\begin{aligned}
{}_a\mathbf{D}_t^p\Big(\lambda f(t) + \mu g(t)\Big) &= \frac{1}{\Gamma(k-p)}\frac{d^k}{dt^k}\int_a^t (t-\tau)^{k-p-1}\Big(\lambda f(\tau) + \mu g(\tau)\Big)d\tau\\
&= \frac{\lambda}{\Gamma(k-p)}\frac{d^k}{dt^k}\int_a^t (t-\tau)^{k-p-1} f(\tau)d\tau\\
&\quad + \frac{\mu}{\Gamma(k-p)}\frac{d^k}{dt^k}\int_a^t (t-\tau)^{k-p-1} g(\tau)d\tau\\
&= \lambda\, {}_a\mathbf{D}_t^p f(t) + \mu\, {}_a\mathbf{D}_t^p g(t).
\end{aligned}$$

2.7.2 The Leibniz Rule for Fractional Derivatives

Let us take two functions, $\varphi(t)$ and $f(t)$, and start with the known Leibniz rule for evaluating the n-th derivative of the product $\varphi(t)f(t)$:

$$\frac{d^n}{dt^n}\Big(\varphi(t)f(t)\Big) = \sum_{k=0}^{n}\binom{n}{k}\varphi^{(k)}(t)f^{(n-k)}(t). \qquad (2.177)$$

Let us now take the right-hand side of formula (2.177) and replace the integer parameter n with the real-valued parameter p. This means that the integer-order derivative $f^{(n-k)}(t)$ will be replaced with the Grünwald–Letnikov fractional-order derivative ${}_aD_t^{p-k}f(t)$. Denoting

$$\Omega_n^p(t) = \sum_{k=0}^{n}\binom{p}{k}\varphi^{(k)}(t)\, {}_aD_t^{p-k}f(t), \qquad (2.178)$$

let us evaluate the sum (2.178).

First, let us suppose that $p = q < 0$. Then we have also $p - k = q - k < 0$ for all k, and according to (2.40)

$$ {}_aD_t^{p-k} f(t) = \frac{1}{\Gamma(-q+k)} \int_a^t (t-\tau)^{-q+k-1} f(\tau) d\tau, \qquad (2.179) $$

which leads to

$$ \Omega_n^q(t) = \sum_{k=0}^{n} \binom{q}{k} \frac{1}{\Gamma(-q+k)} \int_a^t (t-\tau)^{-q+k-1} \varphi^{(k)}(t) f(\tau) d\tau \qquad (2.180) $$

$$ = \int_a^t \left\{ \sum_{k=0}^{k=n} \binom{q}{k} \frac{1}{\Gamma(-q+k)} \varphi^{(k)}(t)(t-\tau)^k \right\} \frac{f(\tau)}{(t-\tau)^{q+1}} d\tau. \qquad (2.181) $$

Taking into account the reflection formula (1.26) for the gamma function, we have

$$ \binom{q}{k} \frac{1}{\Gamma(-q+k)} = \frac{\Gamma(q+1)}{k!\,\Gamma(q-k+1)} \cdot \frac{1}{\Gamma(-q+k)} \qquad (2.182) $$

$$ = \frac{\Gamma(q+1)}{k!} \cdot \frac{\sin(k-q)\pi}{\pi} \qquad (2.183) $$

$$ = (-1)^{k+1} \frac{\Gamma(q+1)}{k!} \frac{\sin(q\pi)}{\pi}, \qquad (2.184) $$

and, therefore, the expression (2.181) takes the form:

$$ \Omega_n^q(t) = -\frac{\sin(q\pi)}{\pi} \Gamma(q+1) \int_a^t \left\{ \sum_{k=0}^{n} \frac{(-1)^k}{k!} \varphi^{(k)}(t)(t-\tau)^k \right\} \frac{f(\tau)}{(t-\tau)^{q+1}} f(\tau) d\tau. \qquad (2.185) $$

Using the Taylor theorem we can write

$$ \sum_{k=0}^{n} \frac{(-1)^k}{k!} \varphi^{(k)}(t)(t-\tau)^k = \varphi(t) + \varphi'(t)(t-\tau) + \ldots + \frac{\varphi^{(n)}(t)}{n!}(t-\tau)^n $$

$$ = \varphi(\tau) + \frac{1}{n!} \int_a^t \varphi^{(n+1)}(\xi)(\tau-\xi)^n d\xi, $$

and therefore we obtain

$$ \Omega_n^q(t) = -\frac{\sin(q\pi)\Gamma(q+1)}{\pi} \int_a^t (t-\tau)^{-q-1} \varphi(\tau) f(\tau) d\tau $$

$$-\frac{\sin(q\pi)\Gamma(q+1)}{\pi\, n!}\int\limits_a^t (t-\tau)^{-q-1} f(\tau)d\tau \int\limits_a^t \varphi^{(n+1)}(\xi)(\tau-\xi)^n d\xi$$

$$= \frac{1}{\Gamma(-q)}\int\limits_a^t (t-\tau)^{-q-1}\varphi(\tau)f(\tau)d\tau$$

$$+\frac{1}{n!\,\Gamma(-q)}\int\limits_a^t (t-\tau)^{-q-1} f(\tau)d\tau \int\limits_\tau^t \varphi^{(n+1)}(\xi)(\tau-\xi)^n d\xi$$

$$= {}_aD_t^q\Big(\varphi(t)f(t)\Big) + R_n^q(t), \tag{2.186}$$

where

$$R_n^q(t) = \frac{1}{n!\,\Gamma(-q)}\int\limits_a^t (t-\tau)^{-q-1} f(\tau)d\tau \int\limits_\tau^t \varphi^{(n+1)}(\xi)(\tau-\xi)^n d\xi. \tag{2.187}$$

Let us now consider the case of $p > 0$. Our first step is to show that the evaluation of $\Omega_n^p(t)$ can be reduced to the evaluation of Ω_n^q for a certain negative q.

Taking into account that $\Gamma(0) = \infty$ we have to put

$$\binom{p-1}{-1} = 0,$$

and using the known property of the binomial coefficients

$$\binom{p}{k} = \binom{p-1}{k} + \binom{p-1}{k-1}$$

we can write

$$\Omega_n^p(t) = \sum_{k=0}^{n}\binom{p-1}{k}\varphi^{(k)}(t)\,{}_aD_t^{p-k}f(t) + \sum_{k=1}^{n}\binom{p-1}{k-1}\varphi^{(k)}(t)\,{}_aD_t^{p-k}f(t). \tag{2.188}$$

Replacing k with $k+1$ in the second sum gives

$$\Omega_n^p(t) = \sum_{k=0}^{n}\binom{p-1}{k}\varphi^{(k)}(t)\frac{d}{dt}\Big({}_aD_t^{p-k-1}f(t)\Big)$$

$$+\sum_{k=0}^{n-1}\binom{p-1}{k}\frac{d\varphi^{(k)}(t)}{dt}\cdot{}_aD_t^{p-k-1}f(t), \tag{2.189}$$

which can be written as

$$\Omega_n^p(t) = \binom{p-1}{n} \varphi^{(n)}(t)\, {}_aD_t^{p-n} f(t) + \frac{d}{dt} \sum_{k=0}^{n-1} \binom{p-1}{k} \varphi^{(k)}(t)\, {}_aD_t^{p-k-1} f(t). \tag{2.190}$$

Adding and subtracting the expression

$$\frac{d}{dt}\left\{ \binom{p-1}{n} \varphi^{(n)}(t)\, {}_aD_t^{p-n-1} f(t) \right\}$$

we obtain

$$\Omega_n^p(t) = \frac{d}{dt} \sum_{k=0}^{n} \binom{p-1}{k} \varphi^{(k)}(t)\, {}_aD_t^{p-k-1} f(t) \tag{2.191}$$

$$- \binom{p-1}{n} \varphi^{(n+1)}\, {}_aD_t^{p-n-1} f(t) \tag{2.192}$$

or

$$\Omega_n^p(t) = \frac{d}{dt} \Omega_n^{p-1}(t) - \binom{p-1}{n} \varphi^{(n+1)}(t)\, {}_aD_t^{p-k-1} f(t). \tag{2.193}$$

The relationship (2.193) says that the evaluation of $\Omega_n^p(t)$ can be reduced to the evaluation of $\Omega_n^{p-1}(t)$. Repeating this procedure we can reduce the evaluation of $\Omega_n^p(t)$ $(p > 0)$ to the evaluation of $\Omega_n^q(t)$ $(q < 0)$.

Let us suppose that $0 < p < 1$. Then $p - 1 < 0$, and according to (2.186) we have

$$\Omega_n^{p-1}(t) = {}_aD_t^{p-1}\Big(\varphi(t) f(t)\Big) + R_n^{p-1}(t). \tag{2.194}$$

To combine (2.194) and (2.193), we have to differentiate (2.194) with respect to t. Taking into account that

$$\frac{d}{dt} R_n^{p-1}(t) = \frac{-p}{n!\,\Gamma(-p+1)} \int_a^t (t-\tau)^{-p-1} f(\tau) d\tau \int_\tau^t \varphi^{n+1}(\xi)(\tau - \xi)^n d\xi$$

$$+ \frac{(-1)^n \varphi^{n+1}(t)}{n!\,\Gamma(-p+1)} \int_a^t (t-\tau)^{-p+n} f(\tau) d\tau \tag{2.195}$$

and that

$$\int_a^t (t-\tau)^{-p+n} f(\tau) d\tau = \Gamma(-p+n+1)\, {}_aD_t^{p-n-1} f(t) \tag{2.196}$$

(since $n - p > 0$), we obtain:

$$\begin{aligned}
\frac{d}{dt}\Omega_n^{p-1}(t) &= {}_aD_t^p\Big(\varphi(t)f(t)\Big) \\
&\quad + \frac{(-1)^n\Gamma(-p+n+1)\varphi^{(n+1)}(t)}{n!\,\Gamma(-p+1)} \cdot {}_aD_t^{p-n-1}f(t) + R_n^p(t) \\
&= {}_aD_t^p\Big(\varphi(t)f(t)\Big) \\
&\quad + \binom{p-1}{n}\varphi^{(n+1)}(t)\,{}_aD_t^{p-n-1}f(t) + R_n^p(t),
\end{aligned} \tag{2.197}$$

and the substitution of this expression into (2.193) gives

$$\Omega_n^p(t) = {}_aD_t^p\Big(\varphi(t)f(t)\Big) + R_n^p(t), \tag{2.198}$$

which has the same form as (2.186).

Using mathematical induction we can prove that the relationship (2.198) holds for all p such that $p + 1 < n$.

Obviously, the relationship (2.198) gives, in fact, the rule for the fractional differentiation of the product of two functions. This rule is a generalization of the Leibniz rule for integer-order differentiation, so it is convenient to preserve Leibniz's name also in the case of fractional differentiation.

The Leibniz rule for fractional differentiation is the following. If $f(\tau)$ is continuous in $[a, t]$ and $\varphi(\tau)$ has $n+1$ continuous derivatives in $[a, t]$, then the fractional derivative of the product $\varphi(t)f(t)$ is given by

$${}_aD_t^p\Big(\varphi(t)f(t)\Big) = \sum_{k=0}^{n}\binom{p}{k}\varphi^{(k)}(t)\,{}_aD_t^{p-k}f(t) - R_n^p(t) \tag{2.199}$$

where $n \geq p + 1$ and

$$R_n^p(t) = \frac{1}{n!\,\Gamma(-p)}\int_a^t (t-\tau)^{-p-1}f(\tau)d\tau \int_\tau^t \varphi^{(n+1)}(\xi)(\tau-\xi)^n d\xi. \tag{2.200}$$

The sum in (2.199) can be considered as a partial sum of an infinite series and $R_n^p(t)$ as a remainder of that series.

Performing two subsequent changes of integration variables, first $\xi = \tau + \zeta(t-\tau)$ and then $\tau = a + \eta(t-a)$ we obtain the following expression for $R_n^p(t)$:

$$R_n^p(t) = \frac{(-1)^n}{n!\,\Gamma(-p)} \int_a^t (t-\tau)^{n-p} f(\tau) d\tau \int_0^1 \varphi^{(n+1)}(\tau + \zeta(t-\tau))\zeta^n d\zeta$$

$$= \frac{(-1)^n (t-a)^{n-p+1}}{n!\,\Gamma(-p)} \int_0^1 \int_0^1 F_a(t,\zeta,\eta) d\eta d\zeta, \qquad (2.201)$$

$$F_a(t,\zeta,\eta) = f(a+\eta(t-a))\varphi^{(n+1)}(a+(t-a)(\zeta+\eta-\zeta\eta)),$$

from which it directly follows that

$$\lim_{n\to\infty} R_n^p(t) = 0$$

if $f(\tau)$ and $\varphi(\tau)$ along with all its derivatives are continuous in $[a,\, t]$. Under this condition the Leibniz rule for fractional differentiation takes the form:

$${}_aD_t^p\Big(\varphi(t)f(t)\Big) = \sum_{k=0}^{\infty} \binom{p}{k} \varphi^{(k)}(t)\, {}_aD_t^{p-k} f(t). \qquad (2.202)$$

The Leibniz rule (2.202) is especially useful for the evaluation of fractional derivatives of a function which is a product of a polynomial and a function with known fractional derivative.

To justify the above operations on $R_n^p(t)$ we have to show that $R_n^p(t)$ has a finite value for $p > 0$.

The function

$$\frac{f(\tau) \int_\tau^t \varphi^{(n+1)}(\xi)(\tau-\xi)^n d\xi}{(t-\tau)^{p+1}} \qquad (2.203)$$

gives an indefinite expression $\frac{0}{0}$ for $\tau = t$. To find the limit we can use the l'Hospital rule. Differentiating the numerator and the denominator with respect to τ we obtain

$$\frac{f'(\tau) \int_\tau^t \varphi^{(n+1)}(\xi)(\tau-\xi)^n d\xi + n f(\tau) \int_\tau^t \varphi^{(n+1)}(\xi)(\tau-\xi)^{n-1} d\xi}{-(p+1)(t-\tau)^p}, \qquad (2.204)$$

which again gives an indefinite expression $\frac{0}{0}$ for $\tau = t$. However, if $m < p \le m+1$, then applying the l'Hospital rule $m+2$ times we will obtain $(t-\tau)^{p-m-1}$ in the denominator (giving infinity for $\tau = t$), while the numerator will consist of the terms containing the multipliers of the form

$$\int_{\tau}^{t} \varphi^{(n+1)}(\xi)(\tau - \xi)^{n-k} d\xi \tag{2.205}$$

which vanish as $\tau \to t$ if $n > k$. Obviously, k cannot be greater than $m+2$, so we can take $n \geq m+2$ and the function (2.203) will tend to 0 for $\tau \to t$. This means that the integral in (2.200) exists in the classical sense even for $p > -1$.

Taking into account the link between the Grünwald–Letnikov fractional derivatives and the Riemann–Liouville ones we see that under the above conditions on $f(t)$ and $\varphi(t)$ the Leibniz rule (2.202) holds also for the Riemann–Liouville derivatives.

2.7.3 Fractional Derivative of a Composite Function

One of the useful consequences of the Leibniz rule for the fractional derivative of a product is a rule for evaluating the fractional derivative of a composite function.

Let us take an analytic function $\varphi(t)$ and $f(t) = H(t-a)$, where $H(t)$ is the Heaviside function. Using the Leibniz rule (2.202) and the formula for the fractional differentiation of the Heaviside function (2.158) we can write:

$$\begin{aligned} {}_aD_t^p \varphi(t) &= \sum_{k=0}^{\infty} \binom{p}{k} \varphi^{(k)}(t)\, {}_aD_t^{p-k} H(t-a) \\ &= \frac{(t-a)^{-p}}{\Gamma(1-p)} \varphi(t) + \sum_{k=1}^{\infty} \binom{p}{k} \frac{(t-a)^{k-p}}{\Gamma(k-p+1)} \varphi^{(k)}(t). \end{aligned} \tag{2.206}$$

Now let us suppose that $\varphi(t)$ is a composite function:

$$\varphi(t) = F(h(t)). \tag{2.207}$$

The k-th order derivative of $\varphi(t)$ is evaluated with the help of the Faà di Bruno formula [2, Chapter 24, §24.1.2]:

$$\frac{d^k}{dt^k} F(h(t)) = k! \sum_{m=1}^{k} F^{(m)}(h(t)) \sum \prod_{r=1}^{k} \frac{1}{a_r!} \left(\frac{h^{(r)}(t)}{r!} \right)^{a_r}, \tag{2.208}$$

where the sum $\sum$ extends over all combinations of non-negative integer values of a_1, a_2, ..., a_k such that

$$\sum_{r=1}^{k} r a_r = k \quad \text{and} \quad \sum_{r}^{k} a_r = m.$$

Substituting (2.207) and (2.208) into (2.206) we obtain the formula for the evaluation of the fractional derivative of a composite function:

$$\begin{aligned} {}_aD_t^p F(h(t)) &= \frac{(t-a)^{-p}}{\Gamma(1-p)}\varphi(t) \\ &+ \sum_{k=1}^{\infty} \binom{p}{k} \frac{k!(t-a)^{k-p}}{\Gamma(k-p+1)} \sum_{m=1}^{k} F^{(m)}(h(t)) \sum \prod_{r=1}^{k} \frac{1}{a_r!} \left(\frac{h^{(r)}(t)}{r!} \right)^{a_r}, \end{aligned} \tag{2.209}$$

where the sum $\sum$ and coefficients a_r have the meaning explained above.

2.7.4 Riemann–Liouville Fractional Differentiation of an Integral Depending on a Parameter

The well-known rule for the differentiation of an integral depending on a parameter with the upper limit depending on the same parameter, namely [68]

$$\frac{d}{dt} \int_0^t F(t,\tau)d\tau = \int_0^t \frac{\partial F(t,\tau)}{\partial t} d\tau + F(t, t-0), \tag{2.210}$$

has its analogue for fractional-order differentiation.

The rule for Riemann–Liouville fractional differentiation of an integral depending on a parameter, when the upper limit also depends on the parameter, is the following:

$${}_0D_t^\alpha \int_0^t K(t,\tau)d\tau = \int_0^t {}_\tau D_t^\alpha K(t,\tau)d\tau + \lim_{\tau \to t-0} {}_\tau D_t^{\alpha-1} K(t,\tau), \tag{2.211}$$

$$(0 < \alpha < 1).$$

Indeed, using (2.210) we have

$$
\begin{aligned}
{}_0D_t^\alpha \int_0^t K(t,\tau)d\tau &= \frac{1}{\Gamma(1-\alpha)}\frac{d}{dt}\int_0^t \frac{d\eta}{(t-\eta)^\alpha}\int_0^\eta K(\eta,\tau)d\tau \\
&= \frac{1}{\Gamma(1-\alpha)}\frac{d}{dt}\int_0^t d\tau \int_\tau^t \frac{K(\eta,\tau)d\eta}{(t-\eta)^\alpha} \\
&= \frac{d}{dt}\int_0^t \widetilde{K}(t,\tau)d\tau \\
&= \int_0^t \frac{\partial}{\partial t}\widetilde{K}(t,\tau)d\tau + \lim_{\tau\to t-0}\widetilde{K}(t,\tau) \\
&= \int_0^t {}_\tau D_t^\alpha K(t,\tau)d\tau + \lim_{\tau\to t-0} {}_\tau D_t^{\alpha-1}K(t,\tau), \qquad (2.212)
\end{aligned}
$$

where

$$
\widetilde{K}(t,\xi) = \frac{1}{\Gamma(1-\alpha)}\int_\xi^t \frac{K(\eta,\xi)d\eta}{(t-\eta)^\alpha}.
$$

The following important particular case must be mentioned. If we have $K(t-\tau)f(\tau)$ instead of $K(t,\tau)$, then relationship (2.211) takes the form:

$$
{}_0D_t^\alpha \int_0^t K(t-\tau)f(\tau)d\tau = \int_0^t {}_0D_\tau^\alpha K(\tau)f(t-\tau)d\tau + \lim_{\tau\to +0} f(t-\tau)\, {}_0D_\tau^{\alpha-1}K(\tau). \qquad (2.213)
$$

It is worth noting that while in the right-hand side of the general formula (2.211) we have fractional derivatives with moving lower terminal τ, all fractional derivatives in (2.213) have the same lower terminal, namely 0. This significant simplification can be very useful in solving applied problems where the fractional differentiation of a convolution integral must be performed.

2.7.5 Behaviour near the Lower Terminal

We have shown in Section 2.3.7 that the Grünwald–Letnikov derivative ${}_aD_t^p f(t)$ and the Riemann–Liouville derivative ${}_a\mathbf{D}_t^p f(t)$ coincide if $f(t)$ is continuous and has a sufficient number of continuous derivatives in the closed interval $[a, t]$.

To study the behaviour of the fractional derivatives at the lower terminal, i.e. for $t \to a+0$, let us suppose that the function $f(t)$ is analytic at least in the interval $[a, \epsilon]$ for some small positive ϵ and, therefore, can be represented by the Taylor series

$$f(t) = \sum_{k=0}^{\infty} \frac{f^{(k)}(a)}{k!}(t-a)^k \tag{2.214}$$

in this interval.

Term-by-term fractional differentiation of (2.214) using the formula for the fractional differentiation of the power function (2.117) gives

$${}_aD_t^p f(t) = {}_a\mathbf{D}_t^p f(t) = \sum_{k=0}^{\infty} \frac{f^{(k)}(a)}{\Gamma(k-p+1)}(t-a)^{k-p}, \tag{2.215}$$

from which it follows that if $f(t)$ has the form (2.214) then

$${}_aD_t^p f(t) = {}_a\mathbf{D}_t^p f(t) \sim \frac{f(a)}{\Gamma(1-p)}(t-a)^{-p}, \quad (t \to a+0), \tag{2.216}$$

and

$$\lim_{t\to a+0} {}_aD_t^p f(t) = \lim_{t\to a+0} {}_a\mathbf{D}_t^p f(t) = \begin{cases} 0, & (p<0) \\ f(a), & (p=0) \\ \infty, & (p>0). \end{cases} \tag{2.217}$$

If we allow $f(t)$ to have an integrable singularity at $t = a$, then it can be written in the form $f(t) = (t-a)^q f_*(t)$, where $f_*(a) \neq 0$ and $q > -1$. Supposing that $f_*(t)$ can be represented by its Taylor series, we can write

$$f(t) = (t-a)^q f_*(t) = (t-a)^q \sum_{k=0}^{\infty} \frac{f_*^{(k)}(a)}{k!}(t-a)^k \tag{2.218}$$

$$= \sum_{k=0}^{\infty} \frac{f_*^{(k)}(a)}{k!}(t-a)^{q+k} \tag{2.219}$$

Performing the term-by-term Riemann–Liouville fractional differentiation of the series (2.219), we obtain

$${}_a\mathbf{D}_t^p f(t) = \sum_{k=0}^{\infty} \frac{f_*^{(k)}(a)}{k!} \frac{\Gamma(q+k+1)}{\Gamma(q+k-p+1)}(t-a)^{q+k-p}, \tag{2.220}$$

from which it follows that

$$ {}_a\mathbf{D}_t^p f(t) \approx \frac{f_*(a)\Gamma(q+1)}{\Gamma(q-p+1)}(t-a)^{q-p}, \quad (t \to a+0), \tag{2.221} $$

and

$$ \lim_{t \to a+0} {}_a\mathbf{D}_t^p f(t) = \begin{cases} 0, & (p < q) \\ \dfrac{f_*(a)\Gamma(q+1)}{\Gamma(q-p+1)}, & (p = q) \\ \infty, & (p > q). \end{cases} \tag{2.222} $$

2.7.6 Behaviour far from the Lower Terminal

To study the behaviour of the fractional derivative far from the lower terminal, i.e. for $t \to \infty$, let us start with the formula obtained for an analytic function $\varphi(t)$ in Section 2.7.3:

$$ {}_aD_t^p\varphi(t) = \sum_{k=0}^{\infty} \binom{p}{k} \frac{(t-a)^{k-p}}{\Gamma(k-p+1)} \varphi^{(k)}(t). \tag{2.223} $$

Using the definition of the binomial coefficients and the reflection formula for the gamma function (1.26) we can write the relationship (2.223) as

$$ \begin{aligned} {}_aD_t^p\varphi(t) &= \sum_{k=0}^{\infty} \frac{\Gamma(p+1)}{\Gamma(k+1)\Gamma(p-k+1)} \frac{(t-a)^{k-p}}{\Gamma(k-p+1)} \varphi^{(k)}(t) \\ &= \frac{\Gamma(p+1)\sin(p\pi)}{\pi} \sum_{k=0}^{\infty} \frac{(-1)^k (t-a)^{k-p}}{(p-k)\, k!} \varphi^{(k)}(t). \end{aligned} \tag{2.224} $$

Now let us suppose that t is far from the lower terminal a, i.e. that $|t| \gg |a|$. Then we can write

$$ (t-a)^{k-p} = t^{k-p}\left(1 - \frac{a}{t}\right)^{k-p} = t^{k-p}\left(1 - \frac{(k-p)a}{t} + O\left(\frac{a^2}{t^2}\right)\right) \tag{2.225} $$

and therefore

$$ (t-a)^{k-p} \approx t^{k-p} + \frac{(p-k)at^k}{t^{p+1}}, \quad (|t| \gg |a|). \tag{2.226} $$

Substituting (2.226) into (2.224) we obtain

$$ \begin{aligned} {}_aD_t^p\varphi(t) \;\approx\; & \frac{\Gamma(p+1)\sin(p\pi)}{\pi} \left\{ \sum_{k=0}^{\infty} \frac{(-1)^k t^{k-p}}{(p-k)\, k!} \varphi^{(k)}(t) \right. \\ & \left. + \frac{a}{t^{p+1}} \sum_{k=0}^{\infty} \frac{(-1)^k t^k \varphi^{(k)}(t)}{k!} \right\} \end{aligned} \tag{2.227} $$

and using (2.223) gives

$$ {}_aD_t^p\varphi(t) \approx {}_0D_t^p\varphi(t) + \frac{a\Gamma(p+1)\sin(p\pi)\varphi(0)}{\pi t^{p+1}}, \quad (|t| \gg |a|). \tag{2.228} $$

Taking $t \to \infty$ we conclude that for large t

$$ {}_aD_t^p\varphi(t) \approx {}_0D_t^p\varphi(t). \tag{2.229} $$

This means that the impact of the instant at which the dynamical process $\varphi(t)$ started (and therefore the impact of the transient effects) vanishes as $t \to \infty$, and therefore for large t the fractional derivative with the lower terminal $t = a$ can be replaced, for example, with the fractional derivative with the lower terminal $t = 0$.

Another way of making the interval between the lower terminal and the upper terminal larger is considering $a \to -\infty$ for a fixed value of t. In this case we have $|a| \gg |t|$ and therefore

$$ (t-a)^{k-p} = a^{k-p}\left(1 - \frac{t}{a}\right)^{k-p} = a^{k-p}\left(1 - \frac{(k-p)t}{a} + O\left(\frac{t^2}{a^2}\right)\right), \tag{2.230} $$

from which it follows that

$$ (t-a)^{k-p} \approx a^{k-p} + \frac{(p-k)ta^k}{a^{p+1}}, \quad (|t| \gg |a|) \tag{2.231} $$

Substitution of (2.231) into (2.224) gives

$$ \begin{aligned} {}_aD_t^p\varphi(t) \approx \frac{\Gamma(p+1)\sin(p\pi)}{\pi} & \left\{ \sum_{k=0}^{\infty} \frac{(-1)^k (t-(t-a))^{k-p}}{(p-k)\,k!} \varphi^{(k)}(t) \right. \\ & \left. + \frac{t}{a^{p+1}} \sum_{k=0}^{\infty} \frac{(-1)^k (t-(t-a))^k \varphi^{(k)}(t)}{k!} \right\} \end{aligned} \tag{2.232} $$

and using (2.223) we obtain

$$ {}_aD_t^p\varphi(t) \approx {}_{t-a}D_t^p\varphi(t) + \frac{t\Gamma(p+1)\sin(p\pi)\varphi(t-a)}{\pi a^{p+1}}, \quad (|a| \gg |t|). \tag{2.233} $$

Therefore, we may conclude that, under certain conditions on $\varphi(t)$, for large negative values of a the fractional derivative with a fixed lower terminal can be replaced with the fractional derivative with a moving lower terminal:

$$ {}_aD_t^p\varphi(t) \approx {}_{t-a}D_t^p\varphi(t). \tag{2.234} $$

2.8 Laplace Transforms of Fractional Derivatives

2.8.1 Basic Facts on the Laplace Transform

Let us recall some basic facts about the Laplace transform.

The function $F(s)$ of the complex variable s defined by

$$F(s) = L\{f(t); s\} = \int_0^\infty e^{-st} f(t) dt \tag{2.235}$$

is called the Laplace transform of the function $f(t)$, which is called the original. For the existence of the integral (2.235) the function $f(t)$ must be of exponential order α, which means that there exist positive constants M and T such that

$$e^{-\alpha t}|f(t)| \leq M \quad \text{for all} \quad t > T.$$

In other words, the function $f(t)$ must not grow faster then a certain exponential function when $t \to \infty$.

We will denote the Laplace transforms by uppercase letters and the originals by lowercase letters.

The original $f(t)$ can be restored from the Laplace tranform $F(s)$ with the help of the inverse Laplace transform

$$f(t) = L^{-1}\{F(s); t\} = \int_{c-\imath\infty}^{c+\imath\infty} e^{st} F(s) ds, \quad c = Re(s) > c_0, \tag{2.236}$$

where c_0 lies in the right half plane of the absolute convergence of the Laplace integral (2.235).

The direct evaluation of the inverse Laplace transform using the formula (2.236) is often complicated; however, sometime it gives useful information on the behaviour of the unknown original $f(t)$ which we look for.

The Laplace transform of the convolution

$$f(t) * g(t) = \int_0^t f(t-\tau) g(\tau) d\tau = \int_0^t f(\tau) g(t-\tau) d\tau \tag{2.237}$$

of the two functions $f(t)$ and $g(t)$, which are equal to zero for $t < 0$, is equal to the product of the Laplace transform of those function:

$$L\{f(t) * g(t);\, s\} = F(s)\, G(s) \tag{2.238}$$

under the assumption that both $F(s)$ and $G(s)$ exist. We will use the property (2.238) for the evaluation of the Laplace transform of the Riemann–Liouville fractional integral.

Another useful property which we need is the formula for the Laplace transform of the derivative of an integer order n of the function $f(t)$:

$$L\{f^{n}(t);\, s\} = s^{n}F(s) - \sum_{k=0}^{n-1} s^{n-k-1} f^{(k)}(0) = s^{n}F(s) - \sum_{k=0}^{n-1} s^{k} f^{(n-k-1)}(0), \tag{2.239}$$

which can be obtained from the definition (2.235) by integrating by parts under the assumption that the corresponding integrals exist.

In the following sections on the Laplace transforms of fractional derivatives we consider the lower terminal $a = 0$.

2.8.2 Laplace Transform of the Riemann–Liouville Fractional Derivative

We will start with the Laplace transform of the Riemann–Liouville and Grünwald–Letnikov fractional integral of order $p > 0$ defined by (2.88), which we can write as a convolution of the functions $g(t) = t^{p-1}$ and $f(t)$:

$$_{0}\mathbf{D}_{t}^{-p} f(t) = {}_{0}D_{t}^{-p} f(t) = \frac{1}{\Gamma(p)} \int_{0}^{t} (t-\tau)^{p-1} f(\tau) d\tau = t^{p-1} * f(t). \tag{2.240}$$

The Laplace transform of the function t^{p-1} is [62]

$$G(s) = L\{t^{p-1};\, s\} = \Gamma(p) s^{-p}. \tag{2.241}$$

Therefore, using the formula for the Laplace transform of the convolution (2.238) we obtain the Laplace transform of the Riemann–Liouville and the Grünwald–Letnikov fractional integral:

$$L\{{}_{0}\mathbf{D}_{t}^{-p} f(t);\, s\} = L\{{}_{0}D_{t}^{-p} f(t);\, s\} = s^{-p} F(s). \tag{2.242}$$

Now let us turn to the evaluation of the Laplace transform of the Riemann–Liouville fractional derivative, which for this purpose we write in the form:

$$ {}_0\mathbf{D}_t^p f(t) = g^{(n)}(t), \tag{2.243}$$

$$ g(t) = {}_0\mathbf{D}_t^{-(n-p)} f(t) \frac{1}{\Gamma(k-p)} \int_0^t (t-\tau)^{n-p-1} f(\tau) d\tau, \tag{2.244}$$

$$ (n-1 \leq p < n). $$

The use of the formula for the Laplace transform of an integer-order derivative (2.239) leads to

$$ L\{ {}_0\mathbf{D}_t^p f(t); \, s\} = s^n G(s) - \sum_{k=0}^{n-1} s^k g^{(n-k-1)}(0). \tag{2.245}$$

The Laplace transform of the function $g(t)$ is evaluated by (2.242):

$$ G(s) = s^{-(n-p)} F(s). \tag{2.246}$$

Additionally, from the definition of the Riemann–Liouville fractional derivative (2.103) it follows that

$$ g^{(n-k-1)}(t) = \frac{d^{n-k-1}}{dt^{n-k-1}} {}_0\mathbf{D}_t^{-(n-p)} f(t) = {}_0\mathbf{D}_t^{p-k-1} f(t). \tag{2.247}$$

Substituting (2.246) and (2.247) into (2.245) we obtain the following final expression for the Laplace transform of the Riemann–Liouville fractional derivative of order $p > 0$:

$$ L\{ {}_0\mathbf{D}_t^p f(t); \, s\} = s^p F(s) - \sum_{k=0}^{n-1} s^k \Big[{}_0\mathbf{D}_t^{p-k-1} f(t) \Big]_{t=0}. \tag{2.248}$$

$$ (n-1 \leq p < n). $$

This Laplace transform of the Riemann–Liouville fractional derivative is well known (see, for example, [179] or [153]). However, its practical applicability is limited by the absense of the physical interpretation of the limit values of fractional derivatives at the lower terminal $t = 0$. At the time of writing, such an interpretation is not known.

2.8.3 Laplace Transform of the Caputo Derivative

To establish the Laplace transform formula for the Caputo fractional derivative let us write the Caputo derivative (2.138) in the form:

$$ {}_0^C D_t^p f(t) = {}_0\mathbf{D}_t^{-(n-p)} g(t), \qquad g(t) = f^{(n)}(t), \tag{2.249}$$

$$(n-1 < p \leq n). \tag{2.250}$$

Using the formula (2.242) for the Laplace transform of the Riemann–Liouville fractional integral gives

$$L\{ {}_0^C D_t^p f(t);\, s\} = s^{-(n-p)} G(s), \tag{2.251}$$

where, according to (2.239),

$$G(s) = s^n F(s) - \sum_{k=0}^{n-1} s^{n-k-1} f^{(k)}(0) = s^n F(s) - \sum_{k=0}^{n-1} s^k f^{(n-k-1)}(0). \tag{2.252}$$

Introducing (2.252) into (2.251) we arrive at the Laplace transform formula for the Caputo fractional derivative:

$$L\{ {}_0^C D_t^p f(t)\} = s^p F(s) - \sum_{k=0}^{n-1} s^{p-k-1} f^{(k)}(0), \tag{2.253}$$

$$(n-1 < p \leq n).$$

Since this formula for the Laplace transform of the Caputo derivative involves the values of the function $f(t)$ and its derivatives at the lower terminal $t = 0$, for which a certain physical interpretation exists (for example, $f(0)$ is the initial position, $f'(0)$ is the initial velocity, $f''(0)$ is the initial acceleration), we can expect that it can be useful for solving applied problems leading to linear fractional differential equations with constant coefficients with accompanying initial conditions in traditional form.

2.8.4 Laplace Transform of the Grünwald–Letnikov Fractional Derivative

First let us consider the case of $0 \leq p < 1$, when the Grünwald–Letnikov fractional derivative (2.54) with the lower terminal $a = 0$ of the function

$f(t)$, which is bounded at $t = 0$, can be written in the following form:

$$ {}_0D_t^p f(t) = \frac{f(0)t^{-p}}{\Gamma(1-p)} + \frac{1}{\Gamma(1-p)} \int_0^t (t-\tau)^{-p} f'(\tau) d\tau. \tag{2.254} $$

Using the Laplace transform of the power function (2.241), the formula for the Laplace transform of the convolution (2.238) and the Laplace transform of the integer-order derivative (2.239) we obtain:

$$ L\{ {}_0D_t^p f(t); \, s\} = \frac{f(0)}{s^{1-p}} + \frac{1}{s^{1-p}} \Big(sF(s) - f(0) \Big) = s^p F(s). \tag{2.255} $$

An example of an application of the formula (2.255) is given in [75].

The Laplace transform of the Grünwald–Letnikov fractional derivative of order $p > 1$ does not exist in the classical sense, because in such a case we have non-integrable functions in the sum in the formula (2.54). The Laplace transforms of such functions are given by divergent integrals. However, the Laplace tranform of the power function (2.241) allows analytic continuation with respect to the parameter p. This approach is equivalent to the generalized functions (distributions) approach [76]. Divergent integrals in such a sense are called finite-part integrals. In this way, assuming that $m < p < m+1$, and using the Laplace transform of the power function (2.241), the formula for the Laplace transform of the convolution (2.238) and the Laplace transform of the integer-order derivative (2.239), we obtain:

$$ \begin{aligned} L\{ {}_0D_t^p f(t); \, s\} &= \sum_{k=0}^{m} f^{(k)}(0) L\{\frac{t^{-p+k}}{\Gamma(-p+m+1)}; \, s\} \\ &\quad + L\{\frac{t^{m-p}}{\Gamma(-p+m+1)} * f^{(m+1)}(t); \, s\} \\ &= \sum_{k=0}^{m} f^{(k)}(0) s^{p-k-1} \\ &\quad + s^{p-m-1} \Big(s^{m+1} F(s) - \sum_{k=0}^{m} f^{(k)}(0) s^{m-k} \Big) \\ &= s^p F(s). \end{aligned} \tag{2.256} $$

We arrived at the same formula as (2.255).

In applications it is necessary to keep in mind that the formula (2.256) holds in the classical sense only for $0 < p < 1$; for $p > 1$ it holds in the sense of generalized functions (distributions) and, therefore, the formulation of an applied problem must also be done using the language of generalized functions, as well as interpretation of the results obtained in this way.

2.8.5 Laplace Transform of the Miller–Ross Sequential Fractional Derivative

Let us introduce the following notation for the Miller–Ross sequential derivative:

$$ {}_a\mathcal{D}_t^{\sigma_m} \equiv {}_aD_t^{\alpha_m}\, {}_aD_t^{\alpha_{m-1}} \ldots {}_aD_t^{\alpha_1}; \tag{2.257}$$

$$ {}_a\mathcal{D}_t^{\sigma_m - 1} \equiv {}_aD_t^{\alpha_m - 1}\, {}_aD_t^{\alpha_{m-1}} \ldots {}_aD_t^{\alpha_1}; \tag{2.258}$$

$$ \sigma_m = \sum_{j=1}^{m} \alpha_j, \qquad 0 < \alpha_j \leq 1, \quad (j = 1, 2, \ldots, m). $$

We can establish the following formula for the Laplace transform of the sequential derivative (2.257):

$$ L\Big\{ {}_0\mathcal{D}_t^{\sigma_m} f(t);\, s \Big\} = s^{\sigma_m} F(s) - \sum_{k=0}^{m-1} s^{\sigma_m - \sigma_{m-k}} \Big[{}_0\mathcal{D}_t^{\sigma_{m-k}-1} f(t) \Big]_{t=0}, \tag{2.259}$$

$$ {}_a\mathcal{D}_t^{\sigma_{m-k}-1} \equiv {}_aD_t^{\alpha_{m-k}-1}\, {}_aD_t^{\alpha_{m-k-1}} \ldots {}_aD_t^{\alpha_1}, $$

$$ (k = 0, 1, \ldots, m-1). $$

The particular case of (2.259) for $f(t)$ m-times differentiable, $\alpha_m = \mu$, $\alpha_k = 1$, $(k = 1, 2, \ldots, m-1)$ was obtained by Caputo [24, p. 41] much earlier. Taking $\alpha_1 = \mu$, $\alpha_k = 1$, $(k = 2, 3, \ldots, m)$ leads under obvious assumptions to the classical formula (2.248).

To prove the formula (2.259) let us first recall the Laplace transform formula for the Riemann–Liouville fractional derivative (2.248), which in the case of $0 < \alpha \leq 1$ takes the form:

$$ L\{ {}_0\mathbf{D}_t^{\alpha} f(t);\, s\} = s^{\alpha} F(s) - \Big[{}_0\mathbf{D}_t^{\alpha-1} f(t) \Big]_{t=0}, \tag{2.260}$$

and then use the formula (2.260) subsequently m times:

$$ L\Big\{ {}_0\mathcal{D}_t^{\sigma_m} f(t);\, s \Big\} = L\Big\{ {}_0D_t^{\alpha_m}\, {}_0\mathcal{D}_t^{\sigma_{m-1}} f(t);\, s \Big\} $$

$$= p^{\alpha_m} L\Big\{ {}_0\mathcal{D}_t^{\sigma_{m-1}} f(t);\, s\Big\} - \Big[{}_0D_t^{\alpha_m - 1}\, {}_0\mathcal{D}_t^{\sigma_{m-1}} f(t)\Big]_{t=0}$$

$$= p^{\alpha_m} L\Big\{ {}_0\mathcal{D}_t^{\sigma_{m-1}} f(t);\, s\Big\} - \Big[{}_0\mathcal{D}_t^{\sigma_m - 1} f(t)\Big]_{t=0}$$

$$= p^{\alpha_m + \alpha_{m-1}} L\Big\{ {}_0\mathcal{D}_t^{\sigma_{m-2}} f(t);\, s\Big\} - p^{\alpha_m}\Big[{}_0\mathcal{D}_t^{\sigma_{m-1} - 1} f(t)\Big]_{t=0} - \Big[{}_0\mathcal{D}_t^{\sigma_m - 1} f(t);\Big]_{t=0}$$

$$\ldots \qquad \ldots \qquad \ldots$$

$$= s^{\sigma_m} F(s) - \sum_{k=0}^{m-1} s^{\sigma_m - \sigma_{m-k}} \Big[{}_0\mathcal{D}_t^{\sigma_{m-k} - 1} f(t)\Big]_{t=0}.$$

2.9 Fourier Transforms of Fractional Derivatives

2.9.1 Basic Facts on the Fourier Transform

The exponential Fourier transform of a continuous function $h(t)$ absolutely integrable in $(-\infty, \infty)$ is defined by

$$F_e\{h(t);\, \omega\} = \int_{-\infty}^{\infty} e^{i\omega t} h(t)\, dt, \tag{2.261}$$

and the original $h(t)$ can be restored from its Fourier transform $H_e(t)$ with the help of the inverse Fourier transform:

$$h(t) = \frac{1}{2\pi} \int_{-\infty}^{\infty} H_e(\omega) e^{-i\omega t}\, d\omega. \tag{2.262}$$

As above, we will denote originals by lowercase letters, and their transforms by uppercase letters.

The Fourier transform of the convolution

$$h(t) * g(t) = \int_{-\infty}^{\infty} h(t-\tau) g(\tau)\, d\tau = \int_{-\infty}^{\infty} h(\tau) g(t-\tau)\, d\tau \tag{2.263}$$

of the two functions $h(t)$ and $g(t)$, which are defined in $(-\infty, \infty)$, is equal to the product of their Fourier transforms:

$$F_e\{h(t) * g(t);\, \omega\} = H_e(\omega)\, G_e(\omega) \tag{2.264}$$

under the assumption that both $H_e(\omega)$ and $G_e(\omega)$ exist. We will use the property (2.264) for the evaluation of the Fourier transforms of the Riemann–Liouville fractional integral and Fourier transforms of fractional derivatives.

Another useful property of the Fourier transform, which is frequently used in solving applied problems, is the Fourier transform of derivatives of $h(t)$. Namely, if $h(t)$, $h'(t)$, ..., $h^{(n-1)}(t)$ vanish for $t \to \pm\infty$, then the Fourier transform of the n-th derivative of $h(t)$ is

$$F_e\{h^{(n)}(t);\, \omega\} = (-i\omega)^n H_e(\omega). \tag{2.265}$$

The Fourier transform is a powerful tool for frequency domain analysis of linear dynamical systems.

2.9.2 Fourier Transform of Fractional Integrals

First we will evaluate the Fourier transform of the Riemann–Liouville fractional integral with the lower terminal $a = -\infty$, i.e. of

$${}_{-\infty}\mathbf{D}_t^{-\alpha} g(t) = \frac{1}{\Gamma(\alpha)} \int\limits_{-\infty}^{t} (t-\tau)^{\alpha-1} g(\tau)\, d\tau, \tag{2.266}$$

where we assume $0 < \alpha < 1$.

Let us start with the Laplace transform of the function

$$h(t) = \frac{t^{\alpha-1}}{\Gamma(\alpha)}$$

(see formula (2.241)), which can be written as

$$\frac{1}{\Gamma(\alpha)} \int\limits_{0}^{\infty} t^{\alpha-1} e^{-st}\, dt = s^{-\alpha}. \tag{2.267}$$

Let us take $s = -i\omega$, where ω is real. It follows from the Dirichlet theorem [68, p. 564] that in such a case the integral (2.267) converges if $0 < \alpha < 1$. Therefore, we immediately obtain the Fourier transform of the function

$$h_+(t) = \begin{cases} \dfrac{t^{\alpha-1}}{\Gamma(\alpha)}, & (t > 0) \\ 0, & (t \le 0) \end{cases}$$

in the form

$$F_e\{h_+(t);\, \omega\} = (-i\omega)^{-\alpha}. \qquad (2.268)$$

Now we can find the Fourier transform of the Riemann–Liouville fractional integral (2.266), which can be written as a convolution (2.263) of the functions $h_+(t)$ and $g(t)$:

$${}_{-\infty}\mathbf{D}_t^{-\alpha} f(t) = h_+(t) * g(t). \qquad (2.269)$$

Using the rule (2.264) we obtain:

$$F_e\left\{{}_{-\infty}\mathbf{D}_t^{-\alpha} g(t);\, \omega\right\} = (i\omega)^{-\alpha} G(\omega), \qquad (2.270)$$

where $G(\omega)$ is the Fourier transform of the function $g(t)$.

The formula (2.270) gives also the Fourier transform of the Grünwald–Letnikov fractional integral ${}_{-\infty}D_t^{-\alpha} g(t)$ and the Caputo fractional integral ${}_{-\infty}^{\ C}D_t^{-\alpha} g(t)$, because in this case they coincide with the Riemann–Liouville fractional integral.

2.9.3 Fourier Transform of Fractional Derivatives

Let us now evaluate the Fourier transform of fractional derivatives.

Considering the lower terminal $a = -\infty$ and requiring the resonable behaviour of $g(t)$ and its derivatives for $t \to -\infty$ we can perform integration by parts and write the Riemann–Liouville, the Grünwald–Letnikov and the Caputo definition in the same form:

$$\left.\begin{matrix} {}_{-\infty}\mathbf{D}_t^{\alpha} g(t) \\ {}_{-\infty}D_t^{\alpha} g(t) \\ {}_{-\infty}^{\ C}D_t^{\alpha} g(t) \end{matrix}\right\} = \frac{1}{\Gamma(n-\alpha)} \int_{-\infty}^{t} \frac{g^{(n)}(\tau)\, d\tau}{(t-\tau)^{\alpha+1-n}} = {}_{-\infty}\mathbf{D}_t^{\alpha-n} g^{(n)}(t), \qquad (2.271)$$

$$(n-1 < \alpha < n).$$

The Fourier transform of (2.271) with the use of the Fourier transform of the Riemann–Liouville fractional integral (2.270) and then the

Fourier transform of an integer-order derivative (2.265) gives the following formula for the exponential Fourier transform of the Riemann–Liouville, Grünwald–Letnikov and Caputo fractional derivatives with the lower terminal $a = -\infty$:

$$\begin{aligned} F_e\{D^\alpha g(t);\, \omega\} &= (-i\omega)^{\alpha-n} F_e\left\{g^{(n)}(t);\, \omega\right\} \\ &= (-i\omega)^{\alpha-n}(-i\omega)^n G(\omega) \\ &= (-i\omega)^\alpha G(\omega), \end{aligned} \tag{2.272}$$

where the symbol D^α denotes any of the mentioned fractional differentiations (Riemann–Liouville ${}_{-\infty}\mathbf{D}_t^\alpha$, Grünwald–Letnikov ${}_{-\infty}D_t^\alpha g(t)$ or Caputo ${}_{-\infty}^{\;C}D_t^\alpha g(t)$).

The Fourier transform of fractional derivatives has been used, for example, by H. Beyer and S. Kempfle [19] for analysing the oscillation equation with a fractional-order damping term:

$$y''(t) + a\ {}_{-\infty}D_t^\alpha y(t) + by(t) = f(t), \tag{2.273}$$

by S. Kempfle and L. Gaul [115] for constructing global solutions of linear fractional differential equations with constant coefficients, and implicitly by R. R. Nigmatullin and Ya. E. Ryabov [166] for studying relaxation processes in insulators.

2.10 Mellin Transforms of Fractional Derivatives

2.10.1 Basic Facts on the Mellin Transform

The Mellin integral transform $F(s)$ of a function $f(t)$, which is defined in the interval $(0,\ \infty)$ is

$$F(s) = \mathcal{M}\left\{f(t);\ s\right\} = \int_0^\infty f(t) t^{s-1} dt, \tag{2.274}$$

where s is complex, such as

$$\gamma_1 < Re(s) < \gamma_2.$$

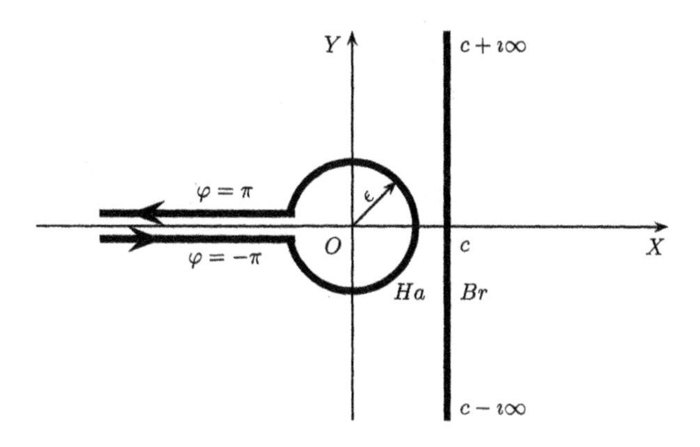

Figure 2.2: *The Bromwich (Br) and the Hankel (Ha) contours.*

The Mellin transform (2.274) exists if the function $f(t)$ is piecewise continuous in every closed interval $[a, b] \subset (0, \infty)$ and

$$\int_0^1 |f(t)| t^{\gamma_1 - 1} dt < \infty, \qquad \int_1^\infty |f(t)| t^{\gamma_2 - 1} dt < \infty. \tag{2.275}$$

If the function $f(t)$ also satisfies the Dirichlet conditions in every closed interval $[a, b] \subset (0, \infty)$, then the function $f(t)$ can be restored using the inverse Mellin transform formula

$$f(t) = \frac{1}{2\pi i} \int_{\gamma - i\infty}^{\gamma + i\infty} F(s) t^{-s} ds, \quad (0 < t < \infty), \tag{2.276}$$

in which $\gamma_1 < \gamma < \gamma_2$. The integration contour in (2.276) is the Bromwich contour (contour Br in Fig. 2.2).

It follows from the definition (2.274) that

$$\mathcal{M}\{t^\alpha f(t);\ s\} = \mathcal{M}\{f(t);\ s + \alpha\} = F(s + \alpha). \tag{2.277}$$

The Mellin transform of the Mellin convolution

$$f(t) * g(t) = \int_0^\infty f(t\tau) g(\tau) d\tau \tag{2.278}$$

of functions $f(t)$ and $g(t)$, the Mellin transforms of which are $F(s)$ and $G(s)$, is given by the formula (see, for example, [249]):

$$\mathcal{M}\Big\{\int_0^\infty f(t\tau)g(\tau)d\tau\,;\ s\Big\} = F(s)\,G(1-s), \tag{2.279}$$

and combining (2.277) and (2.279) gives

$$\mathcal{M}\Big\{t^\lambda\int_0^\infty \tau^\mu f(t\tau)g(\tau)d\tau\,;\ s\Big\} = F(s+\lambda)\,G(1-s-\lambda+\mu). \tag{2.280}$$

Integrating repeatedly by parts, we have the following relationship for the Mellin transform of an integer-order derivative:

$$\begin{aligned}
\mathcal{M}\Big\{f^{(n)}(t);\ s\Big\} &= \int_0^\infty f^{(n)}(t)\,t^{s-1}dt \\
&= \Big[f^{(n-1)}(t)\,t^{s-1}\Big]_0^\infty - (s-1)\int_0^\infty f^{(n-1)}(t)\,t^{s-2}dt \\
&= \Big[f^{(n-1)}(t)\,t^{s-1}\Big]_0^\infty - (s-1)\mathcal{M}\Big\{f^{(n-1)}(t);\ s-1\Big\} \\
&= \dots \\
&= \sum_{k=0}^{n-1}(-1)^k\frac{\Gamma(s)}{\Gamma(s-k)}\Big[f^{(n-k-1)}(t)\,t^{s-k-1}\Big]_0^\infty \\
&\qquad\qquad + (-1)^n\frac{\Gamma(s)}{\Gamma(s-n)}F(s-n) \\
&= \sum_{k=0}^{n-1}\frac{\Gamma(1-s+k)}{\Gamma(1-s)}\Big[f^{(n-k-1)}(t)\,t^{s-k-1}\Big]_0^\infty \\
&\qquad\qquad + \frac{\Gamma(1-s+n)}{\Gamma(1-s)}F(s-n)
\end{aligned} \tag{2.281}$$

where $F(s)$ is the Mellin transform of $f(t)$.

If $f(t)$ and $Re(s)$ are such that all substitutions of the limits $t=0$ and $t=\infty$ give zero, then the formula (2.281) takes the simplest form:

$$\mathcal{M}\Big\{f^{(n)}(t);\ s\Big\} = \frac{\Gamma(1-s+n)}{\Gamma(1-s)}F(s-n). \tag{2.282}$$

2.10.2 Mellin Transform of the Riemann–Liouville Fractional Integrals

Let us evaluate the Mellin transform of the Riemann–Liouville fractional integral ${}_0D_t^{-\alpha}f(t)$, $(\alpha > 0)$. Using the substitution $\tau = t\xi$ we can write

$$\begin{aligned} {}_0D_t^{-\alpha}f(t) &= \frac{1}{\Gamma(\alpha)}\int_0^t (t-\tau)^{\alpha-1}f(\tau)d\tau \\ &= \frac{t^\alpha}{\Gamma(\alpha)}\int_0^1 (1-\xi)^{\alpha-1}f(t\xi)d\xi \\ &= \frac{t^\alpha}{\Gamma(\alpha)}\int_0^\infty f(t\xi)g(\xi)d\xi \end{aligned} \tag{2.283}$$

where

$$g(t) = \begin{cases} (1-t)^{\alpha-1}, & (0 \le t < 1) \\ 0, & (t \ge 1). \end{cases}$$

The Mellin transform of the function $g(t)$ gives simply the Euler beta function (1.20),

$$\mathcal{M}\Big\{g(t);\ s\Big\} = B(\alpha,\ s) = \frac{\Gamma(\alpha)\,\Gamma(s)}{\Gamma(\alpha+s)}. \tag{2.284}$$

Then using the formulas (2.280), (2.283), and (2.284), we obtain:

$$\mathcal{M}\Big\{\ {}_0D_t^{-\alpha}f(t);\ s\Big\} = \frac{1}{\Gamma(\alpha)}\,F(s+\alpha)\,B(\alpha,\ 1-s-\alpha),$$

or

$$\mathcal{M}\Big\{\ {}_0D_t^{-\alpha}f(t);\ s\Big\} = \frac{\Gamma(1-s-\alpha)}{\Gamma(1-s)}\,F(s+\alpha), \tag{2.285}$$

where $F(s)$ is the Mellin transform of the function $f(t)$.

The obtained formula (2.285) reminds us of the particular case of the Mellin transform of the n-th derivative of $f(t)$ (2.282), which can be formally obtained from (2.285) by putting $\alpha = -n$.

2.10.3 Mellin Transform of the Riemann–Liouville Fractional Derivative

Let us take $0 \le n-1 < \alpha < n$. According to the definition of the Riemann–Liouville fractional derivative, we can write

$${}_0D_t^{\alpha}f(t) = \frac{d^n}{dt^n}\,{}_0D_t^{-(n-\alpha)}f(t).$$

Temporarily denoting $g(t) = {}_0D_t^{-(n-\alpha)} f(t)$, and using formulas (2.281) and (2.285), we have:

$$\begin{aligned}
\mathcal{M}\Big\{ {}_0D_t^{\alpha} f(t);\ s\Big\} &= \mathcal{M}\Big\{ \frac{d^n}{dt^n}\, {}_0D_t^{\alpha-n} f(t);\ s\Big\} = \mathcal{M}\Big\{ g^{(n)};\ s\Big\} \\
&= \sum_{k=0}^{n-1} \frac{\Gamma(1-s+k)}{\Gamma(1-s)} \Big[g^{(n-k-1)}(t)\, t^{s-k-1} \Big]_0^{\infty} \\
&\quad + \frac{\Gamma(1-s+n)}{\Gamma(1-s)} G(s-n) \\
&= \sum_{k=0}^{n-1} \frac{\Gamma(1-s+k)}{\Gamma(1-s)} \left[\frac{d^{n-k-1}}{dt^{n-k-1}}\, {}_0D_t^{\alpha-n} f(t)\, t^{s-k-1} \right]_0^{\infty} \\
&\quad + \frac{\Gamma(1-s+n)}{\Gamma(1-s)} \, \frac{\Gamma(1-(s-n)-(n-\alpha))}{\Gamma(1-(s-n))} \\
&\quad \times F((s-n)+(n-\alpha)),
\end{aligned} \tag{2.286}$$

or

$$\begin{aligned}
\mathcal{M}\Big\{ {}_0D_t^{\alpha} f(t);\ s\Big\} &= \sum_{k=0}^{n-1} \frac{\Gamma(1-s+k)}{\Gamma(1-s)} \Big[{}_0D_t^{\alpha-k-1} f(t)\, t^{s-k-1} \Big]_0^{\infty} \\
&\quad + \frac{\Gamma(1-s+\alpha)}{\Gamma(1-s)}\, F(s-\alpha).
\end{aligned} \tag{2.287}$$

If $0 < \alpha < 1$, then (2.287) takes on the form:

$$\mathcal{M}\Big\{ {}_0D_t^{\alpha} f(t);\ s\Big\} = \Big[{}_0D_t^{\alpha-1} f(t)\, t^{s-1} \Big]_0^{\infty} + \frac{\Gamma(1-s+\alpha)}{\Gamma(1-s)}\, F(s-\alpha). \tag{2.288}$$

If the function $f(t)$ and $Re(s)$ are such that all substitutions of the limits $t = 0$ and $t = \infty$ in the formula (2.287) give zero, then it takes on the simplest form:

$$\mathcal{M}\Big\{ {}_0D_t^{\alpha} f(t);\ s\Big\} = \frac{\Gamma(1-s+\alpha)}{\Gamma(1-s)}\, F(s-\alpha). \tag{2.289}$$

2.10.4 Mellin Transform of the Caputo Fractional Derivative

Let us take $0 \leq n-1 \leq \alpha < n$. Temporarily denoting $h(t) = f^{(n)}(t)$ and using the formulas (2.285) and (2.281), we have:

$$\mathcal{M}\Big\{ {}_0^C D_t^{\alpha} f(t);\ s\Big\} = \mathcal{M}\Big\{ {}_0D_t^{-(n-\alpha)} f^{(n)}(t);\ s\Big\}$$

$$
\begin{aligned}
&= \mathcal{M}\Big\{ {}_0D_t^{-(n-\alpha)} h(t);\ s \Big\} \\
&= \frac{1-s-(n-\alpha)}{\Gamma(1-s)} H(s+(n-\alpha)) \\
&= \frac{\Gamma(1-s-n+\alpha)}{\Gamma(1-s)} \\
&\quad \times \Bigg\{ \sum_{k=0}^{n-1} \frac{\Gamma(1-(s+n-\alpha)+k)}{\Gamma(1-(s+n-\alpha))} \Big[f^{(n-k-1)}(t)\, t^{(s+n-\alpha)-k-1} \Big]_0^\infty \\
&\qquad + \frac{\Gamma(1-(s+n-\alpha)+n)}{\Gamma(s+n-\alpha-n)} F((s+n-\alpha)-n) \Bigg\} \\
&= \sum_{k=0}^{n-1} \frac{\Gamma(1-s-n+\alpha+k)}{\Gamma(1-s)} \Big[f^{(n-k-1)}(t)\, t^{s+n-\alpha-k-1} \Big]_0^\infty \\
&\qquad + \frac{\Gamma(1-s-\alpha)}{\Gamma(1-s)} F(s-\alpha)
\end{aligned}
\tag{2.290}
$$

or

$$
\begin{aligned}
\mathcal{M}\Big\{ {}_0^CD_t^\alpha f(t);\ s \Big\} &= \sum_{k=0}^{n-1} \frac{\Gamma(\alpha-m-s)}{\Gamma(1-s)} \Big[f^{(k)}\, t^{s-\alpha+k} \Big]_0^\infty \\
&\quad + \frac{\Gamma(1-s-\alpha)}{\Gamma(1-s)} F(s-\alpha).
\end{aligned}
\tag{2.291}
$$

For $0 < \alpha < 1$ the formula (2.291) takes on the form:

$$
\mathcal{M}\Big\{ {}_0^CD_t^\alpha f(t);\ s \Big\} = \frac{\Gamma(\alpha-s)}{\Gamma(1-s)} \Big[f(t)\, t^{s-\alpha} \Big]_0^\infty + \frac{\Gamma(1-s+\alpha)}{\Gamma(1-s)} F(s-\alpha).
\tag{2.292}
$$

If the function $f(t)$ and $Re(s)$ are such that all substitutions of the limits $t = 0$ and $t = \infty$ in the formula (2.291) give zero, then it takes on the simplest form:

$$
\mathcal{M}\Big\{ {}_0^CD_t^\alpha f(t);\ s \Big\} = \frac{\Gamma(1-s+\alpha)}{\Gamma(1-s)} F(s-\alpha).
\tag{2.293}
$$

2.10.5 Mellin Transform of the Miller–Ross Fractional Derivative

Let us recall the following notation for the Miller–Ross sequential fractional derivative defined by (2.257):

$$
{}_a\mathcal{D}_t^{\sigma_m} \equiv {}_aD_t^{\alpha_m}\, {}_aD_t^{\alpha_{m-1}} \ldots {}_aD_t^{\alpha_1};
$$

$$ {}_a\mathcal{D}_t^{\sigma_m - 1} \equiv {}_aD_t^{\alpha_m - 1}\, {}_aD_t^{\alpha_{m-1}} \ldots {}_aD_t^{\alpha_1}; $$

$$ \sigma_m = \sum_{j=1}^{m} \alpha_j, \qquad 0 < \alpha_j \le 1, \quad (j = 1, 2, \ldots, m). $$

Let us start with $m = 2$. Temporarily denoting $g(t) = {}_0D_t^{\beta}(t)$ and using the formula (2.287), we have:

$$ \begin{aligned} \mathcal{M}\Big\{ {}_0\mathcal{D}_t^{\sigma_2} f(t);\ s \Big\} &= \mathcal{M}\Big\{ {}_0D_t^{\alpha_2} g(t);\ s \Big\} \\ &= \Big[{}_0D_t^{\alpha_2 - 1} g(t)\, t^{s-1} \Big]_0^{\infty} + \frac{\Gamma(1 - s + \alpha_2)}{\Gamma(1 - s)} G(s - \alpha_2) \\ &= \Big[{}_0D_t^{\sigma_2 - 1} f(t)\, t^{s-1} \Big]_0^{\infty} \\ &\quad + \frac{\Gamma(1 - s + \alpha_2)}{\Gamma(1 - s)} \Big\{ \Big[{}_0D_t^{\alpha_1 - 1} f(t)\, t^{s - \alpha_2 - 1} \Big]_0^{\infty} \\ &\quad + \frac{\Gamma(1 - (s - \alpha_2) + \alpha_1)}{\Gamma(1 - (s - \alpha_2))} F((s - \alpha_2) - \alpha_1) \Big\} \\ &= \Big[{}_0\mathcal{D}_t^{\sigma_2 - 1} f(t)\, t^{s-1} \Big]_0^{\infty} \\ &\quad + \frac{\Gamma(1 - s + \alpha_2)}{\Gamma(1 - s)} \Big[{}_0\mathcal{D}_t^{\sigma_1 - 1} f(t)\, t^{s - \alpha_2 - 1} \Big]_0^{\infty} \\ &\quad + \frac{\Gamma(1 - s + \sigma_2)}{\Gamma(1 - s)} F(s - \sigma_2). \end{aligned} \tag{2.294} $$

It can be shown by induction that in the general case the Mellin transform of the Miller–Ross sequential fractional derivative is given by the following expression:

$$ \begin{aligned} \mathcal{M}\Big\{ {}_0\mathcal{D}_t^{\sigma_n} f(t);\ s \Big\} &= \sum_{k=1}^{n} \frac{\Gamma(1 - s + \sigma_n - \sigma_k)}{\Gamma(1 - s)} \Big[{}_0\mathcal{D}_t^{\sigma_k - 1} f(t)\, t^{s - \sigma_n + \sigma_k - 1} \Big]_0^{\infty} \\ &\quad + \frac{\Gamma(1 - s + \sigma_n)}{\Gamma(1 - s)} F(s - \sigma_n). \end{aligned} \tag{2.295} $$

If the function $f(t)$ and $Re(s)$ are such that all substitutions of the limits $t = 0$ and $t = \infty$ in the formula (2.295) give zero, then it takes on the simplest form:

$$ \mathcal{M}\Big\{ {}_0\mathcal{D}_t^{\sigma_n} f(t);\ s \Big\} = \frac{\Gamma(1 - s + \sigma_n)}{\Gamma(1 - s)} F(s - \sigma_n), \tag{2.296} $$

which is the same as expressions (2.287) and (2.291) for the Riemann–Liouville derivative and the Caputo derivative. Therefore, for functions with suitable behaviour for $t \to 0$ and $t \to \infty$ the Mellin transform of the Riemann–Liouville, Caputo, and Miller–Ross fractional derivative may coincide. This is similar to what we also observe in the case of the Laplace and Fourier transforms.

Under the conditions of coincidence, using of (2.277) gives

$$\mathcal{M}\Big\{t^{\alpha}D^{\alpha}f(t);\ s\Big\} = \frac{\Gamma(1-s)}{\Gamma(1-s-\alpha)}F(s), \tag{2.297}$$

and

$$\begin{aligned}\mathcal{M}\Big\{\sum_{k=0}^{n} a_k t^{\alpha+k} D^{\alpha+k} f(t);\ s\Big\} &= F(s)\Gamma(1-s)\sum_{k=0}^{n}\frac{a_k}{\Gamma(1-s-\alpha-k)} \\ &= \frac{\Gamma(s)\Gamma(1-s)}{\Gamma(1-s-\alpha)}\sum_{k=0}^{n}(-1)^k a_k \prod_{j=0}^{k-1}(s+\alpha+1),\end{aligned} \tag{2.298}$$

where D^{α} denotes the Riemann–Liouville, or Caputo, or Miller–Ross fractional derivative.

In particular, we have

$$\mathcal{M}\Big\{t^{\alpha+1}D^{\alpha+1}f(t) + t^{\alpha}D^{\alpha}f(t);\ s\Big\} = \frac{\Gamma(1-s)(1-s-\alpha)}{\Gamma(1-s-\alpha)}F(s), \tag{2.299}$$

and putting $\alpha = 1$ gives the well-known property of the Mellin transform:

$$\mathcal{M}\Big\{t^2 f''(t) + t f'(t);\ s\Big\} = s^2 F(s), \tag{2.300}$$

which is often used in applied problems.

Chapter 3

Existence and Uniqueness Theorems

In this chapter we consider the question of the existence and uniqueness of solutions of initial-value problems for fractional-order differential equations. All the results are given for equations in terms of the Miller–Ross sequential fractional derivatives. This allows direct application of the obtained results to fractional differential equations with the Riemann–Liouville, the Grünwald–Letnikov, and the Caputo fractional derivatives, which can be considered as particular cases of the Miller–Ross sequential fractional derivative.

First we consider the case of linear fractional differential equations with continuous coefficients and prove the existence and uniqueness theorems for a one-term fractional differential equation and for the n-term fractional differential equation.

Then we give the existence and uniqueness theorem for a general fractional differential equation of general form. We also demonstrate on examples that the method of the proof can sometimes be used directly as a method of solution of initial-value problems for fractional differential equations.

Finally, we study the dependence of solutions of a fractional differential equation of a general form on initial conditions and show that small changes of initial conditions may cause only small changes of solution in the intervals not containing the starting point of the interval (the lower terminal of fractional derivatives appearing in the considered equation).

3.1 Linear Fractional Differential Equations

In this section the existence and uniqueness of solutions of initial-value problems for linear fractional differential equations with sequential derivatives are discussed.

Let us consider the following initial-value problem:

$$ {}_0\mathcal{D}_t^{\sigma_n} y(t) + \sum_{j=1}^{n-1} p_j(t)\, {}_0\mathcal{D}_t^{\sigma_{n-j}} y(t) + p_n(t) y(t) = f(t), \qquad (3.1) $$

$$ (0 < t < T < \infty) $$

$$ \Big[{}_0\mathcal{D}_t^{\sigma_k - 1} y(t) \Big]_{t=0} = b_k, \qquad k = 1, \ldots, , n, \qquad (3.2) $$

where

$$ {}_a\mathcal{D}_t^{\sigma_k} \equiv {}_aD_t^{\alpha_k}\, {}_aD_t^{\alpha_{k-1}} \ldots {}_aD_t^{\alpha_1}; $$

$$ {}_a\mathcal{D}_t^{\sigma_k - 1} \equiv {}_aD_t^{\alpha_k - 1}\, {}_aD_t^{\alpha_{k-1}} \ldots {}_aD_t^{\alpha_1}; $$

$$ \sigma_k = \sum_{j=1}^{k} \alpha_j, \quad (k = 1, 2, \ldots, n); $$

$$ 0 < \alpha_j \leq 1, \quad (j = 1, 2, \ldots, n), $$

and $f(t) \in L_1(0, T)$, i.e.

$$ \int_0^T |f(t)| dt < \infty. $$

For simplicity of notation, in the following we assume $f(t) \equiv 0$ for $t > T$.

As the first step, let us consider the case of $p_k(t) \equiv 0$, $(k = 1, \ldots, n)$.

THEOREM 3.1 ∘ *If $f(t) \in L_1(0, T)$, then the equation*

$$ {}_0\mathcal{D}_t^{\sigma_n} y(t) = f(t) \qquad (3.3) $$

has the unique solution $y(t) \in L_1(0, T)$, which satisfies the initial conditions (3.2) •

Proof. Let us construct a solution of the considered problem. Application of the Laplace transform formula of a sequential fractional derivative (2.259) to equation (3.3) gives

$$s^{\sigma_n} Y(s) - \sum_{k=0}^{n-1} s^{\sigma_n - \sigma_{n-k}} \left[{}_0\mathcal{D}_t^{\sigma_{n-k}-1} y(t) \right]_{t=0} = F(s), \qquad (3.4)$$

where $Y(s)$ and $F(s)$ denote the Laplace transforms of $y(t)$ and $f(t)$. Using the initial conditions (3.2), we can write

$$Y(s) = s^{-\sigma_n} F(s) + \sum_{k=0}^{n-1} b_{n-k} s^{-\sigma_{n-k}}, \qquad (3.5)$$

and the inverse Laplace transform gives

$$y(t) = \frac{1}{\Gamma(\sigma_n)} \int_0^t (t-\tau)^{\sigma_n - 1} f(\tau)\, d\tau + \sum_{k=0}^{n-1} \frac{b_{n-k}}{\Gamma(\sigma_{n-k})} t^{\sigma_{n-k}-1}, \qquad (3.6)$$

or, putting $i = n - k$,

$$y(t) = \frac{1}{\Gamma(\sigma_n)} \int_0^t (t-\tau)^{\sigma_n - 1} f(\tau)\, d\tau + \sum_{i=1}^{n} \frac{b_i}{\Gamma(\sigma_i)} t^{\sigma_i - 1}. \qquad (3.7)$$

Using the rule for the Riemann–Liouville fractional differentiation of the power function (2.117), and taking into account that

$$\frac{1}{\Gamma(-m)} = 0, \qquad m = 0, 1, 2, \dots,$$

we easily obtain that

$${}_0\mathcal{D}_t^{\sigma_k} \left(\frac{t^{\sigma_i - 1}}{\Gamma(\sigma_i)} \right) = \begin{cases} \dfrac{t^{\sigma_i - \sigma_k - 1}}{\Gamma(\sigma_i - \sigma_k)}, & (k < i) \\ 0, & (k \geq i) \end{cases} \qquad (3.8)$$

$${}_0\mathcal{D}_t^{\sigma_k - 1} \left(\frac{t^{\sigma_i - 1}}{\Gamma(\sigma_i)} \right) = \begin{cases} \dfrac{t^{\sigma_i - \sigma_k}}{\Gamma(1 + \sigma_i - \sigma_k)}, & (k < i) \\ 1, & (k = i) \\ 0, & (k > i) \end{cases} \qquad (3.9)$$

where $k = 1, 2, \ldots, n$, and $i = 1, 2, \ldots, n$.

It follows from (3.7) that $y(t) \in L_1(0, T)$. Using (3.8) and (3.9), the direct substitution of the function $y(t)$ defined by the expression (3.7) in the equation (3.3) and initial conditions (3.2) shows that $y(t)$ satisfies them, and therefore, the existence of the solution is proved.

The uniqueness follows from the linearity of fractional differentiation and the properties of the Laplace transform. Indeed, if there exist two solutions, $y_1(t)$ and $y_2(t)$, of the considered problem, then the function $z(t) = y_1(t) - y_2(t)$ must satisfy the equation ${}_0\mathcal{D}_t^{\sigma_n} z(t) = 0$ and the zero initial conditions. Then the Laplace transform of $z(t)$ is $Z(s) = 0$, and therefore $z(t) = 0$ almost everywhere in the considered interval, which proves that the solution in $L_1(0, T)$ is unique.

Now we can prove the existence and uniqueness of the solution of the problem (3.1)–(3.2).

THEOREM 3.2 ∘ *If $f(t) \in L_1(0, T)$, and $p_j(t)$ ($j = 1, \ldots, n$) are continuous functions in the closed interval $[0, T]$, then the initial-value problem (3.1)–(3.2) has a unique solution $y(t) \in L_1(0, T)$* •

Proof. The method of proof of this theorem uses the basic idea found in the paper by M. M. Dzhrbashyan and A. B. Nersesyan [50].

Let us assume that the problem (3.1)–(3.2) has a solution $y(t)$, and denote

$$ {}_0\mathcal{D}_t^{\sigma_n} y(t) = \varphi(t). \tag{3.10} $$

Using Theorem 3.1 we can write

$$ y(t) = \frac{1}{\Gamma(\sigma_n)} \int_0^t (t-\tau)^{\sigma_n - 1} \varphi(t)\, dt + \sum_{i=1}^{n} b_i \frac{t^{\sigma_i - 1}}{\Gamma(\sigma_i)}. \tag{3.11} $$

Substituting (3.11) into equation (3.1) written in the form

$$ {}_0\mathcal{D}_t^{\sigma_n} y(t) + \sum_{k=1}^{n-1} p_{n-k}(t)\, {}_0\mathcal{D}_t^{\sigma_k} y(t) + p_n(t) y(t) = f(t), $$

and using (3.8), we obtain the Volterra integral equation of the second kind for the function $\varphi(t)$:

$$ \varphi(t) + \int_0^t K(t,\tau)\, \varphi(\tau)\, dt = g(t), \tag{3.12} $$

where

$$K(t,\tau) = p_n(t)\frac{(t-\tau)^{\sigma_n-1}}{\Gamma(\sigma_n)} + \sum_{k=1}^{n-1} p_{n-k}(t)\frac{(t-\tau)^{\sigma_n-\sigma_k-1}}{\Gamma(\sigma_n-\sigma_k)},$$

$$g(t) = f(t) - p_n(t)\sum_{i=1}^{n} b_i \frac{t^{\sigma_i-1}}{\Gamma(\sigma_i)} - \sum_{k=1}^{n-1} p_{n-k}(t) \sum_{i=k+1}^{n} b_i \frac{t^{\sigma_i-\sigma_k-1}}{\Gamma(\sigma_i-\sigma_k)}.$$

Since the functions $p_j(t)$ $(j = 1, \ldots, n)$ are continuous in $[0, T]$, then the kernel $K(t,\tau)$ can be written in the form of a weakly singular kernel

$$K(t,\tau) = \frac{K^*(t,\tau)}{(t-\tau)^{1-\mu}}, \tag{3.13}$$

where $K^*(t,\tau)$ is continuous for $0 \le t \le T, \quad 0 \le \tau \le T,$ and

$$\mu = \min\{\sigma_n, \sigma_n - \sigma_{n-1}, \sigma_n - \sigma_{n-2}, \ldots, \sigma_n - \sigma_1,\} = \min\{\sigma_n, \alpha_n\}.$$

Similarly, $g(t)$ can be written in the form

$$g(t) = \frac{g^*(t)}{t^{1-\nu}}, \tag{3.14}$$

where $g^*(t)$ is continuous in $[0, T]$, and

$$\begin{aligned} \nu = \min\{&\sigma_1, \ldots, \sigma_n;\ \sigma_2 - \sigma_1, \ldots, \sigma_n - \sigma_1; \\ &\sigma_3 - \sigma_2, \ldots, \sigma_n - \sigma_2;\ \ldots;\ \sigma_n - \sigma_{n-1}\} \\ = \min\{&\sigma_1, \ldots, \sigma_n; \alpha_2, \ldots, \alpha_n\} = \min\{\alpha_1, \alpha_2, \ldots, \alpha_n\}. \end{aligned}$$

Obviously, $0 < \mu \le 1$, $0 < \nu \le 1$. It is known (for example, [220]) that the equation (3.12) with the weakly singular kernel (3.13) and the right-hand side $g(t) \in L_1(0, T)$ has a unique solution $\varphi(t) \in L_1(0, T)$. Then, according to Theorem 3.1, the unique solution $y(t) \in L_1(0, T)$ of the problem (3.10), (3.2), which is at the same time the solution of the problem (3.1)–(3.2), can be determined using the formula (3.11). This ends the proof of Theorem 3.2.

In many applied problems, which are considered in this book, the zero initial conditions on the function $y(t)$ and its integer-order derivatives are used. There are three main reasons for this:

- our physical interpretation of fractional derivatives (see Section 2.6), from which it follows that zero initial conditions mean the absolute beginning of the process represented by the function $y(t)$,

- difficulties with numerical approximation of initial conditions of the type (3.2),

- the coincidence of the Riemann–Liouville, Grünwald–Letnikov, Caputo, and Miller–Ross derivative in the case of a proper number of zero initial conditions on the function $y(t)$ and its integer-order derivatives; this coincidence prevents misinterpretation of the problem formulation and solution.

Because of this, we consider this particular case of Theorem 3.2 separately.

Let us suppose that $m-1 \leq \sigma_n < m$, and that

$$y^{(j)}(0) = 0, \qquad (j = 0, 1, \ldots, m-1). \tag{3.15}$$

In such a case, using the composition rule for the Riemann–Liouville derivatives (see Section 2.3.6), we can replace all sequential fractional derivatives in equation (3.1) by the Riemann–Liouville fractional derivatives of the same order σ_k, which gives:

$${}_0D_t^{\sigma_n} y(t) + \sum_{j=1}^{n-1} p_j(t)\, {}_0D_t^{\sigma_{n-j}} y(t) + p_n(t) y(t) = f(t). \tag{3.16}$$

Recalling Section 2.3.7, we note that it follows from the zero initial conditions (3.15) that all the conditions (3.2) are zero, i.e., $b_k = 0$, $k = 1, 2, \ldots, n$. Moreover, $f(t)$ can be taken to be continuous, and the following statement holds:

THEOREM 3.3 ∘ *If $f(t)$ and $p_j(t)$ $(j = 1, \ldots, n)$ are continuous functions in the closed interval $[0, T]$, then the initial-value problem (3.16), (3.15), where $m-1 \leq \sigma_n < m$, $\sigma_n > \sigma_{n-1} > \sigma_{n-2} > \ldots > \sigma_2 > \sigma_1 > 0$, has a unique solution $y(t)$, which is continuous in $[0, T]$* •

3.2 Fractional Differential Equation of a General Form

Besides linear fractional differential equations, non-linear equations also appear in applications. Because of this, in this section we discuss the existence and uniqueness of a solution of an initial-value problem for the fractional differential equation of a general form in terms of the

Miller–Ross sequential fractional derivatives. Due to the link between the Miller–Ross, the Riemann–Liouville, the Grünwald–Letnikov, and the Caputo fractional derivatives, the results given below can be used for all these mutations of fractional differentiation.

Let us consider the initial-value problem

$$ {}_0\mathcal{D}_t^{\sigma_n} y(t) = f(t, y), \tag{3.17} $$

$$ \left[{}_0\mathcal{D}_t^{\sigma_k - 1} y(t) \right]_{t=0} = b_k, \qquad k = 1, \ldots, n, \tag{3.18} $$

where, as in the previous section,

$$ {}_a\mathcal{D}_t^{\sigma_k} \equiv {}_aD_t^{\alpha_k}\, {}_aD_t^{\alpha_{k-1}} \ldots {}_aD_t^{\alpha_1}; $$

$$ {}_a\mathcal{D}_t^{\sigma_k - 1} \equiv {}_aD_t^{\alpha_k - 1}\, {}_aD_t^{\alpha_{k-1}} \ldots {}_aD_t^{\alpha_1}; $$

$$ \sigma_k = \sum_{j=1}^{k} \alpha_j, \qquad (k = 1, 2, \ldots, n); $$

$$ 0 < \alpha_j \leq 1, \qquad (j = 1, 2, \ldots, n). $$

Let us suppose that $f(t, y)$ is defined in a domain G of a plane (t, y), and define a region $R(h, K) \subset G$ as a set of points $(t, y) \in G$, which satisfy the following inequalities:

$$ 0 < t < h, \qquad \left| t^{1-\sigma_1} y(t) - \sum_{i=1}^{n} b_i \frac{t^{\sigma_i - \sigma_1}}{\Gamma(\sigma_i)} \right| \leq K, \tag{3.19} $$

where h and K are constant.

THEOREM 3.4 ∘ *Let $f(t, y)$ be a real-valued continuous function, defined in the domain G, satisfying in G the Lipschitz condition with respect to y, i.e.,*

$$ |f(t, y_1) - f(t, y_2)| \leq A |y_1 - y_2|, $$

such that

$$ |f(t, y)| \leq M < \infty \qquad \textit{for all} \qquad (t, y) \in G. $$

Let also

$$ K \geq \frac{M h^{\sigma_n - \sigma_1 + 1}}{\Gamma(1 + \sigma_n)}. $$

Then there exists in the region $R(h, K)$ a unique and continuous solution $y(t)$ of the problem (3.17)–(3.18) •

Proof. The method of proof of this theorem is based on the ideas due to E. Pitcher and W. E. Sewell [188] and M. A. Al-Bassam [4].

First, let us reduce the problem (3.17)–(3.18) to an equivalent fractional integral equation.

Using the formula (3.7), or performing subsequently the fractional integration of order α_n, α_{n-1}, ..., α_1 with the help of the composition rule (2.108), we obtain

$$y(t) = \sum_{i=1}^{n} \frac{b_i}{\Gamma(\sigma_i)} t^{\sigma_i - 1} + \frac{1}{\Gamma(\sigma_n)} \int_0^t (t-\tau)^{\sigma_n - 1} f(\tau, y(\tau))\, d\tau. \qquad (3.20)$$

We see that if $y(t)$ satisfies (3.17)–(3.18), then it also satisfies the equation (3.20).

On the other hand, if $y(t)$ is a solution of (3.20), then applying to (3.20) the sequential fractional derivative operator ${}_0\mathcal{D}_t^{\sigma_n}$ and the formula (3.8) we obtain for $y(t)$ the fractional differential equation (3.17). The use of (3.9) shows that if $y(t)$ satisfies (3.20), then it satisfies the conditions (3.18). Therefore, the equation (3.20) is equivalent to the initial-value problem (3.17)–(3.18).

Now let us define the sequence of functions $y_0(t)$, $y_1(t)$, $y_2(t)$, ..., by the following relationships:

$$y_0(t) = \sum_{i=1}^{n} \frac{b_i}{\Gamma(\sigma_i)} t^{\sigma_i - 1}, \qquad (3.21)$$

$$y_m(t) = \sum_{i=1}^{n} \frac{b_i}{\Gamma(\sigma_i)} t^{\sigma_i - 1} + \frac{1}{\Gamma(\sigma_n)} \int_0^t (t-\tau)^{\sigma_n - 1} f(\tau, y_{m-1}(\tau))\, d\tau, \quad (3.22)$$

$$m = 1, 2, 3, \ldots$$

We will show that $\lim_{m\to\infty} y_m(t)$ exists and gives the required solution $y(t)$ of the equation (3.20).

First, it can be shown by induction that for $0 < t \leq h$ we have $y_m(t) \in R(h, K)$ for all m. Indeed,

$$\left| t^{1-\sigma_1} y_m(t) - \sum_{i=1}^{n} b_i \frac{t^{\sigma_i - \sigma_1}}{\Gamma(\sigma_i)} \right| = \left| \frac{t^{1-\sigma_1}}{\Gamma(\sigma_n)} \int_0^t (t-\tau)^{\sigma_n - 1} f(\tau, y_{m-1}(\tau)) d\tau \right|$$

$$\leq \frac{M t^{\sigma_n - \sigma_1 + 1}}{\Gamma(1+\sigma_n)} \leq \frac{M h^{\sigma_n - \sigma_1 + 1}}{\Gamma(1+\sigma_n)} \leq K, \qquad (3.23)$$

and for the same reasons we have the same inequality for $y_1(t)$:

$$\left| t^{1-\sigma_1} y_1(t) - \sum_{i=1}^{n} b_i \frac{t^{\sigma_i - \sigma_1}}{\Gamma(\sigma_i)} \right| \leq \frac{M h^{\sigma_n - \sigma_1 + 1}}{\Gamma(1+\sigma_n)} \leq K.$$

Further, it can be shown by induction that for all m

$$|y_m(t) - y_{m-1}(t)| \leq \frac{M A^{m-1} t^{m\sigma_n}}{\Gamma(1+m\sigma_n)}. \qquad (3.24)$$

Indeed, using (3.23), we have for $m = 1$:

$$|y_1(t) - y_0(t)| \leq \frac{M t^{\sigma_n}}{\Gamma(1+\sigma_n)}, \qquad (0 < t \leq h). \qquad (3.25)$$

Let us suppose that

$$|y_{m-1}(t) - y_{m-2}(t)| \leq \frac{M A^{m-2} t^{(m-1)\sigma_n}}{\Gamma(1+(m-1)\sigma_n)}, (0 < t \leq h). \qquad (3.26)$$

Then, using (3.22) and (3.26), and recalling the Riemann–Liouville fractional derivative of the power function (2.117), we have

$$|y_m(t) - y_{m-1}(t)| \leq \frac{A}{\Gamma(\sigma_n)} \int_0^t (t-\tau)^{\sigma_n} |y_{m-1}(\tau) - y_{m-2}(\tau)| d\tau$$

$$\leq \frac{M A^{m-1}}{\Gamma(1+(m-1)\sigma_n)} \frac{1}{\Gamma(\sigma_n)} \int_0^t (t-\tau)^{\sigma_n - 1} \tau^{(m-1)\sigma_n} d\tau$$

$$= \frac{M A^{m-1}}{\Gamma(1+(m-1)\sigma_n)} \, {}_0D_t^{-\sigma_n} t^{(m-1)\sigma_n}$$

$$= \frac{M A^{m-1}}{\Gamma(1+(m-1)\sigma_n)} \frac{\Gamma(1+(m-1)\sigma_n) t^{(m-1)\sigma_n + \sigma_n}}{\Gamma(1+(m-1)\sigma_n + \sigma_n)}$$

$$= \frac{M A^{m-1} t^{m\sigma_n}}{\Gamma(1+m\sigma_n)}. \qquad (3.27)$$

This means that (3.24) holds for all m.

Now let us consider the series

$$y^*(t) = \lim_{m\to\infty}\Big(y_m(t) - y_0(t)\Big) = \sum_{j=1}^{\infty}\Big(y_j(t) - y_{j-1}(t)\Big). \tag{3.28}$$

According to the estimate (3.24), for $0 \le t \le h$ the absolute value of its terms is less than the corresponding terms of the convergent numeric series

$$M\sum_{j=1}^{\infty}\frac{MA^{j-1}h^{j\sigma_n}}{\Gamma(1+j\sigma_n)} = \frac{M}{A}\Big(E_{\sigma_n,1}(Ah_n^{\sigma}) - 1\Big), \tag{3.29}$$

where $E_{\lambda,\mu}(z)$ is the Mittag-Leffler function (see Section 1.2). This means that the series (3.28) converges uniformly. Obviously, each term $\Big(y_j(t) - y_{j-1}(t)\Big)$ of the series (3.28) is a continuous function of t for $0 \le t \le h$. Therefore, the sum of the series (3.28), $y^*(t)$, is a continuous function for $0 \le t \le h$, and

$$y(t) = \lim_{m\to\infty} y_m(t) = y_0(t) + y^*(t)$$

is a continuous function for $0 < t \le h$.

The uniform convergence of the sequence of $y_m(t)$ allows us to take $m \to \infty$ in the relationship (3.22). This gives the equation (3.20), showing that $y(t)$, the limit function of the process defined by (3.21) and (3.22), is the solution of (3.20).

Finally, let us prove the uniqueness of the solution. Let us suppose that $\tilde{y}(t)$ is another solution of the equation (3.20), which is continuous in the interval $0 < t \le h$. Then it follows from (3.20) that the function $z(t) = y(t) - \tilde{y}(t)$ satisfies the equation

$$z(t) = \frac{1}{\Gamma(\sigma_n)}\int_0^t (t-\tau)^{\sigma_n - 1} f(\tau, z(\tau))d\tau, \tag{3.30}$$

from which it follows that $z(0) = 0$. Therefore, $z(t)$ is continuous for $0 \le t \le h$. Then $|z(t)| < B$ for $0 \le t \le h$, where B is constant, and we obtain from the equation (3.30) that

$$|z(t)| \le \frac{ABt^{\sigma_n}}{\Gamma(1+\sigma_n)}, \qquad (0 \le t \le h). \tag{3.31}$$

Repeating this estimates j times, we obtain

$$|z(t)| \le \frac{A^j\, B\, t^{j\sigma_n}}{\Gamma(\sigma_n)}, \qquad j = 1,\, 2,\, \dots \tag{3.32}$$

In the right-hand side we recognize — up to the constant multiplier B — the general term of the series for the Mittag-Leffler function $E_{\sigma_n,1}(At^{\sigma}_n)$, and therefore for all t

$$\lim_{j\to\infty} \frac{A^j\, t^{j\sigma_n}}{\Gamma(\sigma_n)} = 0.$$

Taking the limit of (3.32) as $j \to \infty$, we conclude that $z(t) \equiv 0$, and $\tilde{y}(t) \equiv y(t)$ for $0 < t \le h$. This ends the proof of Theorem 3.4.

3.3 Existence and Uniqueness Theorem as a Method of Solution

In some cases, Theorem 3.4 can be used directly as a method for the solution of fractional differential equations. We will illustrate this below on two examples.

Example 3.1. Let us consider the initial-value problem in terms of sequential fractional derivatives (the notation is the same as in Theorem 3.4):

$${}_0\mathcal{D}_t^{\sigma_n} y(t) = \lambda y(t) \tag{3.33}$$

$$\left[{}_0\mathcal{D}_t^{\sigma_k - 1} y(t)\right]_{t=0} = b_k, \qquad k = 1,\, \dots,\, ,\, n. \tag{3.34}$$

In this case we have $f(t,y) = \lambda y$. In accordance with the proof of Theorem 3.4, let us take

$$y_0(t) = \sum_{i=1}^{n} b_i\, \frac{t^{\sigma_i - 1}}{\Gamma(\sigma_i)}, \tag{3.35}$$

$$\begin{aligned} y_m(t) &= y_0(t) + \frac{\lambda}{\Gamma(\sigma_n)} \int_0^t (t-\tau)^{\sigma_n - 1} y_{m-1}(\tau)\, d\tau \\ &= y_0(t) + \lambda\, {}_0D_t^{-\sigma_n} y_{m-1}(t), \end{aligned} \tag{3.36}$$

$$m = 1,\, 2,\, 3,\, \dots$$

Using (3.35) and (3.36), and applying the formula for the fractional differentiation of the power function (2.117), we obtain:

$$\begin{aligned} y_1(t) &= y_0(t) + \lambda\, {}_0D_t^{-\sigma_n} \left\{ \sum_{i=1}^{n} b_i \frac{t^{\sigma_i - 1}}{\Gamma(\sigma_i)} \right\} \\ &= y_0(t) + \lambda \sum_{i=1}^{n} b_i \frac{t^{\sigma_n + \sigma_i - 1}}{\Gamma(\sigma_n + \sigma_i)}, \\ y_2(t) &= y_0(t) + \lambda\, {}_0D_t^{-\sigma_n} y_1(t) \\ &= y_0(t) + \lambda\, {}_0D_t^{-\sigma_n} \left\{ y_0(t) + \lambda \sum_{i=1}^{n} b_i \frac{t^{\sigma_n + \sigma_i - 1}}{\Gamma(\sigma_n + \sigma_i)} \right\} \\ &= y_0(t) + \lambda \sum_{i=1}^{n} b_i \frac{t^{\sigma_n + \sigma_i - 1}}{\Gamma(\sigma_n + \sigma_i)} + \lambda^2 \sum_{i=1}^{n} b_i \frac{t^{2\sigma_n + \sigma_i - 1}}{\Gamma(2\sigma_n + \sigma_i)} \\ &= \sum_{i=1}^{n} b_i \sum_{k=0}^{2} \frac{\lambda^k\, t^{k\sigma_n + \sigma_i - 1}}{\Gamma(k\sigma_n + \sigma_i)}, \end{aligned}$$

and it can be shown by induction that

$$y_m(t) = \sum_{i=1}^{n} b_i \sum_{k=0}^{m} \frac{\lambda^k\, t^{k\sigma_n + \sigma_i - 1}}{\Gamma(k\sigma_n + \sigma_i)}, \qquad m = 1, 2, 3, \ldots \tag{3.37}$$

Taking the limit of (3.37) as $m \to \infty$, we obtain the solution of the problem (3.33)–(3.34):

$$y(t) = \sum_{i=1}^{n} b_i \sum_{k=0}^{\infty} \frac{\lambda^k\, t^{k\sigma_n + \sigma_i - 1}}{\Gamma(k\sigma_n + \sigma_i)} = \sum_{i=1}^{n} b_i\, t^{\sigma_i - 1} E_{\sigma_n, \sigma_i}(\lambda t^{\sigma_n}), \tag{3.38}$$

where $E_{\alpha,\beta}(z)$ is the Mittag-Leffler function (see Section 1.2). In this particular example, the solution (3.38) can also be obtained by the Laplace transform method.

If $n = 1$ and $\alpha_1 = 1$, then the initial-value problem (3.33)–(3.34) takes on the form

$$y'(t) = \lambda y(t), \qquad y(0) = b_1, \tag{3.39}$$

and, taking into account the relationship (1.57), the formula (3.38) gives the classical solution of the problem (3.39):

$$y(t) = b_1 E_{1,1}(\lambda t) = e^{\lambda t}.$$

Example 3.2. [4] Let us consider the following initial-value problem in terms of the Riemann–Liouville fractional derivatives:

$$ {}_0D_t^{\alpha} y(t) = t^{\alpha} y(t), \qquad \left[{}_0D_t^{\alpha-1} y(t) \right]_{t=0} = b, \tag{3.40} $$

where $0 < \alpha < 1$.

In this case $f(t, y) = t^{\alpha} y$. In accordance with the proof of Theorem 3.4, let us take

$$ y_0(t) = b \frac{t^{\alpha-1}}{\Gamma(\alpha)}, \tag{3.41} $$

$$ y_m(t) = b \frac{t^{\alpha-1}}{\Gamma(\alpha)} + \frac{1}{\Gamma(\alpha)} \int_0^t (t-\tau)^{\alpha-1} \tau^{\alpha} y_{m-1}(\tau)\, d\tau \tag{3.42} $$

$$ m = 1, 2, 3, \ldots $$

Using (3.35) and (3.36), and applying the formula for the fractional differentiation of the power function (2.117), it can be shown by induction that

$$ y_m(t) = b \frac{t^{\alpha-1}}{\Gamma(\alpha)} + b \sum_{k=1}^{m} \frac{\Gamma(2\alpha)\,\Gamma(4\alpha) \cdots \Gamma(2k\alpha)}{\Gamma(\alpha)\,\Gamma(3\alpha) \cdots \Gamma(2k\alpha + \alpha)} t^{\alpha(2k+1)-1}, $$

$$ m = 1, 2, 3, \ldots $$

and taking the limit as $m \to \infty$ gives the solution:

$$ y(t) = b \frac{t^{\alpha-1}}{\Gamma(\alpha)} + b \sum_{k=1}^{\infty} \frac{\prod_{j=1}^{k} \Gamma(2j\alpha)}{\prod_{j=0}^{k} \Gamma(2j\alpha + \alpha)} t^{\alpha(2k+1)-1}. \tag{3.43} $$

3.4 Dependence of a Solution on Initial Conditions

In this section we consider the changes in the solution which are caused by small changes in initial conditions.

Let us introduce small changes in the initial conditions (3.18):

$$ \left[{}_0\mathcal{D}_t^{\sigma_k - 1} y(t) \right]_{t=0} = b_k + \delta_k, \qquad k = 1, \ldots, n, \tag{3.44} $$

where δ_k $(k = 1, \ldots, n)$ are arbitrary constants.

The following theorem is a generalization of Al-Bassam's result [4].

THEOREM 3.5 ∘ *Let the assumptions of Theorem 3.4 hold. If $y(t)$ is a solution of the equation (3.17) satisfying the initial conditions (3.18), and $\tilde{y}(t)$ is a solution of the same equations satisfying the initial conditions (3.44), then for $0 < t \leq h$ the following holds:*

$$|y(t) - \tilde{y}(t)| \leq \sum_{i=1}^{n} |\delta_i| \, t^{\sigma_i - 1} \, E_{\sigma_n, \sigma_i}(A t_n^{\sigma}), \tag{3.45}$$

where $E_{\alpha,\beta}(z)$ is the Mittag-Leffler function. •

Proof. In accordance with Theorem 3.4 we have:

$$y(t) = \lim_{m \to \infty} y_m(t),$$

$$y_0(t) = \sum_{i=1}^{n} b_i \frac{t^{\sigma_i - 1}}{\Gamma(\sigma_i)}, \tag{3.46}$$

$$y_m(t) = y_0(t) + \frac{1}{\Gamma(\sigma_n)} \int_0^t (t-\tau)^{\sigma_n - 1} f(\tau, y_{m-1}(\tau)) d\tau, \tag{3.47}$$

$$m = 1, 2, \ldots,$$

and

$$\tilde{y}(t) = \lim_{m \to \infty} \tilde{y}_m(t),$$

$$\tilde{y}_0(t) = \sum_{i=1}^{n} (b_i + \delta_i) \frac{t^{\sigma_i - 1}}{\Gamma(\sigma_i)}, \tag{3.48}$$

$$\tilde{y}_m(t) = \tilde{y}_0(t) + \frac{1}{\Gamma(\sigma_n)} \int_0^t (t-\tau)^{\sigma_n - 1} f(\tau, \tilde{y}_{m-1}(\tau)) d\tau, \tag{3.49}$$

$$m = 1, 2, \ldots$$

From (3.46) and (3.48) it directly follows that

$$|y_0(t) - \tilde{y}_0(t)| \leq \sum_{i=1}^{n} |\delta_i| \frac{t^{\sigma_i - 1}}{\Gamma(\sigma_i)}. \tag{3.50}$$

Using subsequently the relationships (3.47) and (3.49), the Lipschitz condition for the function $f(t, y)$, the inequality (3.50), and the rule for the Riemann–Liouville fractional differentiation of the power function (2.117), we obtain:

$$
\begin{aligned}
|y_1(t) - \tilde{y}_1(t)| &= \Bigg| \sum_{i=1}^{n} \delta_i \frac{t^{\sigma_i - 1}}{\Gamma(\sigma_i)} \\
&\quad + \frac{1}{\Gamma(\sigma_n)} \int_0^t (t-\tau)^{\sigma_n - 1} \{ f(\tau, y_0(\tau)) - f(\tau, \tilde{y}_0(\tau)) \} \, d\tau \Bigg| \\
&\le \sum_{i=1}^{n} |\delta_i| \frac{t^{\sigma_i - 1}}{\Gamma(\sigma_i)} + \frac{A}{\Gamma(\sigma_n)} \int_0^t (t-\tau)^{\sigma_n - 1} |y_0(\tau) - \tilde{y}_0(\tau)| \, d\tau \\
&\le \sum_{i=1}^{n} |\delta_i| \frac{t^{\sigma_i - 1}}{\Gamma(\sigma_i)} + \frac{A}{\Gamma(\sigma_n)} \int_0^t (t-\tau)^{\sigma_n - 1} \left\{ \sum_{i=1}^{n} |\delta_i| \frac{\tau^{\sigma_i - 1}}{\Gamma(\sigma_i)} \right\} d\tau \\
&\le \sum_{i=1}^{n} |\delta_i| \frac{t^{\sigma_i - 1}}{\Gamma(\sigma_i)} + A \, {}_0D_t^{\sigma_n} \left\{ \sum_{i=1}^{n} |\delta_i| \frac{t^{\sigma_i - 1}}{\Gamma(\sigma_i)} \right\} \\
&= \sum_{i=1}^{n} |\delta_i| \frac{t^{\sigma_i - 1}}{\Gamma(\sigma_i)} + A \sum_{i=1}^{n} |\delta_i| \frac{t^{\sigma_n + \sigma_i - 1}}{\Gamma(\sigma_n + \sigma_i)} \\
&= \sum_{i=1}^{n} |\delta_i| \, t^{\sigma_i - 1} \sum_{k=0}^{1} \frac{A^k \, t^{k\sigma_n}}{\Gamma(k\sigma_n + \sigma_i)}.
\end{aligned}
$$

Similarly, we have

$$
\begin{aligned}
|y_2(t) - \tilde{y}_2(t)| &\le \sum_{i=1}^{n} |\delta_i| \frac{t^{\sigma_i - 1}}{\Gamma(\sigma_i)} + \frac{A}{\Gamma(\sigma_n)} \int_0^t (t-\tau)^{\sigma_n - 1} |y_1(\tau) - \tilde{y}_1(\tau)| \, d\tau \\
&\le \sum_{i=1}^{n} |\delta_i| \frac{t^{\sigma_i - 1}}{\Gamma(\sigma_i)} \\
&\quad + A \, {}_0D_t^{-\sigma_n} \left\{ \sum_{i=1}^{n} |\delta_i| \frac{t^{\sigma_i - 1}}{\Gamma(\sigma_i)} + A \sum_{i=1}^{n} |\delta_i| \frac{t^{\sigma_n + \sigma_i - 1}}{\Gamma(\sigma_n + \sigma_i)} \right\} \\
&= \sum_{i=1}^{n} |\delta_i| \frac{t^{\sigma_i - 1}}{\Gamma(\sigma_i)} \\
&\quad + A \sum_{i=1}^{n} |\delta_i| \frac{t^{\sigma_n + \sigma_i - 1}}{\Gamma(\sigma_n + \sigma_i)} + A^2 \sum_{i=1}^{n} |\delta_i| \frac{t^{2\sigma_n + \sigma_i - 1}}{\Gamma(2\sigma_n + \sigma_i)} \\
&= \sum_{i=1}^{n} |\delta_i| \, t^{\sigma_i - 1} \sum_{k=0}^{2} \frac{A^k \, t^{k\sigma_n}}{\Gamma(k\sigma_n + \sigma_i)},
\end{aligned}
$$

and by induction

$$|y_m(t) - \tilde{y}_m(t)| \le \sum_{i=1}^{n} |\delta_i| \, t^{\sigma_i - 1} \sum_{k=0}^{m} \frac{A^k \, t^{k\sigma_n}}{\Gamma(k\sigma_n + \sigma_i)}, \tag{3.51}$$

$$m = 1, \, 2, \, \ldots$$

Taking the limit of (3.51) as $m \to \infty$, we obtain

$$\begin{aligned} |y(t) - \tilde{y}(t)| &\le \sum_{i=1}^{n} |\delta_i| \, t^{\sigma_i - 1} \sum_{k=0}^{\infty} \frac{A^k \, t^{k\sigma_n}}{\Gamma(k\sigma_n + \sigma_i)} \\ &= \sum_{i=1}^{n} |\delta_i| \, t^{\sigma_i - 1} E_{\sigma_n, \sigma_i}(A t^{\sigma_n}), \end{aligned}$$

which ends the proof of Theorem 3.5.

It follows from this theorem that for every ε between 0 and h small changes in initial conditions (3.18) cause only small changes of the solution in the closed interval $[\varepsilon, \, h]$ (which does not contain zero).

On the other hand, the solution may change significantly in $[0, \, \varepsilon]$. Indeed, if the non-disturbed initial conditions (3.18) are zero (i.e., $b_k = 0$, $k = 1, 2, \ldots, n$), than the non-disturbed solution $y(t)$ is continuous in $[0, \varepsilon]$, and therefore bounded. However, the solution $\tilde{y}(t)$, corresponding to the disturbed initial conditions, may contain terms of the form $\delta_i \, t^{\sigma_i - 1}/\Gamma(\sigma_i)$, which for $\sigma_i < 1$ will make the disturbed solution unbounded at $t = 0$.

Chapter 4

The Laplace Transform Method

Differential equations of fractional order appear more and more frequently in various research areas and engineering applications. An effective and easy-to-use method for solving such equations is needed.

However, known methods have certain disadvantages. Methods, described in detail in [179, 153, 13] for fractional differential equations of rational order, do not work in the case of arbitrary real order. On the other hand, there is an iteration method described in [232], which allows solution of fractional differential equations of arbitrary real order but it works effectively only for relatively simple equations, as well as the series method [179, 70]. Other authors (e.g. [13, 29]) used in their investigations the one-parameter Mittag-Leffler function $E_\alpha(z) = \sum_{k=0}^{\infty} \frac{z^k}{\Gamma(\alpha k+1)}$. Still other authors [235, 80] prefer the Fox H-function [69], which seems to be too general to be frequently used in applications.

Instead of this variety of different methods, we introduce here a method which is free of these disadvantages and suitable for a wide class of initial value problems for fractional differential equations. The method uses the Laplace transform technique and is based on the formula of the Laplace transform of the Mittag-Leffler function in two parameters $E_{\alpha,\beta}(z)$. We hope that this method could be useful for obtaining solutions of different applied problems appearing in physics, chemistry, electrochemistry, engineering, etc.

This chapter deals with the solution of fractional linear differential equations with constant coefficients.

In Section 4.1 we give solutions to some initial-value problems for "standard" fractional differential equations. Some of them were solved by other authors earlier by other methods, and the comparison in such cases just underlines the simplicity and the power of the Laplace transform method.

In Section 4.2 we extend the proposed method for the case of so-called "sequential" fractional differential equations, i.e. equations in terms of the Miller–Ross sequential derivatives). For this purpose, we use the Laplace transform for the Miller–Ross sequential fractional derivative given by formula (2.259). The "sequential" analogues of the problems solved in Section 4.1 are considered. Naturally, we arrive at solutions which are different from those obtained in the Section 4.1.

The operational calculus, which can be applied to the fractional differential equations considered in this chapter, has been developed in the papers by Yu. F. Luchko and H. M. Srivastava [128], and by S. B. Hadid and Yu. Luchko [100]. R. Gorenflo and Yu. Luchko also developed an operational method for solving generalized Abel integral equations of the second kind [86].

4.1 Standard Fractional Differential Equations

The following examples illustrate the use of (1.80) for solving fractional-order differential equations with constant coefficients. In this chapter we use the classical formula for the Laplace transform of the fractional derivative, as given, e.g., in [179, p. 134] or [153, p. 123]:

$$\int\limits_0^\infty e^{-st}\,{}_0D_t^\alpha f(t)\,dt = s^\alpha F(s) - \sum_{k=0}^{n-1} s^k \left[{}_0D_t^{\alpha-k-1} f(t)\right]_{t=0}, \qquad (4.1)$$

$$(n-1 < \alpha \leq n).$$

4.1.1 Ordinary Linear Fractional Differential Equations

In this section we give some examples of the solution of ordinary linear differential equations of fractional order.

Example 4.1. A slight generalization of an equation solved in [179, p. 157]:

$${}_0D_t^{1/2} f(t) + af(t) = 0, \quad (t > 0); \qquad \left[{}_0D_t^{-1/2} f(t)\right]_{t=0} = C. \qquad (4.2)$$

Applying the Laplace transform we obtain

$$F(s) = \frac{C}{s^{1/2} + a}, \qquad C = \left[{}_0D_t^{-1/2} f(t) \right]_{t=0}$$

and the inverse transform with the help of (1.81) gives the solution of (4.2):

$$f(t) = Ct^{-1/2} E_{\frac{1}{2},\frac{1}{2}}(-a\sqrt{t}). \tag{4.3}$$

Using the series expansion (1.56) of $E_{\alpha,\beta}(t)$, it is easy to check that for $a = 1$ the solution (4.3) is identical to the solution

$$f(t) = C(\frac{1}{\sqrt{\pi t}} - e^t \mathrm{erfc}(\sqrt{t})),$$

obtained in [179] in a more complex way.

Example 4.2. Let us consider the following equation:

$${}_0D_t^Q f(t) + {}_0D_t^q f(t) = h(t), \tag{4.4}$$

which "encounters very great difficulties except when the difference $q - Q$ is integer or half-integer" [179, p.156].

Suppose that $0 < q < Q < 1$. The Laplace transform of equation (4.4) leads to

$$(s^Q + s^q)F(s) = C + H(s), \tag{4.5}$$

$$C = \left[{}_0D_t^{q-1} f(t) + {}_0D_t^{Q-1} f(t) \right]_{t=0},$$

and then

$$F(s) = \frac{C + H(s)}{s^Q + s^q} = \frac{C + H(s)}{s^q(s^{Q-q} + 1)} = \Big(C + H(s)\Big) \frac{s^{-q}}{s^{Q-q} + 1}. \tag{4.6}$$

After inversion with the help of (1.80) for $\alpha = Q - q$ and $\beta = Q$, we obtain the solution:

$$f(t) = C\,G(t) + \int_0^t G(t - \tau) h(\tau) d\tau, \tag{4.7}$$

$$C = \left[{}_0D_t^{q-1} f(t) + {}_0D_t^{Q-1} f(t) \right]_{t=0}, \quad G(t) = t^{Q-1} E_{Q-q,Q}(-t^{Q-q}).$$

The case $0 < q < Q < n$ (for example, the equation obtained in [184]) can be solved similarly.

Example 4.3. Let us consider the following initial value problem for a non-homogeneous fractional differential equation under non-zero initial conditions:

$$ {}_0D_t^{\alpha} y(t) - \lambda y(t) = h(t), \quad (t > 0); \tag{4.8} $$

$$ \left[{}_0D_t^{\alpha-k} y(t) \right]_{t=0} = b_k, \quad (k = 1, 2, \ldots, n), \tag{4.9} $$

where $n - 1 < \alpha < n$. The problem (4.8) was solved in [232] by the iteration method. With the help of the Laplace transform and the formula (1.80) we obtain the same solution directly and easily.

Indeed, taking into account the initial conditions (4.9), the Laplace transform of equation (4.8) yields

$$ s^{\alpha} Y(s) - \lambda Y(s) = H(s) + \sum_{k=1}^{n} b_k s^{k-1}, $$

from which

$$ Y(s) = \frac{H(s)}{s^{\alpha} - \lambda} + \sum_{k=1}^{n} b_k \frac{s^{k-1}}{s^{\alpha} - \lambda}, \tag{4.10} $$

and the inverse Laplace transform using (1.80) gives the solution:

$$ y(t) = \sum_{k=1}^{n} b_k t^{\alpha-k} E_{\alpha,\alpha-k+1}(\lambda t^{\alpha}) + \int_0^t (t-\tau)^{\alpha-1} E_{\alpha,\alpha}(\lambda (t-\tau)^{\alpha}) h(\tau) d\tau. \tag{4.11} $$

4.1.2 Partial Linear Fractional Differential Equations

The proposed approach can be successfully used for solving partial linear differential equations of fractional order.

Example 4.4. Nigmatullin's fractional diffusion equation

Let us consider the following initial boundary value problem for the fractional diffusion equation in one space dimension:

$$ {}_0D_t^{\alpha} u(x,t) = \lambda^2 \frac{\partial^2 u(x,t)}{\partial x^2}, \quad (t > 0, \ -\infty < x < \infty); \tag{4.12} $$

$$ \lim_{x \to \pm\infty} u(x,t) = 0; \quad \left[{}_0D_t^{\alpha-1} u(x,t) \right]_{t=0} = \varphi(x). \tag{4.13} $$

We assume here $0 < \alpha < 1$. An equation of the type (4.12) was deduced by Nigmatullin [164] and by Westerlund [253] and studied by Mainardi [131]). We will give a simple solution of problem (4.12) demonstrating once again the advantage of using the Mittag-Leffler function in two parameters (1.56).

Taking into account the boundary conditions (4.13), the Fourier transform with respect to variable x gives:

$$ {}_0D_t^{\alpha}\overline{u}(\beta,t) + \lambda^2\beta^2\overline{u}(\beta,t) = 0 \tag{4.14}$$

$$ \left[{}_0D_t^{\alpha-1}\overline{u}(x,t)\right]_{t=0} = \overline{\varphi}(\beta), \tag{4.15}$$

where β is the Fourier transform parameter. Applying the Laplace transform to (4.14) and using the initial condition (4.15) we obtain

$$ \overline{U}(\beta,s) = \frac{\varphi(\beta)}{s^{\alpha}+\lambda^2\beta^2}. \tag{4.16}$$

The inverse Laplace transform of (4.16) using (1.80) gives

$$ \overline{u}(\beta,t) = \varphi(\beta)t^{\alpha-1}E_{\alpha,\alpha}(-\lambda^2\beta^2t^{\alpha}), \tag{4.17}$$

and then the inverse Fourier transform produces the solution of the initial-value problem (4.12)–(4.13):

$$ u(x,t) = \int_{-\infty}^{\infty} G(x-\xi,t)\varphi(\xi)d\xi, \tag{4.18}$$

$$ G(x,t) = \frac{1}{\pi}\int_0^{\infty} t^{\alpha-1}E_{\alpha,\alpha}(-\lambda^2\beta^2t^{\alpha})\cos\beta x d\beta. \tag{4.19}$$

Let us evaluate integral (4.19). The Laplace transform of (4.19) and formula 1.2(11) from [62] produce

$$ g(x,s) = \frac{1}{\pi}\int_0^{\infty}\frac{\cos(\beta x)\,d\beta}{\lambda^2\beta^2+s^{\alpha}} = \frac{1}{2\lambda}s^{-\alpha/2}e^{-|x|\lambda^{-1}s^{\alpha/2}}, \tag{4.20}$$

and the inverse Laplace transform gives

$$ G(x,t) = \frac{1}{4\lambda\pi i}\int_{Br} e^{st}s^{-\frac{\alpha}{2}}\exp(-|x|\lambda^{-1}s^{\alpha/2})ds. \tag{4.21}$$

Performing the substitutions $\sigma = st$ and $z = |x|\lambda^{-1}t^{-\rho}$ ($\rho = \alpha/2$) and transforming the Bromwich contour Br to the Hankel contour Ha (see Fig. 2.2), as was done in a similar case by Mainardi [131], we obtain

$$G(x,t) = \frac{t^{1-\rho}}{2\lambda}\frac{1}{2\pi i}\int\limits_{Ha} e^{\sigma - z\sigma^{\rho}}\frac{d\sigma}{\sigma^{\rho}} = \frac{1}{2\lambda}t^{\rho-1}W(-z,-\rho,\rho), \qquad z = \frac{|x|}{\lambda t^{\rho}} \tag{4.22}$$

where $W(z,\lambda,\mu)$ is the Wright function (1.156). We would like to note that, in fact, we have just evaluated the Fourier cosine-transform of the function $u_1(\beta) = t^{\alpha-1}E_{\alpha,\alpha}(-\lambda^2\beta^2t^{\alpha})$.

It is easy to check that for $\alpha = 1$ (the traditional diffusion equation) the fractional Green function (4.22) reduces to the classical expression

$$G(x,t) = \frac{1}{2\lambda\sqrt{\pi t}}\exp(-\frac{x^2}{4\lambda^2 t}). \tag{4.23}$$

Example 4.5. The Schneider–Wyss fractional diffusion equation

The following example shows that the proposed method can be effectively applied also to fractional integral equations. Let us consider the Schneider–Wyss type formulation of the diffusion equation [235] (for simplicity and comparison with the previous example — in one spatial dimension):

$$u(x,t) = \varphi(x) + \lambda^2 {}_0D_t^{-\alpha}\frac{\partial^2 u(x,t)}{\partial x^2}, \qquad (-\infty < x < \infty, \quad t > 0); \tag{4.24}$$

$$\lim_{x\to\pm\infty} u(x,t) = 0, \quad u(x,0) = \varphi(x). \tag{4.25}$$

Applying the Fourier transform with respect to the spatial variable x and the Laplace transform with respect to time t, we obtain:

$$\overline{U}(\beta,s) = \frac{\varphi(\beta)s^{\alpha-1}}{s^{\alpha} + \lambda^2\beta^2}, \tag{4.26}$$

where $\overline{U}(\beta,p)$ is the Fourier–Laplace transform of $u(x,t)$, β is the Fourier transfrom parameter and p is the Laplace transform parameter.

Inverting Laplace and Fourier transforms as was done in the previous Example 4.4, we obtain the solution of problem (4.24):

$$u(x,t) = \int\limits_{-\infty}^{\infty} G(x-\xi,t)\varphi(\xi)d\xi, \tag{4.27}$$

$$G(x,t) = \frac{1}{\pi}\int_0^\infty E_{\alpha,1}(-\lambda^2\beta^2 t^\alpha)\cos\beta x\, d\beta. \tag{4.28}$$

Let us evaluate integral (4.28). The Laplace transform of (4.28) and formula 1.2(11) from [62] produce

$$g(x,s) = \frac{s^{\alpha-1}}{\pi}\int_0^\infty \frac{\cos(\beta x)d\beta}{s^\alpha + \lambda^2\beta^2} = \frac{1}{2\lambda}s^{\frac{\alpha}{2}-1}\exp(-|x|\lambda^{-1}s^{\alpha/2}), \tag{4.29}$$

and the inverse Laplace transform gives:

$$G(x,t) = \frac{1}{4\lambda\pi i}\int_{Br} e^{st}s^{\frac{\alpha}{2}-1}\exp(-|x|\lambda^{-1}s^{\alpha/2})ds. \tag{4.30}$$

Performing the substitutions $\sigma = st$ and $z = |x|\lambda^{-1}t^{-\rho}$ ($\rho = \alpha/2$) and transforming the Bromwich contour Br to the Hankel contour Ha (see Fig. 2.2), as was done in a similar case by Mainardi [131], we obtain

$$G(x,t) = \frac{t^{-\rho}}{2\lambda}\frac{1}{2\pi i}\int_{Ha} e^{\sigma - z\sigma^\rho}\frac{d\sigma}{\sigma^{1-\rho}} = \frac{1}{2\lambda}t^{-\rho}M(z,\rho), \quad z = \frac{|x|}{\lambda t^\rho} \tag{4.31}$$

where $M(z,\rho) = W(-z,-\rho,1-\rho)$ is the Mainardi function (1.160).

The last expression is identical to the expression which was obtained by Mainardi [131] in another way.

We would like to note at this point, as in the previous example, that we have just evaluated the Fourier cosine-transform of the function $u_2(\beta) = E_{\alpha,1}(-\lambda^2\beta^2 t^\alpha)$.

For $\alpha = 1$ the fractional Green's function (4.31) also reduces to the classical expression (4.23). The case of an arbitrary number of space dimensions can be solved similarly.

For $\alpha = 1$ both generalizations (Nigmatullin's as well as that by Schneider and Wyss) of the diffusion problem give the standard diffusion problem, and the solutions reduce to the classical solution. However, it is obvious that the asymptotic behaviour of (4.18) and (4.27) for $t \to 0$, and $t \to \infty$ is different (see also the discussion in [80] on two different generalizations of the standard relaxation equation and the discussion in [72] on two fractional models — one based on fractional derivatives and the other based on fractional integrals — for mechanical stress relaxation).

This difference was caused by initial conditions of different types. The class of solutions is determined by the number and the type of initial conditions.

4.2 Sequential Fractional Differential Equations

Let us consider initial value problems of the form:

$$ {}_0\mathcal{L}_t y(t) = f(t); \qquad {}_0\mathcal{D}_t^{\sigma_k - 1} y(t)\Big|_{t=0} = b_k, \quad (k = 1, \dots, n), \tag{4.32} $$

$$ {}_a\mathcal{L}_t y(t) \equiv {}_a\mathcal{D}_t^{\sigma_n} y(t) + \sum_{k=1}^{n-1} p_k(t)\, {}_a\mathcal{D}_t^{\sigma_{n-k}} y(t) + p_n(t) y(t), \tag{4.33} $$

where the following notation is used for the Miller–Ross sequential derivatives:

$$ {}_a\mathcal{D}_t^{\sigma_k} \equiv {}_aD_t^{\alpha_k}\, {}_aD_t^{\alpha_{k-1}} \dots {}_aD_t^{\alpha_1}; $$

$$ {}_a\mathcal{D}_t^{\sigma_k - 1} \equiv {}_aD_t^{\alpha_k - 1}\, {}_aD_t^{\alpha_{k-1}} \dots {}_aD_t^{\alpha_1}; $$

$$ \sigma_k = \sum_{j=1}^{k} \alpha_j, \quad (k = 1, 2, \dots, n); $$

$$ 0 < \alpha_j \le 1, \quad (j = 1, 2, \dots, n). $$

The fractional differential equation (4.32) is a sequential fractional differential equation, according to the terminology used by Miller and Ross [153]. To extend the Laplace transform method using the advantages of (1.80) for such equations with constant coefficients, the formula (2.259) can be used.

4.2.1 Ordinary Linear Fractional Differential Equations

In this section we give solutions of the "sequential" analogues of "standard" linear ordinary fractional differential equations with constant coefficients. Of course, we must take appropriate initial conditions, also in terms of sequential fractional derivatives.

Example 4.6. Let us consider the sequential analogue of Example 4.1:

$$ {}_0D_t^{\alpha} \left({}_0D_t^{\beta} y(t) \right) + a y(t) = 0 \tag{4.34} $$

$$ \left[{}_0D_t^{\alpha - 1} \left({}_0D_t^{\beta} y(t) \right) \right]_{t=0} = b_1, \qquad \left[{}_0D_t^{\beta - 1} y(t) \right]_{t=0} = b_2, \tag{4.35} $$

where $0 < \alpha < 1$, $0 < \beta < 1$, $\alpha + \beta = 1/2$.

The formula (2.259) of the Laplace transform of the sequential fractional derivative allows us to utilize the initial conditions (4.35). To use (2.259), we take $\alpha_1 = \beta$, $\alpha_2 = \alpha$, and $m = 2$. Therefore, $\sigma_1 = \beta$, $\sigma_2 = \alpha+\beta$. Then the Laplace transform (2.259) of equation (4.34) gives:

$$(s^{\alpha+\beta} + a)Y(s) = s^{\alpha} b_2 + b_1, \tag{4.36}$$

from which it follows that

$$Y(s) = b_2 \frac{s^{\alpha}}{s^{\alpha+\beta} + a} + b_1 \frac{1}{s^{\alpha+\beta} + a}, \tag{4.37}$$

and after the Laplace inversion with the help of (1.80) we find the solution to the problem (4.34)–(4.35):

$$y(t) = b_2 t^{\beta-1} E_{\alpha+\beta,\beta}(-at^{\alpha+\beta}) + b_1 t^{\alpha+\beta-1} E_{\alpha+\beta,\alpha+\beta}(-at^{\alpha+\beta}). \tag{4.38}$$

For $\beta = 0$ and $\alpha = 1/2$ (and assuming, of course, $b_2 = 0$), we can obtain from (4.38) the solution of Example 4.1.

Example 4.7. Let us now consider the following sequential analogue for the equation considered in Example 4.2:

$${}_0D_t^{\alpha} \left({}_0D_t^{\beta} y(t) \right) + {}_0D_t^{q} y(t) = h(t), \tag{4.39}$$

where $0 < \alpha < 1$, $0 < \beta < 1$, $0 < q < 1$, $\alpha + \beta = Q > q$.

The Laplace transform (2.259) of equation (4.39) gives:

$$(s^{\alpha+\beta} + s^{q})Y(s) = H(s) + s^{\alpha} b_2 + b_1, \tag{4.40}$$

$$b_1 = \left[{}_0D_t^{\alpha-1} \left({}_0D_t^{\beta} y(t) \right) \right]_{t=0} + \left[{}_0D_t^{q-1} y(t) \right]_{t=0},$$

$$b_2 = \left[{}_0D_t^{\beta-1} y(t) \right]_{t=0}.$$

Writing $Y(s)$ in the form

$$Y(s) = \frac{s^{-q} H(s)}{s^{\alpha+\beta-q} + 1} + b_2 \frac{s^{\alpha-q}}{s^{\alpha+\beta-q} + 1} + b_1 \frac{s^{-q}}{s^{\alpha+\beta-q} + 1} \tag{4.41}$$

and finding the inverse Laplace transform with the help of (1.80), we obtain the solution:

$$\begin{aligned} y(t) &= b_2 t^{\beta-1} E_{\alpha+\beta-q,\beta}(-t^{\alpha+\beta-q}) + b_1 t^{\alpha+\beta-q} E_{\alpha+\beta-q,\alpha+\beta}(-t^{\alpha+\beta-q}) \\ &+ \int_0^t (t-\tau)^{\alpha+\beta-1} E_{\alpha+\beta-q,\alpha+\beta} \left(-(t-\tau)^{\alpha+\beta-q} \right) h(\tau) d\tau. \end{aligned} \tag{4.42}$$

It is easy to see that this solution contains the solution of Example 4.2 as a special case.

Example 4.8. Let us consider the following initial value problem for the sequential fractional differential equation:

$$ {}_0D_t^{\alpha_2}({}_0D_t^{\alpha_1}y(t)) - \lambda y(t) = h(t); \tag{4.43} $$

$$ \left[{}_0D_t^{\alpha_2-1}\left({}_0D_t^{\alpha_1}y(t)\right)\right]_{t=0} = b_1, \quad \left[{}_0D_t^{\alpha_1-1}y(t)\right]_{t=0} = b_2. \tag{4.44} $$

Let us consider $0 < \alpha_1 < 1$, $0 < \alpha_2 < 1$. The Laplace transform (2.259) of equation (4.43) gives

$$ (s^{\alpha_1+\alpha_2} - \lambda)Y(s) = s^{\alpha_2}b_2 + b_1, $$

and after inversion using (1.80) we obtain the solution:

$$ y(t) = b_2 t^{\alpha_1-1}E_{\alpha,\alpha_1}(\lambda t^{\alpha}) + b_1 t^{\alpha-1}E_{\alpha,\alpha}(\lambda t^{\alpha}) + \int_0^t (t-\tau)^{\alpha-1}E_{\alpha,\alpha}(\lambda(t-\tau)^{\alpha})h(\tau)d\tau, \tag{4.45} $$

$$ (\alpha = \alpha_1 + \alpha_2). $$

Let us take α the same as in Example 4.3. Using (1.56), (1.82) and (2.213), it is easy to verify that (4.45) is the solution of (4.43). It is also worthwhile to note that if $b_1 \neq 0$, $b_2 \neq 0$, then (4.45) is *not* a solution of the equation ${}_0D_t^{\alpha}y(t) - \lambda y(t) = h(t)$ from Example 4.3; also (4.11) is *not* a solution of equation (4.43). On the other hand, equations (4.8) and (4.43) are very close to one another: the *fractional Green's function* in both cases is $G(t) = (t)^{\alpha-1}E_{\alpha,\alpha}(\lambda t^{\alpha})$. We will return to this observation in Chapter 5.

4.2.2 Partial Linear Fractional Differential Equations

Example 4.9. Let us consider Mainardi's [131] initial value problem for the fractional diffusion–wave equation:

$$ {}_0D_t^{\alpha}u(x,t) = \lambda^2 \frac{\partial^2 u(x,t)}{\partial x^2}, \qquad (|x| < \infty, t > 0) \tag{4.46} $$

$$ u(x,0) = f(x), \qquad (|x| < \infty) \tag{4.47} $$

$$\lim_{x \pm \infty} u(x,t) = 0, \qquad (t > 0) \tag{4.48}$$

where $0 < \alpha < 1$.

The type of the initial condition (4.47) suggests that the fractional derivative in equation (4.46) must be interpreted as a properly chosen sequential fractional derivative ${}_0\mathcal{D}_t^{\alpha} = {}_0D_t^{\alpha_2}\,{}_0D_t^{\alpha_1}$. The Laplace transform formula (2.259) for $\alpha_2 = \alpha - 1$, $\alpha_1 = 1$, and $k = 2$ (this gives Caputo's formula [24]), i.e.,

$$\mathcal{L}\{\,{}_0\mathcal{D}_t^{\alpha} y(t);\, s\} = s^{\alpha} Y(s) - s^{\alpha-1} y(0), \tag{4.49}$$

applied to the problem (4.46)–(4.48), yields:

$$s^{\alpha}\overline{u}(x,s) - s^{\alpha-1} f(x) = \lambda^2 \frac{\partial^2 \overline{u}(x,s)}{\partial x^2}, \qquad (|x| < \infty) \tag{4.50}$$

$$\lim_{x \pm \infty} \overline{u}(x,s) = 0, \qquad (t > 0). \tag{4.51}$$

Applying now the Fourier exponential transform to equation (4.50) and utilizing the boundary conditions (4.51), we obtain:

$$U(\beta,s) = \frac{s^{\alpha-1}}{s^{\alpha} + \lambda^2\beta^2} F(\beta), \tag{4.52}$$

where $U(\beta,p)$ and $F(\beta)$ are the Fourier transforms of $\overline{u}(x,s)$ and $f(x)$.

The inverse Laplace transform of the fraction $s^{\alpha-1}/(s^{\alpha} + \lambda^2\beta^2)$ is $E_{\alpha,1}(-\lambda^2\beta^2 t^{\alpha})$ (where $E_{\lambda,\mu}(z)$ is the Mittag-Leffler function in two parameters). Therefore, the inversion of the Fourier and the Laplace transform gives the solution in the following form:

$$u(x,t) = \int_{-\infty}^{\infty} G(x-\xi,t) f(\xi)\, d\xi, \tag{4.53}$$

$$\begin{aligned} G(x,t) &= \frac{1}{\pi}\int_{0}^{\infty} E_{\alpha,1}(-\lambda^2\beta^2 t^{\alpha}) \cos(\beta x)\, d\beta \\ &= \frac{1}{2\lambda} t^{-\rho} W(-z, -\rho, 1-\rho), \end{aligned} \tag{4.54}$$

where $W(z,\lambda,\mu)$ is the Wright function (1.156). This solution is identical to the solution of the Schneider–Wyss fractional (integro-differential) diffusion equation (4.27).

Chapter 5

Fractional Green's Function

There is something common in solutions of the corresponding "standard" and "sequential" fractional differential equations considered in Chapter 4: they both have the same *fractional Green's function.*

It seems that the notion of Green's function of a fractional differential equation appeared for the first time in the book by S. I. Meshkov [149], namely for the equation of the form (5.19), which is considered below.

The definition of fractional Green's function suggested and used by K. S. Miller and B. Ross [153, Chapter V] applies to fractional differential equations containing only derivatives of order $k\alpha$, where k is integer.

In this chapter, which is based on the papers [201, 208], we give a more general definition of the fractional Green's function and discuss some of its properties, necessary for constructing solutions of initial-value problems for fractional linear differential equations with constant coefficients.

We give the solution of the initial-value problem for the ordinary fractional linear differential equation with constant coefficients using only its Green's function. Due to this result, the solution of such initial value problems reduces to finding the fractional Green's functions. We obtained explicit expressions for the fractional Green's function for some special cases (one- , two- , three- and four-term equations).

The explicit expression for an arbitrary fractional linear ordinary differential equation with constant coefficients ends this chapter.

5.1 Definition and Some Properties

We consider here equation (4.32) under homogeneous initial conditions $b_k = 0$, $(k = 1, \ldots, n)$, i.e.

$$ {}_0\mathcal{L}_t y(t) = f(t); \qquad \left[{}_0\mathcal{D}_t^{\sigma_k - 1} y(t)\right]_{t=0} = 0, \quad (k = 1, \ldots, n) \tag{5.1} $$

$$ {}_a\mathcal{L}_t y(t) \equiv {}_a\mathcal{D}_t^{\sigma_n} y(t) + \sum_{k=1}^{n-1} p_k(t)\, {}_a\mathcal{D}_t^{\sigma_{n-k}} y(t) + p_n(t) y(t), $$

$$ {}_a\mathcal{D}_t^{\sigma_k} \equiv {}_aD_t^{\alpha_k}\, {}_aD_t^{\alpha_{k-1}} \ldots {}_aD_t^{\alpha_1}; \quad {}_a\mathcal{D}_t^{\sigma_k - 1} \equiv {}_aD_t^{\alpha_k - 1}\, {}_aD_t^{\alpha_{k-1}} \ldots {}_aD_t^{\alpha_1}; $$

$$ \sigma_k = \sum_{j=1}^{k} \alpha_j, \quad (k = 1, 2, \ldots, n); \qquad 0 \le \alpha_j \le 1, \quad (j = 1, 2, \ldots, n). $$

The following definition is a "fractionalization" of the definition given in [160].

5.1.1 Definition

The function $G(t, \tau)$ satisfying the following conditions

a) ${}_\tau\mathcal{L}_t G(t, \tau) = 0$ for every $\tau \in (0, t)$;

b) $\lim_{\tau \to t-0} \left({}_\tau\mathcal{D}_t^{\sigma_k - 1} G(t, \tau)\right) = \delta_{k,n}$, $\quad k = 0, 1, \ldots, n$,
($\delta_{k,n}$ is Kronecker's delta);

c) $\lim\limits_{\substack{\tau, t \to +0 \\ \tau < t}} \left({}_\tau\mathcal{D}_t^{\sigma_k} G(t, \tau)\right) = 0, \quad k = 0, 1, \ldots, n-1$

is called the Green's function of equation (5.1).

5.1.2 Properties

1. Using (2.211), it can be shown that $y(t) = \int_0^t G(t, \tau) f(\tau) d\tau$ is the solution of problem (5.1).

Let us outline the proof of this statement. Evaluating

$$ {}_0\mathcal{D}_t^{\sigma_1} y(t), \quad {}_0\mathcal{D}_t^{\sigma_2} y(t), \quad \ldots, \quad {}_0\mathcal{D}_t^{\sigma_n} y(t) $$

using the rule (2.211) and condition (b) of the definition of the Green's function, we obtain:

$$ {}_0\mathcal{D}_t^{\sigma_1} y(t) = {}_0D_t^{\alpha_1} \int_0^t G(t, \tau) f(\tau) d\tau $$

$$= \int_0^t {}_{\tau}D_t^{\alpha_1} G(t,\tau) f(\tau) d\tau + \lim_{\tau \to t-0} {}_{\tau}D_t^{\alpha_1 - 1} G(t,\tau) f(\tau)$$

$$= \int_0^t {}_{\tau}\mathcal{D}_t^{\sigma_1} G(t,\tau) f(\tau) d\tau \qquad (5.2)$$

$$\begin{aligned} {}_0\mathcal{D}_t^{\sigma_2} y(t) &= {}_0D_t^{\alpha_2}({}_0D_t^{\alpha_1} y(t)) = {}_0D_t^{\alpha_2} \int_0^t {}_{\tau}D_t^{\alpha_1} G(t,\tau) f(\tau) d\tau \\ &= \int_0^t {}_{\tau}D_t^{\alpha_2}({}_{\tau}D_t^{\alpha_1} G(t,\tau)) f(\tau) d\tau \\ &\quad + \lim_{\tau \to t-0} {}_{\tau}D_t^{\alpha_2 - 1}({}_{\tau}D_t^{\alpha_1} G(t,\tau)) f(\tau) \\ &= \int_0^t {}_{\tau}\mathcal{D}_t^{\sigma_2} G(t,\tau) f(\tau) d\tau \end{aligned} \qquad (5.3)$$

$$\dots \quad \dots \quad \dots$$

$$\begin{aligned} {}_0\mathcal{D}_t^{\sigma_{n-1}} y(t) &= {}_0D_t^{\alpha_{n-1}}({}_0D_t^{\sigma_{n-2}} y(t)) = {}_0D_t^{\alpha_{n-1}} \int_0^t {}_{\tau}\mathcal{D}_t^{\sigma_{n-2}} G(t,\tau) f(\tau) d\tau \\ &= \int_0^t {}_{\tau}D_t^{\alpha_{n-1}}({}_{\tau}\mathcal{D}_t^{\sigma_{n-2}} G(t,\tau)) f(\tau) d\tau \\ &\quad + \lim_{\tau \to t-0} {}_{\tau}D_t^{\alpha_{n-1} - 1}({}_{\tau}\mathcal{D}_t^{\sigma_{n-2}} G(t,\tau)) f(\tau) \\ &= \int_0^t {}_{\tau}\mathcal{D}_t^{\sigma_{n-1}} G(t,\tau) f(\tau) d\tau \end{aligned} \qquad (5.4)$$

$$\begin{aligned} {}_0\mathcal{D}_t^{\sigma_n} y(t) &= {}_0D_t^{\alpha_n}({}_0\mathcal{D}_t^{\sigma_{n-1}} y(t)) = {}_0D_t^{\alpha_n} \int_0^t {}_{\tau}\mathcal{D}_t^{\sigma_{n-1}} G(t,\tau) f(\tau) d\tau \\ &= \int_0^t {}_{\tau}D_t^{\alpha_n}({}_{\tau}\mathcal{D}_t^{\sigma_{n-1}} G(t,\tau)) f(\tau) d\tau \end{aligned}$$

$$+ \lim_{\tau \to t-0} {}_{\tau}D_t^{\alpha_n - 1}({}_{\tau}\mathcal{D}_t^{\sigma_{n-1}} G(t,\tau)) f(\tau)$$

$$= \int_0^t {}_{\tau}\mathcal{D}_t^{\sigma_n} G(t,\tau) f(\tau) d\tau + f(t). \tag{5.5}$$

Multiplying these equations by the corresponding coefficients and summarizing, we obtain

$${}_0\mathcal{L}_t y(t) = \int_0^t {}_{\tau}\mathcal{L}_t G(t,\tau) f(\tau) d\tau + f(t) = f(t), \tag{5.6}$$

which completes the proof.

2. For fractional differential equations with constant coefficients we have

$$G(t,\tau) \equiv G(t-\tau).$$

This is obvious because in such a case the Green's function can be obtained by the Laplace transform method.

The type (standard or sequential) of the equation is not important for determining the Green's function, because due to condition (b) in the Green's function definition all non-integral addends vanish.

3. Appropriate derivatives of the Green's function $G(t,\tau)$ form a set of linearly independent solutions of a homogeneous ($f(t) \equiv 0$) equation (4.32) (for a simple illustration, see Examples 4.3 and 4.8).

Let us demonstrate this for the case of the linear fractional differential equations with constant coefficients, which are the main subject for study in this work and for which we have $G(t,\tau) \equiv G(t-\tau)$.

Let us take $0 < \lambda < \sigma_n$. First, the function

$$y_{\lambda}(t) = {}_0D_t^{\lambda} G(t) \tag{5.7}$$

is a solution of the corresponding homogeneous equation. Indeed,

$${}_0\mathcal{L}_t y_{\lambda}(t) = {}_0\mathcal{L}_t({}_0D_t^{\lambda} G(t)) = {}_0D_t^{\lambda}({}_0\mathcal{L}_t G(t)) = 0. \tag{5.8}$$

We used here the fact that ${}_0\mathcal{L}_t \, {}_0D_t^{\lambda} = {}_0D_t^{\lambda} \, {}_0\mathcal{L}_t$, which follows from condition (c) in the definition of the fractional Green's function.

Second,

$${}_0D_t^{\sigma_n - \lambda - 1} y_{\lambda}(t) \Big|_{t=0} = 1. \tag{5.9}$$

In fact,

$$\begin{aligned}\left[{}_0D_t^{\sigma_n-\lambda-1}y_\lambda(t)\right]_{t=0} &= \left[{}_0D_t^{\sigma_n-\lambda-1}\left({}_0D_t^{\lambda}G(t)\right)\right]_{t=0}\\ &= \left[{}_0D_t^{\sigma_n-1}G(t)\right]_{t=0} = 1.\end{aligned} \tag{5.10}$$

Here we used the relationship

$${}_0D_t^{\sigma_n-\lambda-1}\,{}_0D_t^{\lambda}G(t) = {}_0D_t^{(\sigma_n-\lambda-1)+\lambda}G(t), \tag{5.11}$$

which follows from condition (c) of the definition of the Green's function, and then condition (b).

We see that having the fractional Green's function of equation (5.1), we can determine particular solutions of the corresponding homogeneous equation, which are necessary for satisfying inhomogeneous initial conditions.

Therefore, the solution of linear fractional differential equations with constant coefficients reduces to finding the fractional Green's function. After that, we can immediately write the solution of the inhomogeneous equation satisfying given inhomogeneous initial conditions.

This solution has the form

$$y(t) = \sum_{k=1}^{n} b_k\psi_k(t) + \int_0^t G(t-\tau)f(\tau)d\tau, \tag{5.12}$$

$$b_k = \left[{}_0\mathcal{D}_t^{\sigma_k-1}y(t)\right]_{t=0} \tag{5.13}$$

$$\psi_k(t) = {}_0\mathcal{D}_t^{\sigma_n-\sigma_k}G(t), \qquad {}_0\mathcal{D}_t^{\sigma_n-\sigma_k} \equiv {}_aD_t^{\alpha_n}\,{}_aD_t^{\alpha_{n-1}}\dots{}_aD_t^{\alpha_{k+1}}. \tag{5.14}$$

Because of this, in the following sections we find some explicit expressions for fractional Green's functions, including a general linear fractional differential equation.

5.2 One-term Equation

The fractional Green's function $G_1(t)$ for the one-term fractional-order differential equation with constant coefficients

$$a\,{}_0D_t^{\alpha}y(t) = f(t), \tag{5.15}$$

where the derivative can be either "classical" (i.e., considered in the book by Oldham and Spanier) or "sequential" (Miller and Ross), is found by the inverse Laplace transform of the following expression:

$$g_1(p) = \frac{1}{ap^{\alpha}}. \tag{5.16}$$

The inverse Laplace transform then gives

$$G_1(t) = \frac{1}{a} \frac{t^{\alpha-1}}{\Gamma(\alpha)}. \tag{5.17}$$

The solution of equation (5.15) under homogeneous initial conditions is

$$y(t) = \frac{1}{a\Gamma(\alpha)} \int_0^t \frac{f(\tau)d\tau}{(t-\tau)^{1-\alpha}} = \frac{1}{a}\, {}_0D_t^{-\alpha} f(t). \tag{5.18}$$

Using [188, lemma 3.3], we can easily verify that expression (5.18) gives the solution of equation (5.15), if $f(x)$ is continuous in $[0, \infty)$.

5.3 Two-term Equation

The fractional Green's function $G_2(t)$ for the two-term fractional-order differential equation with constant coefficients

$$a\, {}_0D_t^{\alpha} y(t) + by(t) = f(t), \tag{5.19}$$

where the derivative can be either "classical" (i.e., considered in the book by Oldham and Spanier) or "sequential" (Miller and Ross), is found by the inverse Laplace transform of the following expression:

$$g_2(p) = \frac{1}{ap^{\alpha} + b} = \frac{1}{a} \frac{1}{p^{\alpha} + \frac{b}{a}}, \tag{5.20}$$

which leads to

$$G_2(t) = \frac{1}{a} t^{\alpha-1} E_{\alpha,\alpha}(-\frac{b}{a} t^{\alpha}). \tag{5.21}$$

For instance, the function $G_2(t)$ plays a key role in the solution of Example 4.1 and 4.3.

Taking in (5.21) $b = 0$ and using the definition of the Mittag-Leffler function (1.56), we obtain the Green's function $G_1(t)$ for the one-term equation.

5.4 Three-term Equation

The fractional Green's function $G_3(t)$ for the three-term fractional-order differential equation with constant coefficients

$$a\, {}_0D_t^{\beta} y(t) + b\, {}_0D_t^{\alpha} y(t) + c\, y(t) = f(t), \tag{5.22}$$

where the derivatives can be either "classical" (i.e., considered in the book by Oldham and Spanier) or "sequential" (Miller and Ross), is found by the inverse Laplace transform of the following expression:

$$g_3(p) = \frac{1}{ap^{\beta} + bp^{\alpha} + c} \tag{5.23}$$

Assuming $\beta > \alpha$, we can write $g_3(p)$ in the form

$$\begin{aligned} g_3(p) &= \frac{1}{c}\, \frac{cp^{-\alpha}}{ap^{\beta-\alpha} + b}\, \frac{1}{1 + \dfrac{cp^{-\alpha}}{ap^{\beta-\alpha} + b}} \\ &= \frac{1}{c} \sum_{k=0}^{\infty} (-1)^k \left(\frac{c}{a}\right)^{k+1} \frac{p^{-\alpha k - \alpha}}{\left(p^{\beta-\alpha} + \frac{b}{a}\right)^{k+1}}. \end{aligned} \tag{5.24}$$

The term-by-term inversion, based on the general expansion theorem for the Laplace transform given in [42, §22], using (1.80) produces

$$G_3(t) = \frac{1}{a} \sum_{k=0}^{\infty} \frac{(-1)^k}{k!} \left(\frac{c}{a}\right)^k t^{\beta(k+1)-1} E^{(k)}_{\beta-\alpha,\beta+\alpha k}\left(-\frac{b}{a} t^{\beta-\alpha}\right), \tag{5.25}$$

where $E_{\lambda,\mu}(z)$ is the Mittag-Leffler function in two parameters,

$$E^{(k)}_{\lambda,\mu}(y) \equiv \frac{d^k}{dy^k} E_{\lambda,\mu}(y) = \sum_{j=0}^{\infty} \frac{(j+k)!\, y^j}{j!\, \Gamma(\lambda j + \lambda k + \mu)}, \qquad (k = 0, 1, 2, \ldots). \tag{5.26}$$

We assume in this solution that $a \neq 0$, because otherwise we have the two-term equation (5.19). We can also assume $c \neq 0$, because for $c = 0$

$$g_3(p) = \frac{1}{ap^{\beta} + bp^{\alpha}} = \frac{p^{-\alpha}}{ap^{\beta-\alpha} + b},$$

and the Laplace inversion can be done in the same way as in the case of the two-term equation.

Two special cases of equation (5.22) were considered by Bagley and Torvik [16] (for $\beta = 2$ and $\alpha = 3/2$) and by Caputo [24] (for $\beta = 2$ and

$0 < \alpha < 1$). It is easy to show that our solution (5.25) contains Caputo's solution as a particular case.

Indeed, substituting (5.26) into (5.25) and changing the order of summation, we obtain:

$$G_3(t) = \frac{1}{c}\sum_{k=0}^{\infty}(-1)^k\left(\frac{c}{a}\right)^{k+1}\sum_{j=0}^{\infty}(-1)^j\left(\frac{b}{a}\right)^j\frac{(j+k)!\, t^{\beta(j+k)+\beta-1-\alpha j}}{k!\, j!\, \Gamma(\beta(j+k+1)-\alpha j)}$$

$$= \frac{1}{c}\sum_{j=0}^{\infty}\left(\frac{-b}{a}\right)^j\sum_{k=0}^{\infty}(-1)^k\left(\frac{c}{a}\right)^{k+1}\frac{(j+k)!\, t^{\beta(j+k)+\beta-1-\alpha j}}{k!\, j!\, \Gamma(\beta(j+k+1)-\alpha j)} \quad (5.27)$$

$$(5.28)$$

$$(5.29)$$

For $\beta = 2$ this expression is identical with the expression obtained by Caputo [24, formula (2.27)].

5.5 Four-term Equation

The fractional Green's function $G_4(t)$ for the four-term fractional-order differential equation with constant coefficients

$$a\,{}_0D_t^{\gamma}y(t) + b\,{}_0D_t^{\beta}y(t) + c\,{}_0D_t^{\alpha}y(t) + d\,y(t) = f(t), \quad (5.30)$$

where the derivatives can be, as in the previous section, either "classical" or "sequential", is found by the inverse Laplace transform of the following expression:

$$g_4(p) = \frac{1}{ap^{\gamma} + bp^{\beta} + cp^{\alpha} + d}. \quad (5.31)$$

Assuming $\gamma > \beta > \alpha$, we can write $g(p)$ in the form

$$g_4(p) = \frac{1}{ap^{\gamma} + bp^{\beta}}\,\frac{1}{1 + \frac{cp^{\alpha}+d}{ap^{\gamma}+bp^{\beta}}}$$

$$= \frac{a^{-1}p^{-\beta}}{p^{\gamma-\beta} + a^{-1}b}\,\frac{1}{1 + \frac{a^{-1}cp^{\alpha-\beta}+a^{-1}dp^{-\beta}}{p^{\gamma-\beta}+a^{-1}b}}$$

$$= \sum_{m=0}^{\infty}(-1)^m\frac{a^{-1}p^{-\beta}}{(p^{\gamma-\beta}+a^{-1}b)^{m+1}}\left(\frac{c}{a}p^{\alpha-\beta} + \frac{d}{a}p^{-\beta}\right)^m$$

$$= \sum_{m=0}^{\infty}(-1)^m\frac{a^{-1}p^{-\beta}}{(p^{\gamma-\beta}+a^{-1}b)^{m+1}}\sum_{k=0}^{m}\binom{m}{k}\frac{c^k d^{m-k}}{a^m}p^{\alpha k-\beta m}$$

$$= \frac{1}{a}\sum_{m=0}^{\infty}(-1)^m\left(\frac{d}{a}\right)^m\sum_{k=0}^{m}\binom{m}{k}\left(\frac{c}{d}\right)^k\frac{p^{\alpha k-\beta m-\beta}}{(p^{\gamma-\beta}+a^{-1}b)^{m+1}}. \quad (5.32)$$

The term-by-term inversion, based on the general expansion theorem for the Laplace transform given in [42, §22], using (1.80) gives the final expression for the fractional Green's function for equation (5.30):

$$G_4(t) = \frac{1}{a}\sum_{m=0}^{\infty}\frac{1}{m!}\left(\frac{-d}{a}\right)^m\sum_{k=0}^{m}\binom{m}{k}\left(\frac{c}{d}\right)^k$$

$$\times\, t^{\gamma(m+1)-\alpha k-1}E^{(m)}_{\gamma-\beta,\gamma+\beta m-\alpha k}(-\frac{b}{a}t^{\gamma-\beta}). \quad (5.33)$$

We assumed in this solution that $a \neq 0$, because in the opposite case we have the three-term equation (5.22). We can also assume $d \neq 0$, because in the case of $d = 0$, after writing

$$g_4(p) = \frac{p^{-\alpha}}{ap^{\gamma-\alpha}+bp^{\beta-\alpha}+c} \quad (5.34)$$

the Laplace inversion can be done in the same way as in the case of the three-term equation.

5.6 General Case: n-term Equation

The above results can be essentially generalized.

The fractional Green's function $G_n(t)$ for the n-term fractional-order differential equation with constant coefficients

$$a_n\, D^{\beta_n}y(t) + a_{n-1}\, D^{\beta_{n-1}}y(t) + \ldots + a_1\, D^{\beta_1}y(t) + a_0\, D^{\beta_0}y(t) = f(t), \quad (5.35)$$

where the derivatives $D^{\alpha} \equiv {}_0D_t^{\alpha}$ can be, as in the previous sections, either "classical " or "sequential", is found by the inverse Laplace transform of the following expression:

$$g_n(p) = \frac{1}{a_np^{\beta_n}+a_{n-1}p^{\beta_{n-1}}+\ldots+a_1p^{\beta_1}+a_0p^{\beta_0}}. \quad (5.36)$$

Let us assume $\beta_n > \beta_{n-1} > \ldots > \beta_1 > \beta_0$ and write $g_n(p)$ in the form:

$$g_n(p) = \frac{1}{a_np^{\beta_n}+a_{n-1}p^{\beta_{n-1}}}\;\frac{1}{1+\dfrac{\sum_{k=0}^{n-2}a_kp^{\beta_k}}{a_np^{\beta_n}+a_{n-1}p^{\beta_{n-1}}}}$$

$$
\begin{aligned}
&= \frac{a_n^{-1} p^{-\beta_{n-1}}}{p^{\beta_n - \beta_{n-1}} + \frac{a_{n-1}}{a_n}} \frac{1}{1 + \frac{a_n^{-1} p^{-\beta_{n-1}} \sum_{k=0}^{n-2} a_k p^{\beta_k}}{p^{\beta_n - \beta_{n-1}} + \frac{a_{n-1}}{a_n}}} \\
&= \sum_{m=0}^{\infty} \frac{(-1)^m a_n^{-1} p^{-\beta_{n-1}}}{\left(p^{\beta_n - \beta_{n-1}} + \frac{a_{n-1}}{a_n}\right)^{m+1}} \left(\sum_{k=0}^{n-2} \left(\frac{a_k}{a_n}\right) p^{\beta_k - \beta_{n-1}} \right)^m \\
&= \sum_{m=0}^{\infty} \frac{(-1)^m a_n^{-1} p^{-\beta_{n-1}}}{\left(p^{\beta_n - \beta_{n-1}} + \frac{a_{n-1}}{a_n}\right)^{m+1}} \\
&\quad \sum_{\substack{k_0 + k_1 + \ldots + k_{n-2} = m \\ k_0 \geq 0; \ldots, k_{n-2} \geq 0}} (m; k_0, k_1, \ldots, k_{n-2}) \prod_{i=0}^{n-2} \left(\frac{a_i}{a_n}\right)^{k_i} p^{(\beta_i - \beta_{n-1}) k_i} \\
&= \frac{1}{a_n} \sum_{m=0}^{\infty} (-1)^m \sum_{\substack{k_0 + k_1 + \ldots + k_{n-2} = m \\ k_0 \geq 0; \ldots, k_{n-2} \geq 0}} (m; k_0, k_1, \ldots, k_{n-2}) \\
&\quad \prod_{i=0}^{n-2} \left(\frac{a_i}{a_n}\right)^{k_i} \frac{p^{-\beta_{n-1} + \sum_{i=0}^{n-2} (\beta_i - \beta_{n-1}) k_i}}{\left(p^{\beta_n - \beta_{n-1}} + \frac{a_{n-1}}{a_n}\right)^{m+1}}
\end{aligned} \tag{5.37}
$$

where $(m; k_0, k_1, \ldots, k_{n-2})$ are the multinomial coefficients [2].

The term-by-term inversion, based on the general expansion theorem for the Laplace transform given in [42, §22], using (1.80) gives the final expression for the fractional Green's function for equation (5.35):

$$
\begin{aligned}
G_n(t) &= \frac{1}{a_n} \sum_{m=0}^{\infty} \frac{(-1)^m}{m!} \sum_{\substack{k_0 + k_1 + \ldots + k_{n-2} = m \\ k_0 \geq 0; \ldots, k_{n-2} \geq 0}} (m; k_0, k_1, \ldots, k_{n-2}) \\
&\quad \prod_{i=0}^{n-2} \left(\frac{a_i}{a_n}\right)^{k_i} t^{(\beta_n - \beta_{n-1}) m + \beta_n + \sum_{j=0}^{n-2} (\beta_{n-1} - \beta_j) k_j - 1} \\
&\quad \times E^{(m)}_{\beta_n - \beta_{n-1}, + \beta_n + \sum_{j=0}^{n-2} (\beta_{n-1} - \beta_j) k_j} \left(- \frac{a_{n-1}}{a_n} t^{\beta_n - \beta_{n-1}} \right)
\end{aligned} \tag{5.38}
$$

Chapter 6

Other Methods for Solution of Fractional Order Equations

In this chapter some further analytical methods for solving fractional-order integral and differential equations are described, namely the Mellin transform method, the power series method, and Yu. I. Babenko's symbolic method. We also include the method of orthogonal polynomials for the solution of integral equations of fractional order, and give a collection of so-called spectral relationships for various types of kernels. All the methods described in this chapter are also illustrated by examples.

6.1 The Mellin Transform Method

In some cases, solutions of fractional differential equations can be obtained using the Mellin transform (see Sections 2.10.1–2.10.5).

Example 6.1. Let us consider the equation

$$t^{\alpha+1} D^{\alpha+1} y(t) + t^{\alpha} D^{\alpha} y(t) = f(t). \tag{6.1}$$

If we suppose that

$$y(0) = y'(0) = 0, \qquad y(\infty) = y'(\infty) = 0, \tag{6.2}$$

then D^{α} can mean the Riemann–Liouville, or the Caputo, or the Miller–Ross fractional derivative.

Applying the Mellin transform to equation (6.1), and using the formula (2.299), we obtain

$$Y(s) = F(s)\,\frac{\Gamma(1-s-\alpha)}{\Gamma(1-s)(1-s-\alpha)} = F(s)\,G(1-s), \tag{6.3}$$

$$G(s) = \frac{\Gamma(s-\alpha)}{\Gamma(s)(s-\alpha)} = \frac{\Gamma(s-\alpha+1)}{\Gamma(s)(s-\alpha)^2}. \tag{6.4}$$

If the inverse Mellin transform $g(t)$ of the function $G(s)$ is known, then the solution of equation (6.1) is the Mellin convolution (2.278) of the functions $f(t)$ and $g(t)$:

$$y(t) = \int_0^\infty f(t\tau)g(\tau)d\tau. \tag{6.5}$$

It can be shown that $g(t) = 0$ for $t > 1$. Indeed, its Mellin transform $G(s)$ can be written as

$$G(s) = G_1(s)G_2(s), \tag{6.6}$$

$$G_1(s) = \frac{\Gamma(s-\alpha)}{\Gamma(s)}, \qquad G_2(s) = \frac{1}{s-\alpha}.$$

The inverse Mellin transforms of the functions $G_1(s)$ and $G_2(s)$ can be found using formulas 7.3(20) and 7.1(3) from tables [62]:

$$g_1(t) = \begin{cases} \dfrac{t^{-\alpha}(1-t)^{\alpha-1}}{\Gamma(\alpha)} & 0<t<1, \\ 0, & t>1, \end{cases} \qquad g_2(t) = \begin{cases} t^{-\alpha}, & 0<t<1, \\ 0, & t>1, \end{cases} \tag{6.7}$$

and using the formula 6.1(14) from the same tables, we have the inverse transform of $G(s)$:

$$g(t) = \int_0^\infty g_1(t/\tau)\,g_2(\tau)\,\frac{d\tau}{\tau}. \tag{6.8}$$

It follows from the expressions (6.7) for the functions $g_1(t)$ and $g_2(t)$ that $g(t) = 0$ for $t > 1$, and that for $0 < t < 1$

$$g(t) = \int_t^1 g_1(t/\tau)\,g_2(\tau)\,\frac{d\tau}{\tau}. \tag{6.9}$$

An explicit expression for the function $g(t)$ for $0 < t < 1$ can be obtained by evaluating the integral (6.9) or by inverting $G(s)$ given by (6.4). In the second case, we can use the residue theorem.

The function $G(s)$ has a double pole at $s = \alpha$ and ordinary poles at $s = -n + \alpha - 1$ ($n = 0, 1, 2, \dots$), all lying in the left half-plane. Therefore, using the residue theorem gives

$$\begin{aligned} g(t) &= \int\limits_{Br} G(s)\, t^{-s} ds \\ &= \sum \Big[\text{Res}\; G(s)\, t^{-s}\Big]_{s=\alpha} + \sum_{n=0}^{\infty} \Big[\text{Res}\; G(s)\, t^{-s}\Big]_{s=-n+\alpha-1} \\ &= \frac{t^{-\alpha}}{\Gamma(\alpha)} \Big(\ln(t) + \psi(\alpha) - \gamma\Big) \\ &\quad + \sum_{n=0}^{\infty} \frac{(-1)^n t^{n-\alpha+1}}{(n+1)\,\Gamma(n+2)\,\Gamma(-n+\alpha-1)}, \end{aligned} \tag{6.10}$$

where $\gamma = 0.577215\dots$ is the Euler constant, and $\psi(z) = \Gamma'(z)/\Gamma(z)$ is the logarithmic derivative of the gamma function [63, Section 1.7].

Therefore, the solution of equation (6.1), which vanishes at $t = 0$ and $t = \infty$, is

$$y(t) = \int\limits_0^1 f(t\tau) g(\tau) d\tau, \tag{6.11}$$

where the function $g(t)$ is given by the expression (6.10), in which the power series converges for $|t| < 1$.

A fractional differential equation of the form

$$\sum_{k=0}^{n} a_k t^{\alpha+k} D^{\alpha+k} y(t) = f(t)$$

can be solved similarly with the help of formula (2.298).

6.2 Power Series Method

The power series method is the most transparent method of solution of fractional differential and integral equations. The idea of this method is to look for the solution in the form of a power series; the coefficients of the series must be determined.

Sometimes it is possible to find the general expression for the coefficients, at other times it is only possible to find the recurrence relation for the coefficients. In both cases, the solution can be computed approximately as partial sum of the series. This explains why the power series method is frequently used for solving applied problems.

The considerable disadvantage of the power series method is that it requires the power series expansions for all known (given) functions and non-constant coefficients appearing in the fractional-order equation; in real problems, however, known functions and non-constant coefficients are often the results of measurements and in such cases their power series expansions are unavailable.

On the other hand, there are important problems leading to non-linear fractional differential and integral equations or to fractional-order equations with non-constant coefficients, which could be solved at present only with the help of the power series method.

Let us consider several examples of the use of the power series method.

6.2.1 One-term Equation

The first example is the equation which we call the one-term equation, because there is only one term in its left-hand side:

$$ {}_0D_t^{\alpha} y(t) = f(t), \quad (t > 0), \tag{6.12} $$

where we suppose $0 < \alpha < 1$.

a) Let us first take the initial condition

$$ y(0) = 0, \tag{6.13} $$

and assume that the function $f(t)$ can be expanded in the Taylor series converging for $0 \leq t \leq R$, where R is the radius of convergence:

$$ f(t) = \sum_{n=0}^{\infty} \frac{f^{(n)}(0)}{n!} t^n. \tag{6.14} $$

Recalling the rule for the Riemann–Liouville fractional differentiation of the power function (2.117) we can write

$$ {}_0D_t^{\alpha} t^{\nu} = \frac{\Gamma(1+\nu)}{\Gamma(1+\nu-\alpha)} t^{\nu-\alpha}. \tag{6.15} $$

Taking into account the formula (6.15) we note that we can look for the solution of the equation (6.12) in the form of the following power series:

$$y(t) = t^{\alpha} \sum_{n=0}^{\infty} y_n t^n = \sum_{n=0}^{\infty} y_n t^{n+\alpha}. \tag{6.16}$$

Substituting the expressions (6.16) and (6.14) into the equation (6.12) and using (6.15) we obtain:

$$\sum_{n=0}^{\infty} y_n \frac{\Gamma(1+n+\alpha)}{\Gamma(n+1)} t^n = f(t) = \sum_{n=0}^{\infty} \frac{f^{(n)}(0)}{n!} t^n, \tag{6.17}$$

and comparison of the coefficients of both series gives

$$y_n = \frac{f^{(n)}(0)}{\Gamma(1+n+\alpha)}, \quad (n = \overline{1,\infty}). \tag{6.18}$$

Therefore, under the above assumptions the solution of the equation (6.12) is

$$y(t) = t^{\alpha} \sum_{n=0}^{\infty} \frac{f^{(n)}(0)}{\Gamma(1+n+\alpha)} t^n, \quad (0 \leq t \leq R). \tag{6.19}$$

In the case of the simple equation (6.12) we can easily transform the expression (6.19):

$$\begin{aligned} y(t) &= \sum_{n=0}^{\infty} \frac{f^{(n)}(0)}{n!} \frac{\Gamma(n+1)}{\Gamma(1+n+\alpha)} t^{n+\alpha} \\ &= \sum_{n=0}^{\infty} \frac{f^{(n)}(0)}{n!} \, {}_0D_t^{-\alpha} t^n \\ &= {}_0D_t^{-\alpha} \left\{ \sum_{n=0}^{\infty} \frac{f^{(n)}(0)}{n!} t^n \right\} \\ &= {}_0D_t^{-\alpha} f(t). \end{aligned} \tag{6.20}$$

Of course, since we looked for the solution $y(t)$ satisfying the zero initial condition (6.13) we could directly apply α-th-order fractional integration to both sides of the equation (6.12) and an application of the composition law for the Riemann–Liouville fractional derivatives (see Section 2.3.6) would give the expression (6.20). However, the use of the inverse operator is often impossible.

This approach can also be used if the right-hand side of the equation (6.12) has the form

$$f(t) = t^{\beta} g(t), \quad (\beta > -1), \tag{6.21}$$

$$g(t) = \sum_{n=0}^{\infty} \frac{g^{(n)}(0)}{n!} t^n.$$

In such a case we can look for the solution satisfying the initial condition (6.13) in the following form:

$$y(t) = t^{\alpha+\beta} \sum_{n=0}^{\infty} y_n t^n = \sum_{n=0}^{\infty} y_n t^{n+\alpha+\beta}, \tag{6.22}$$

and determine the coefficients y_n in the same manner as above. The result is

$$y_n = \frac{\Gamma(1+n+\beta)\, g^{(n)}(0)}{\Gamma(1+n+\alpha+\beta)\, \Gamma(n+1)}, \quad (n = \overline{1,\infty}). \tag{6.23}$$

b) If we have to solve the equation (6.12) under the non-zero initial condition

$$y(0) = A, \quad (A \neq 0), \tag{6.24}$$

then the solution exists only if (see Section 2.7.5)

$$f(t) \sim \frac{At^{-\alpha}}{\Gamma(1-\alpha)}, \quad (t \to 0). \tag{6.25}$$

Let us suppose that

$$f(t) = \frac{At^{-\alpha}}{\Gamma(1-\alpha)} + \sum_{n=1}^{\infty} f_n t^{n-\alpha}, \tag{6.26}$$

where the coefficients f_n are known.

Then we can look for the solution in the form

$$y(t) = \sum_{n=0}^{\infty} y_n t^n. \tag{6.27}$$

Substituting (6.27) and (6.26) into equation (6.12) we obtain

$$\sum_{n=0}^{\infty} y_n \frac{\Gamma(1+n)}{\Gamma(a+n-\alpha)} t^{n-\alpha} = f(t) = \frac{At^{-\alpha}}{\Gamma(1-\alpha)} + \sum_{n=1}^{\infty} f_n t^{n-\alpha}, \tag{6.28}$$

and therefore

$$y_0 = A; \quad y_n = \frac{\Gamma(1+n-\alpha)}{\Gamma n+1} f_n, \quad (n = \overline{1,\infty}). \tag{6.29}$$

c) Finally, let us consider the following initial condition for the equation (6.12):

$$ {}_0D_t^{\alpha-1} y(t)\Big|_{t=0} = B, \tag{6.30}$$

where B is constant.

In this case we can look for the solution in the form

$$y(t) = t^{\alpha-1} \sum_{n=0}^{\infty} y_n t^n = \sum_{n=0}^{\infty} y_n t^{n+\alpha-1}. \tag{6.31}$$

Let us assume that $f(t)$ can be expanded in the Taylor series (6.14). Substituting (6.31) and (6.14) into equation (6.12), using the derivative of the power function (6.15) and recalling that $1/\Gamma(0) = 0$ (see Section 1.1.2) we obtain after the obvious change of summation index

$$\sum_{n=0}^{\infty} y_{n+1} \frac{\Gamma(n+\alpha+1)}{\Gamma(n+1)} t^n = \sum_{n=0}^{\infty} \frac{f^{(n)}(0)}{\Gamma(n+1)} t^n, \tag{6.32}$$

and the comparison of the coefficients gives

$$y_{n+1} = \frac{f^{(n)}(0)}{\Gamma(n+\alpha+1)}, \quad (n = \overline{0,\infty}). \tag{6.33}$$

The coefficient y_0 must be determined from the initial condition (6.30). Using the formula (6.15) we have:

$$\begin{aligned} {}_0D_t^{\alpha-1} y(t) &= \sum_{n=0}^{\infty} y_n \, {}_0D_t^{\alpha-1} t^{n+\alpha+1} \\ &= \sum_{n=0}^{\infty} y_n \frac{\Gamma(n+\alpha)}{\Gamma(n+1)} t^n, \end{aligned} \tag{6.34}$$

and taking $t \to 0$ we obtain

$$y_0 = \frac{B}{\Gamma(\alpha)}. \tag{6.35}$$

These examples show that even the simplest fractional-order equation, such as (6.12), requires special attention to the formulation of the initial condition. The initial condition and the right-hand side of this equation determine the class of solutions. We can get an idea about the possible form of the solution only after the analysis of the initial condition, the right-hand side and the determination of the class of solutions. In thinking about the form of the series representing the solution the key role belongs to the rule for the differentiation of the power function (2.117) or (6.15).

6.2.2 Equation with Non-constant Coefficients

Let us consider the following initial value problem for the fractional differential equation:

$$\frac{d}{dt}\Big(f(t)\,(y(t)+1)\Big) + \lambda\, {}_0D_t^{1/2} y(t) = 0, \qquad (0 < t < 1) \tag{6.36}$$

$$y(0) = 0, \tag{6.37}$$

where the function $f(t)$ is a given function. For certain particular types of $f(t)$ the problem (6.36)–(6.37) allows us to obtain an analytical solution by the power series method. For example, let us assume that

$$f(t) = \sum_{n=0}^{\infty} f_n t^{n/2}, \quad f_0 = 1. \tag{6.38}$$

Then we can look for the solution $y(t)$ in the form of a similar fractional power series:

$$y(t) = \sum_{n=1}^{\infty} y_n t^{n/2}. \tag{6.39}$$

The substitution of (6.39) and (6.38) into equation (6.36) and comparison of the coefficients of the resulting power series lead to the following recurrence relationships:

$$y_1 = -f_1, \qquad \sum_{k=0}^{n} y_{n+1-k} f_k + \lambda y_n \frac{\Gamma\left(\frac{n}{2}+1\right)}{\Gamma\left(\frac{n+3}{2}\right)} = -f_{n+1}. \tag{6.40}$$

Due to the construction of the solution (6.39), the initial condition (6.37) is satisfied automatically.

If we take, for example,

$$f(t) = 1 - \sqrt{t}, \tag{6.41}$$

then from (6.40) it follows that

$$\begin{aligned} y_1 &= 1, \\ y_2 &= y_1 \left(1 - \lambda \frac{\Gamma(3/2)}{\Gamma(2)}\right), \\ &\dots \\ y_{n+1} &= y_n \left(1 - \lambda \frac{\Gamma\left(\frac{n+2}{2}\right)}{\Gamma\left(\frac{n+3}{2}\right)}\right). \end{aligned} \qquad (6.42)$$

For an arbitrary λ the series (6.39) converges in the interval $0 \leq t < 1$. If

$$\lambda = \frac{\Gamma\left(\frac{n+2}{2}\right)}{\Gamma\left(\frac{n+3}{2}\right)}$$

then the solution $y(t)$ is given by a finite sum, for example,

$$\lambda = \frac{\Gamma(2)}{\Gamma(3/2)} = \frac{2}{\sqrt{\pi}}, \qquad y(t) = \sqrt{t}; \qquad (6.43)$$

$$\lambda = \frac{\Gamma(5/2)}{\Gamma(2)} = \frac{3\sqrt{\pi}}{4}, \qquad y(t) = \sqrt{t} + \left(1 - \frac{3\pi}{8}\right) t. \qquad (6.44)$$

If we take $f(t)$ in the form [11]

$$f(t) = \frac{1 - t^{n+\frac{1}{2}}}{1 + \frac{\Gamma(n+\frac{3}{2}) t^n}{\lambda \Gamma(n+1)}} \qquad (6.45)$$

then the solution is given by

$$y(t) = \frac{\Gamma\left(n + \frac{3}{2}\right) t^n}{\lambda \Gamma(n+1)}. \qquad (6.46)$$

The physical problem leading to the fractional differential equation (6.36) and the numerical solution of the initial-value problem (6.36)–(6.37) are considered in Section 8.3.3.

6.2.3 Two-term Non-linear Equation

Let us consider the following initial-value problem for the two-term non-linear fractional differential equation:

$${}_0D_t^{1/2} y(t) - \lambda \Big(y(t) - y_0\Big)^2 = 0, \quad (t > 0), \qquad (6.47)$$

$$y(0) = 0, \tag{6.48}$$

where λ and y_0 are given constants.

In this case we can look for the solution in the form of the fractional power series

$$y(t) = \sum_{n=1}^{\infty} y_n t^{n/2}, \tag{6.49}$$

based on the observation that both ${}_0D_t^{1/2} y(t)$ and $(y - A)^2$ produce series of the same form. Obviously, the solution (6.49) satisfies the initial condition (6.48) automatically.

Substituting (6.49) into equation (6.47), using the formula (6.15) and comparing the coefficients in the fractional power series, we obtain the following recurrence relations for the coefficients y_n:

$$\begin{aligned}
y_1 &= \lambda\, y_0^2 \frac{\Gamma(1)}{\Gamma(\frac{3}{2})}, \\
y_2 &= 2\,\lambda\, y_0\, y_1 \frac{\Gamma(\frac{3}{2})}{\Gamma(2)}, \\
y_3 &= \lambda\,(y_1^2 + 2\, y_0\, y_2) \frac{\Gamma(2)}{\Gamma(\frac{5}{2})}, \\
\ldots &\quad \ldots \quad \ldots \\
y_n &= \lambda \sum_{k=0}^{n-1} y_k\, y_{n-k-1} \frac{\Gamma(\frac{n+1}{2})}{\Gamma(\frac{n+2}{2})}.
\end{aligned}$$

Finding the convergence interval for the series (6.49), where the coefficients y_n are defined by the above recurrence relationships, is difficult. However, computations show that this series can be used for computing the solution for small t.

6.3 Babenko's Symbolic Calculus Method

In this section we describe the method used by Yu. I. Babenko in his book [11] for solving various types of fractional integral and differential equations. The method itself is close to the Laplace transform method. It can be used in more cases than, for example, the Laplace transform method, but it is always necessary to check for the validity of the formal solutions. In the general case, such checking is not a simple task.

6.3.1 The Idea of the Method

We will explain the idea of Babenko's method on the following example. Let us consider the Abel integral equation of the second kind:

$$y(t) + \frac{\lambda}{\Gamma(\alpha)} \int_0^t (t-\tau)^{\alpha-1} y(\tau) d\tau = f(t), \quad (t > 0) \tag{6.50}$$

where we suppose $\alpha > 0$, λ is constant. Denoting

$$D^{-\alpha} y(t) \equiv {}_0D_t^{-\alpha} y(t) = \frac{1}{\Gamma(\alpha)} \int_0^t (t-\tau)^{\alpha-1} y(\tau) d\tau, \tag{6.51}$$

we can write equation (6.50) in the form

$$(1 + \lambda D^{-\alpha}) y(t) = f(t), \tag{6.52}$$

and the solution can be immediately written in the symbolic form

$$y(t) = (1 + \lambda D^{-\alpha})^{-1} f(t), \tag{6.53}$$

where $(1 + \lambda D^{-\alpha})^{-1}$ denotes the left inverse operator to the operator $(1 + \lambda D^{-\alpha})$.

The expression (6.53) is concise, but not suitable for practical purposes and computation. Using the binomial expansion of $(1 + \lambda D^{-\alpha})^{-1}$, we can write (6.53) as

$$y(t) = \sum_{n=0}^{\infty} (-1)^n \lambda^n D^{-\alpha n} f(t). \tag{6.54}$$

Since for many functions $f(t)$ all the fractional integrals in the right-hand side of the expression (6.54) can be evaluated explicitly, the formula (6.54) gives in those cases the formal solutions in the form of a series, the sum of which can sometimes be evaluated.

For example, let us take $f(tq) = t$. An appplication of the formula (2.117) for the Riemann–Liouville fractional differentiation of the power function then gives

$$D^{-\alpha n} f(t) = D^{-\alpha n} t = \frac{\Gamma(2) t^{\alpha n + 1}}{\Gamma(\alpha n + 2)} = \frac{t^{\alpha n+1}}{\Gamma(\alpha n + 2)},$$

and substituting this expression into (6.54) we obtain the solution:

$$y(t) = \sum_{n=0}^{\infty} \frac{(-1)^n \lambda^n t^{\alpha n+1}}{\Gamma(\alpha n + 2)} = t \sum_{n=0}^{\infty} \frac{(-\lambda t^{\alpha})^n}{\Gamma(\alpha n + 2)} = t E_{\alpha,2}(-\lambda t^{\alpha}), \qquad (6.55)$$

where $E_{\alpha,\beta}(t)$ is the Mittag-Leffler function defined by (1.56). The solution (6.55) can also be obtained by the Laplace transform method described in Chapter 4 with the use of formula (1.82) for the arbitrary-order differentiation and integration of the Mittag-Leffler function.

Moreover, the expression (6.54) can also be used for obtaining the closed-form solution of the equation (6.50). For example, using the definition of the Riemann–Liouville fractional integral (2.88) and the definition of the Mittag-Leffler function (1.56) we can write

$$\begin{aligned}
y(t) &= \sum_{n=0}^{\infty} (-1)^n \lambda^n D^{-\alpha n} f(t) \\
&= \sum_{n=0}^{\infty} (-1)^n \lambda^n \frac{1}{\Gamma(\alpha n)} \int_0^t (t-\tau)^{\alpha n - 1} f(\tau) d\tau \\
&= \sum_{n=0}^{\infty} (-1)^n \lambda^n \frac{1}{\Gamma(\alpha n + 1)} \frac{d}{dt} \int_0^t (t-\tau)^{\alpha n} f(\tau) d\tau \\
&= \frac{d}{dt} \int_0^t \left\{ \sum_{n=0}^{\infty} \frac{(-\lambda)^n (t-\tau)^{\alpha n}}{\Gamma(\alpha n + 1)} \right\} f(\tau) d\tau \\
&= \frac{d}{dt} \int_0^t E_{\alpha,1}\Big(-\lambda(t-\tau^{\alpha})\Big) f(\tau) d\tau. \qquad (6.56)
\end{aligned}$$

The solution obtained in this manner is formal; the interchange of summation and integration requires justification or the same final result must be obtained by some other method. For example, in the considered case the expression (6.56) can also be obtained with the help of the Laplace transform method described in Chapter 4.

6.3.2 Application in Heat and Mass Transfer

As the following example shows, Babenko's method can also be used for solving certain problems related to partial differential equations of heat and mass transfer theory.

Let us consider the following sample heat transfer problem for the half-plane:

$$\Big(\frac{\partial}{\partial t} - \frac{\partial^2}{\partial x^2} + \gamma\Big)T(x,t) = 0; \quad (t > 0; \quad 0 < x < \infty), \tag{6.57}$$

$$T(0,t) = T_s(t), \tag{6.58}$$

$$T(\infty,t) = 0, \tag{6.59}$$

$$T(x,0) = 0, \tag{6.60}$$

where the constant γ and the function $T_s(t)$ are given. Let us look for the heat flux at the boundary, i.e. for

$$q_s(t) = \frac{\partial T}{\partial x}\bigg|_{x=0}.$$

The first step is to suppose that equation (6.57) can be written in the form

$$\Big(L - \frac{\partial}{\partial x}\Big)\Big(L + \frac{\partial}{\partial x}\Big)T(x,t) = 0, \tag{6.61}$$

where the operator L is defined by the following symbolic expression:

$$L = \sqrt{\frac{\partial}{\partial t} + \gamma} = \sqrt{D+\gamma}, \qquad \Big(D \equiv \frac{\partial}{\partial t}\Big). \tag{6.62}$$

Then we can utilize the boundary condition (6.59) for noticing that all decaying solutions of the original equation (6.57) are also the solutions of the equation which correspond to the second operator in (6.61):

$$\Big(L + \frac{\partial}{\partial x}\Big)T(x,t) = 0. \tag{6.63}$$

Putting $x = 0$ in (6.63) we obtain the required heat flux at the boundary:

$$q_s(t) = -\sqrt{D+\gamma}\,T_s(t). \tag{6.64}$$

For computations we must have an interpretation of the operator in (6.64). The first way is to use the formal binomial expansion:

$$q_s(t) = -\sqrt{D+\gamma}\,T_s(t) = -D^{1/2}\sqrt{1+\gamma D^{-1}}\,T_s(t)$$
$$= -\sum_{n=0}^{\infty}\binom{1/2}{n}\gamma^n D^{\frac{1}{2}-n}T_s(t). \tag{6.65}$$

The second way suggested by Yu. I. Babenko is based on the identity

$$L^2 = \sqrt{D+\gamma}\,\sqrt{D+\gamma} = D+\gamma, \tag{6.66}$$

which in some cases allows us to write the operator L in the form

$$L = \sqrt{D+\gamma} = e^{-\gamma t} D^{1/2} e^{\gamma t}. \tag{6.67}$$

Indeed, if $f(0) = f'(0) = 0$ (this is a necessary condition; see Section 2.3.6), then

$$\begin{aligned} L^2 f(t) &= \left(e^{-\gamma t} D^{1/2} e^{\gamma t}\right)\left(e^{-\gamma t} D^{1/2} e^{\gamma t}\right) f(t) \\ &= e^{-\gamma t} D e^{\gamma t} f(t) = e^{-\gamma t}\left\{e^{\gamma t} D f(t) + \gamma e^{\gamma t} f(t)\right\} \\ &= D f(t) + \gamma f(t) = \left(D+\gamma\right) f(t). \end{aligned} \tag{6.68}$$

Therefore, if the given surface temperature $T_s(t)$ satisfies the conditions

$$T_s(0) = T_s'(0) = 0,$$

then the heat flux (6.64) can be written in the form

$$q_s(t) = -e^{-\gamma t} D^{1/2} e^{\gamma t} T_s(t), \tag{6.69}$$

where $D^{1/2}$ denotes the half-order Riemann–Liouville fractional derivative.

In general, the justification of Babenko's approach is not known, and therefore it is necessary to look for such justification for each particular problem. However, it is a powerful tool for determining the possible form of the solution. Numerous examples of the application of this symbolic method for solving integer- and fractional-order differential equations appearing in heat and mass transfer problems are given by Yu. I. Babenko in his book [11].

6.3.3 Link to the Laplace Transform Method

There is a certain link between Babenko's method and the Laplace transform method.

Let us consider the heat conduction problem (6.57)–(6.60) for a half-plane for $\gamma = 0$. In such a case we have

$$L = D^{1/2} = {}_0D_t^{1/2}, \tag{6.70}$$

and the solution for the heat flux at the boundary takes on the form

$$q_s(t) = -\,{}_0D_t^{1/2} T_s(t). \tag{6.71}$$

This result can be obtained by the Laplace transform method as described below in Section 7.7.3.

6.4 Method of Orthogonal Polynomials

It is well known that the solution of the integral equation of the first kind is an ill-posed problem [248]. For example, in the case of the classical Abel's integral equation ($0 < \alpha < 1$)

$$\frac{1}{\Gamma(\alpha)} \int_0^t \frac{y(\tau)}{(t-\tau)^{1-\alpha}} d\tau = f(t), \quad (0 < t < 1) \tag{6.72}$$

the main difficulty is the differentiation which appears in the explicit formulas for the solution of the equation (6.72). In many applied problems the function $f(t)$ is known from measurements (or computed approximately), and differentiation leads to magnification of noise in measured data and to wrong numerical results. Attempts to circumvent this difficulty are described, for example, in [67], [154], and [82]. It is clear that the form of solution which does not require differentiation will be more useful for today's numerous applications in physics, engineering and other fields. The method of orthogonal polynomials developed by G. Ya. Popov [211] and described in detail in his monograph [213] provides a tool for the numerical solution of certain classes of integral equations of the first kind in the presense of noise in $f(t)$. A similar approach was suggested also by R. Gorenflo and Y. Kovetz for the solution of Abel's integral equation [85].

In this section we deal with the application of the method of orthogonal polynomials to the solution of fractional integral equations. We briefly describe the method, present a collection of spectral relationships and give two examples. The special function notation is in accordance with [2].

The following notation is used below:

${}_pF_q\left(\begin{array}{cccc} a_1 & a_2 & \dots & a_p \\ b_1 & b_2 & \dots & b_q \end{array} \middle| z \right)$ is the hypergeometric function;
$\Gamma(z)$ is the Euler gamma function;

$B(x,y)$ is the Euler beta fuction;
$P_m^{\alpha,\beta}(t)$ is the Jacobi polynomial;
$Q_m^{\alpha,\beta}(t) = (1-t)^\alpha(1+t)^\beta P_m^{\alpha,\beta}(t)$.

6.4.1 The Idea of the Method

Let us introduce the idea of the method of orthogonal polynomials on an example of the solution of the so-called characteristic Cauchy singular integral equation of the first kind

$$\frac{1}{\pi}\int_{-1}^{1}\frac{y(\tau)d\tau}{\tau - t} = f(t), \quad (-1 < t < 1), \tag{6.73}$$

where the function $f(t)$ is given and $y(t)$ is unknown.

Equation (6.73) plays an important role in the theory of elasticity, in fracture mechanics, in fluid mechanics, in the theory of electricity, in acoustics and in other applied sciences.

To obtain the solution of equation (6.73) we will use the following three relationships:

$$\frac{1}{\pi}\int_{-1}^{1}\frac{T_n(\tau)d\tau}{(\tau-t)\sqrt{1-\tau^2}} = U_{n-1}(t), \tag{6.74}$$

$$-1 < t < 1, \quad (n = 1, 2, \ldots),$$

$$\frac{1}{\pi}\int_{-1}^{1}\frac{P_n^{\frac{1}{2},-\frac{1}{2}}(\tau)}{(\tau-t)}\sqrt{\frac{1-\tau}{1+\tau}}d\tau = -P_n^{-\frac{1}{2},\frac{1}{2}}(t), \tag{6.75}$$

$$-1 < t < 1, \quad (n = 0, 1, 2, \ldots),$$

$$\frac{1}{\pi}\int_{-1}^{1}\frac{U_n(\tau)\sqrt{1-\tau^2}d\tau}{(\tau-t)} = -T_{n+1}(t), \tag{6.76}$$

$$-1 < t < 1, \quad (n = 0, 1, 2, \ldots),$$

where $T_n(t)$ is the Chebyshev polynomial of the first kind, $U_n(t)$ is the Chebyshev polynomial of the second kind and $P_n^{\lambda,\mu}(t)$ is the Jacobi polynomial.

Relationships (6.74)–(6.76) are called *the spectral relationships*.

The nature of the applied problem, which was reduced to the Cauchy singular integral equation (6.73), determines the class of solutions. In the classical theory of the Cauchy singular integral equations the following three cases are considered:

a) solution unbounded at $t = \pm 1$;

b) solution unbounded at $t = -1$ and bounded at $t = 1$ (or, what is equivalent, bounded at $t = -1$ and unbounded at $t = 1$);

c) solution bounded at $t = \pm 1$.

The method of orthogonal polynomials allows us to obtain a solution in all three cases.

a) Solution Unbounded at $t = \pm 1$

Let us suppose that we look for the solution unbounded at both ends $t = \pm 1$. Then, comparing equation (6.73) and the spectral relationship (6.74), we see that it is possible to look for the solution $y(t)$ in the form of a series of Chebyshev polynomials of the first kind:

$$y(t) = \frac{1}{\sqrt{1-t^2}} \sum_{n=0}^{\infty} y_n T_n(t), \tag{6.77}$$

where the coefficients y_n must be found.

Substituting (6.77) into equation (6.73), using the spectral relationship (6.74) and the known integral

$$\frac{1}{\pi} \int_{-1}^{1} \frac{d\tau}{(\tau - t)\sqrt{1-\tau^2}} = 0$$

we obtain:

$$\sum_{n=1}^{\infty} y_n U_{n-1}(t) = f(t), \qquad (-1 < t < 1). \tag{6.78}$$

To determine y_n, we multiply both sides of the equation (6.78) by $U_{n-1}(t)$ and integrate from $t = -1$ to $t = 1$. Taking into account the orthogonality of the Chebyshev polynomials of the second kind

$$\int_{-1}^{1} U_k(t) U_m(t) \sqrt{1-t^2}\, dt = \begin{cases} \frac{\pi}{2}, & k = m \\ 0, & k \neq m \end{cases}$$

we obtain:

$$y_n = \frac{2}{\pi}\int_{-1}^{1} f(t)U_{n-1}(t)\sqrt{1-t^2}dt, \quad (n = 1, 2, \ldots). \tag{6.79}$$

The constant y_0 can be found from the additional condition providing the uniqueness of the solution. For example, if the additional condition is

$$\int_{-1}^{1} y(t)dt = A, \quad (A = const), \tag{6.80}$$

then

$$\int_{-1}^{1} y(t)dt = \sum_{n=0}^{\infty} y_n \int_{-1}^{1} \frac{T_n(t)dt}{\sqrt{1-t^2}} = \sum_{n=0}^{\infty} y_n \int_{-1}^{1} \frac{T_n(t)T_0(t)dt}{\sqrt{1-t^2}} = \pi y_0, \tag{6.81}$$

and therefore

$$y_0 = \frac{A}{\pi}. \tag{6.82}$$

The formulas (6.77), (6.79) and (6.82) give the solution of the equation (6.73) in the class of unbounded functions.

b) Solution Unbounded at $t = -1$ and Bounded at $t = 1$

Let us now suppose that we look for the solution which is unbounded at $t = -1$ and bounded at $t = 1$. In this case, comparing the equation (6.73) and the spectral relationship (6.74), we see that it is possible to look for the solution $y(t)$ in the form of a series of Jacobi polynomials $P_n^{\frac{1}{2},-\frac{1}{2}}(t)$

$$y(t) = \sqrt{\frac{1-t}{1+t}} \sum_{k=0}^{\infty} y_n P_n^{\frac{1}{2},-\frac{1}{2}}(t), \tag{6.83}$$

where the coefficients y_n must be determined.

Substituting (6.83) into equation (6.73) and applying the spectral relationship (6.75) we obtain:

$$-\sum_{n=0}^{\infty} y_n P_n^{-\frac{1}{2},\frac{1}{2}}(t) = f(t), \qquad (-1 < t < 1). \tag{6.84}$$

To determine y_n, we multiply both sides of equation (6.84) by $P_n^{-\frac{1}{2},\frac{1}{2}}(t)$ and integrate from $t = -1$ to $t = 1$. Taking into account the orthogonality of the Jacobi polynomials $P_n^{-\frac{1}{2},\frac{1}{2}}(t)$

$$\int_{-1}^{1} P_k^{-\frac{1}{2},\frac{1}{2}}(t) P_m^{-\frac{1}{2},\frac{1}{2}}(t) \sqrt{\frac{1+t}{1-t}} dt = \begin{cases} \frac{\Gamma^2(m+\frac{1}{2})}{\Gamma^2(m+1)}, & k = m \\ 0, & k \neq m \end{cases}$$

we obtain:

$$y_n = \frac{\Gamma^2(n+1)}{\Gamma^2(n+\frac{1}{2})} \int_{-1}^{1} f(t) P_n^{-\frac{1}{2},\frac{1}{2}}(t) \sqrt{\frac{1+t}{1-t}} dt, \quad n = 0, 1, 2, \ldots \tag{6.85}$$

The formulas (6.83) and (6.85) give the solution of the equation (6.73) in terms of the class of functions unbounded at $t = -1$ and bounded at $t = 1$. In contrast with the case of an unbounded solution, we do not need any additional condition for determining the unique solution of the equation (6.73), because the selected class of solutions is narrower.

c) Solution Bounded at $t = \pm 1$

Finally, let us suppose that we look for the solution which is bounded at both ends $t = \pm 1$. Then, comparing the equation (6.73) and the spectral relationship (6.76), we see that it is possible to look for the solution $y(t)$ in the form of a series of Chebyshev polynomials of the second kind:

$$y(t) = \sqrt{1-t^2} \sum_{n=0}^{\infty} y_n U_n(t), \tag{6.86}$$

where the coefficients y_n must be determined.

Substituting (6.86) into equation (6.73) and utilizing the spectral relationship (6.76) we obtain:

$$-\sum_{n=0}^{\infty} y_n T_{n+1}(t) = f(t), \qquad (-1 < t < 1). \tag{6.87}$$

To determine y_n, we multiply both sides of equation (6.87) by $T_{n+1}(t)$, $(n = 0, 1, 2, \ldots)$, and integrate from $t = -1$ to $t = 1$. Taking into account the orthogonality of the Chebyshev polynomials of the first kind

$$\int_{-1}^{1} \frac{T_k(t) T_m(t)}{\sqrt{1-t^2}} dt = \begin{cases} \pi, & k = m = 0 \\ \frac{\pi}{2}, & k = m \neq 0 \\ 0, & k \neq m \end{cases}$$

we obtain:

$$y_n = \frac{2}{\pi}\int\limits_{-1}^{1} \frac{f(t)T_{n+1}(t)}{\sqrt{1-t^2}}dt, \quad (n = 0, 1, 2, \ldots). \tag{6.88}$$

Multiplying (6.87) by $T_0(t)$ and integrating from $t = -1$ to $t = 1$ we obtain the condition for the existence of the solution of equation (6.73) in terms of the class of functions bounded at $t = \pm 1$:

$$\int\limits_{-1}^{1} \frac{f(t)dt}{\sqrt{1-t^2}} = 0. \tag{6.89}$$

If the right-hand side of the Cauchy singular integral equation (6.73) satisfies the condition (6.89), then the solution $y(t)$ is given by formulas (6.86) and (6.88); otherwise the solution does not exist.

It is worth noting that the spectral relationships (6.74)–(6.76) allow us also to obtain a solution of the characteristic Cauchy singular integral equation of the second kind

$$y(t) - \frac{\lambda}{\pi}\int\limits_{-1}^{1} \frac{y(\tau)d\tau}{\tau - t} = f(t), \quad (-1 < t < 1), \tag{6.90}$$

and of the so-called complete equations corresponding to the characteristic ones (6.73) and (6.90), namely

$$\frac{1}{\pi}\int\limits_{-1}^{1} \left(\frac{1}{\tau - t} + K(t,\tau)\right) y(\tau)d\tau = f(t), \quad (-1 < t < 1), \tag{6.91}$$

$$y(t) - \frac{\lambda}{\pi}\int\limits_{-1}^{1} \left(\frac{1}{\tau - t} + K(t,\tau)\right) y(\tau)d\tau = f(t), \quad (-1 < t < 1), \tag{6.92}$$

where the kernel $K(t, \tau)$ does not contain singularities. The corresponding integral equation is reduced to an infinite system of linear algebraic equations, whose approximate solution can be found by the method of reduction. The details and the justification of such a solution procedure can be found in [213].

6.4.2 General Scheme of the Method

Now let us briefly describe the general scheme of the method following [213].

Let us consider the integral equation of the first kind

$$\int_a^b K(t,\tau)y(\tau)d\tau = f(t), \quad (a < t < b). \tag{6.93}$$

Suppose there is a *spectral relationship* for the integral operator with the kernel $K(t,\tau)$. This means that there are two families of orthogonal polynomials, $\{p_n^+(t)\}_{n=0}^\infty$ and $\{p_n^-(t)\}_{n=0}^\infty$, and non-zero numbers $\{\sigma_n\}_{n=0}^\infty$ such that

$$\int_a^b K(t,\tau)p_n^+(\tau)p_+(\tau)d\tau = \sigma_n g_+(t)p_n^-(t), \quad (a < t < b, \quad n = \overline{0,\infty}). \tag{6.94}$$

Let us suppose that the polynomials $p_n^+(t)$ and $p_n^-(t)$ are orthonormal in (a, b), i.e.

$$\int_a^b \frac{p_m^\pm(t)p_n^\pm(t)dt}{w_\pm(t)} = \delta_{nm}, \quad w_\pm^{-1} = p_\pm(t)g_\mp(t), \tag{6.95}$$

where δ_{nm} is the Kronecker delta.

Then we can look for the solution of the equation (6.93) in the form of the series

$$y(\tau) = p_+(\tau)\sum_{m=0}^\infty y_m p_m^+(t), \quad (a < t < b), \tag{6.96}$$

where the coefficients y_n must be determined.

Substituting (6.96) into the equation (6.93) and using the spectral relationship (6.94) we obtain

$$g_+(t)\sum_{m=0}^\infty \sigma_m y_m p_m^-(t) = f(t), \quad (a < t < b). \tag{6.97}$$

Multiplying both sides of (6.97) by $p_-(t)p_n^-(t)$ and integrating from a to b with the use of the orthonormality condition (6.95) gives

$$y_n\sigma_n = f_n, \quad f_n = \int_a^b f(t)p_-(t)p_n^-(t)dt, \tag{6.98}$$

from which we find y_n. Therefore, the solution of the equation (6.93) can be written in the explicit form

$$y(t) = p_+(t) \sum_{m=0}^{\infty} \frac{f_m p_m^+(t)}{\sigma_m^+}. \tag{6.99}$$

In the case of the equation of the first kind

$$\int_a^b R(t,\tau) y(\tau) d\tau = f(t), \quad (a < t < b), \tag{6.100}$$

with the kernel $R(t,\tau)$ for which the spectral relationship is unknown, it may often be possible to write it in the form

$$\int_a^b \Big\{ K(t,\tau) + D(t,\tau) \Big\} y(\tau) d\tau = f(t), \quad (a < t < b), \tag{6.101}$$

where $D(t,\tau)$ is a regular kernel. Then substitution of (6.96) into (6.101), the use of the spectral relationship (6.94) and the orthonormality condition (6.95) lead to the infinite system of linear algebraic equations:

$$\sigma_n y_n + \sum_{m=0}^{\infty} d_{nm} y_m = f_n, \quad (n = \overline{0,\infty}), \tag{6.102}$$

$$d_{nm} = \int_a^b \int_a^b D(t,\tau) p_-(t) p_+(\tau) p_m^+(\tau) p_n^-(t) dt d\tau,$$

where f_n is the same as above.

The infinite system (6.102) can be solved approximately by the reduction method. It is worth noting that the solution of the integral equation of the first kind (6.101) is reduced to the solution of the infinite linear algebraic system of the second kind (6.102), which is an advantage from the point of view of approximate numerical solutions. The conditions of convergence of the reduction method for the system (6.102) are given in [112], [111], and [213].

For some kernels $K(t,\tau)$ more than one spectral relationship may be known. For example, in the case of the Cauchy singular kernel $(t-\tau)^{-1}$ we have three spectral relationships for classical Jacobi polynomials (see [63, formulas 10.12(47) and 10.12.(48)] and [213, formula A-12.4, p. 304]), and an infinite number of them for generalized Jacobi polynomials [192] (note also the spectral relationships (6.114), (6.113), and (6.117) below). In such a case, the definite choice of the spectral relationship depends on the character of the problem resulting in the considered integral equation; the character of the problem determines the class of solution. The spectral relationship must be selected after choosing the class of solutions and is accomplished, if necessary, with additional conditions for determining additional constants (i.e. finding the unique solution) or with the conditions (usually for the right-hand side) of solvability (see, for example, [192] and [193]).

The availability of a suitable spectral relationship is a necessary condition for the application of the method of orthogonal polynomials. A wide collection of spectral relationships for various integral operators of the first and the second kinds is given in [213], but it does not contain the relationships given in this paper.

In the subsections following below we present a collection of spectral relationships for the three types of fractional differential operators:

- the Riesz fractional integrals with the kernel $R(t,\tau) = |t-\tau|^{-\alpha}$,
- the left Riemann–Liouville fractional integrals

$$
{}_aD_t^{-\alpha}f(t) = \frac{1}{\Gamma(\alpha)} \int_a^t (t-\tau)^{\alpha-1} f(\tau)d\tau, \qquad (6.103)
$$

- the right fractional Riemann–Liouville integrals (sometimes also called the Weyl fractional integrals)

$$
{}_tD_b^{-\alpha}f(t) = \frac{1}{\Gamma(\alpha)} \int_t^b (\tau-t)^{\alpha-1} f(\tau)d\tau. \qquad (6.104)
$$

6.4.3 Riesz Fractional Potential

We will start by obtaining the spectral relationships for the integral operator with the kernel $|x-t|^{-\nu}$ called the Riesz potential operator [232].

THEOREM 6.1 ∘ *If* $\alpha > -1$, $\beta > -1$, $0 < \nu < 1$ *and* γ *is an arbitrary real number, then for* $-1 < x < 1$ *the following hold*

$$\int_{-1}^{1} \left(\operatorname{sign}(x-t) + \frac{\tan(\pi\gamma)}{\tan\frac{\nu\pi}{2}} \right) \frac{Q_m^{\alpha,\beta}(t)}{|x-t|^\nu} dt$$

$$= \frac{\sin\pi(\gamma - \frac{\nu}{2})\Phi_1(x) + \sin\pi(\gamma + \frac{\nu}{2} - \beta)\Phi_2(x)}{\Gamma(\nu)\sin\frac{\nu\pi}{2}\cos\gamma\pi}, \qquad m = 0,1,2,\ldots, \tag{6.105}$$

where

$$\Phi_1(x) = \frac{\Gamma(m+\alpha+1)\Gamma(m+\nu)\Gamma(\beta-\nu+1)(-1)^m}{2^{-\alpha-\beta+\nu-1}\Gamma(m+\alpha+\beta-\nu+2)\,m!}$$

$$\times\ {}_2F_1\left(\begin{matrix} m+\nu, \quad \nu-m-\alpha-\beta-1 \\ -\beta+\nu \end{matrix} \middle| \frac{1+x}{2} \right) \tag{6.106}$$

$$\Phi_2(x) = \frac{\Gamma(m+\beta+1)\Gamma(\nu-\beta-1)(-1)^{m+1}}{2^{-\alpha}(1+x)^{\nu-\beta-1}}$$

$$\times\ {}_2F_1\left(\begin{matrix} m+\beta+1, \quad -m-\alpha \\ \beta-\nu+2 \end{matrix} \middle| \frac{1+x}{2} \right) \quad \bullet \tag{6.107}$$

Proof. To prove this statement, let us consider the following integral

$$J(x) = \int_{-1}^{1} k(x-t) Q_m^{\alpha,\beta}(t) dt, \qquad |x| \le 1 \tag{6.108}$$

$$Re(\alpha,\beta) > -1, \qquad m = 0,1,2,\ldots$$

where the function $k(z)$ is defined by

$$k(z) = e^{i\pi\gamma} \int_0^\infty s^{\nu-1} e^{izs} ds, \qquad 0 < \nu < 1. \tag{6.109}$$

Substituting (6.109) into equation (6.108), interchanging the order of integration, using the Rodrigue formula for Jacobi polynomials and integrating by parts we obtain:

$$J(x) = (-i)^m B(m+\beta+1, m+\alpha+1) e^{i\pi\gamma} 2^{m+\alpha+\beta+1}(m!)^{-1}$$

$$\times \int_0^\infty s^{m+\nu-1} e^{i(1+x)s}\ {}_1F_1\left(\begin{matrix} m+\beta+1 \\ 2m+\alpha+\beta+2 \end{matrix} \middle| -2is \right) ds. \tag{6.110}$$

Evaluating integral (6.110) with the help of [62, formula 6.9(9))] and then using [63, formula 2.10(2)], we find

$$J(x) = e^{i\pi(\gamma - \frac{\nu}{2})}\Phi_1(x) + e^{i\pi(\gamma + \frac{\nu}{2} - \beta)}\Phi_2(x), \tag{6.111}$$

where $\Phi_1(x)$ and $\Phi_2(x)$ are given by (6.106) and (6.107).

On the other hand, with the help of [62, formulas 6.5(1) and 6.5(21)], integral (6.109) can be expressed in the form

$$k(z) = \Gamma(\nu)|z|^{-\nu} \exp\left\{ i\pi \left(\gamma + \frac{\nu}{2}\mathrm{sign}(z) \right) \right\}. \tag{6.112}$$

Substituting (6.112) into (6.108), taking into account the equation (6.111), and separating the imaginary part, we obtain (6.105), which ends the proof of Theorem 6.1. (Consideration of the real part after replacing γ with $\gamma + \frac{1}{2}$ leads to the same final expression.)

THEOREM 6.2 ∘ *If* $\alpha > -1$, $\beta > -1$ *and* $0 < \nu < 1$, *then for* $-1 < x < 1$ *the following hold*

$$\int_{-1}^{1} \frac{Q_m^{\alpha,\beta}(t)dt}{|x-t|^\nu} = \frac{\cos\frac{\pi\nu}{2}\Phi_1(x) + \cos\pi\left(\frac{\nu}{2} - \beta\right)\Phi_2(x)}{\Gamma(\nu)\cos\frac{\pi\nu}{2}},$$

$$m = 0, 1, 2, \ldots,$$

where $\Phi_1(x)$ *and* $\Phi_2(x)$ *are the same as in Theorem 6.1* •

Proof. This is a particular case of Theorem 6.1 for $\gamma = \frac{1}{2}$.

THEOREM 6.3 ∘ *If* $\alpha > -1$, $\beta > -1$ *and* $0 < \nu < 1$, *then for* $-1 < x < 1$ *the following hold*

$$\int_{-1}^{1} 2\frac{\mathrm{sign}(x-t)}{|x-t|^\nu} Q_m^{\alpha,\beta}(t)dt = \frac{-\sin\frac{\pi\nu}{2}\Phi_1(x) + \sin\pi\left(\frac{\nu}{2} - \beta\right)\Phi_2(x)}{\Gamma(\nu)\sin\frac{\pi\nu}{2}},$$

$$m = 0, 1, 2, \ldots,$$

where $\Phi_1(x)$ *and* $\Phi_2(x)$ *are the same as in Theorem 6.1* •

Proof. This is a particular case of Theorem 6.1 for $\gamma = 0$.

THEOREM 6.4 ∘ *If* $0 < \nu < 1$, γ *is an arbitrary real number and* r *and* k *are integer numbers such that* $r > -1+\gamma-\frac{\nu}{2}$, $k > -1-\gamma-\frac{\nu}{2}$, *then for* $-1 < x < 1$ *the following hold*

$$\int_{-1}^{1}\left(\operatorname{sign}(x-t)+\frac{\tan(\pi\gamma)}{\tan\frac{\nu\pi}{2}}\right)\frac{Q_m^{-\gamma+\frac{\nu}{2}+r,\gamma+\frac{\nu}{2}+k}(t)}{|x-t|^{\nu}}dt$$

$$=\frac{\pi(-1)^{r+k+1}\sin\pi(\gamma-\frac{\nu}{2})2^{r+k+1}\Gamma(m+\nu)}{m!\Gamma(\nu)\sin\frac{\nu\pi}{2}\cos(\gamma\pi)\sin\pi(-\gamma+\frac{\nu}{2}-k)}P_{m+r+k+1}^{\gamma+\frac{\nu}{2}-r-1,-\gamma+\frac{\nu}{2}-k-1}(x), \tag{6.113}$$

$$m+r+k+1\geq 0. \quad \bullet$$

Proof. Indeed, taking in (6.105) $\beta=\gamma+\frac{\nu}{2}+k$ (in this case the second term in (6.105) disappears) and $\alpha=-\gamma+\frac{\nu}{2}+r$ (due to this choice $\Phi_1(x)$ becomes a polynomial), we obtain (6.113), which completes the proof of Theorem 6.4.

Theorem 6.4 is a generalization of Popov's formula [213, formula A-6.3, p. 298].

THEOREM 6.5 (Spectral relationship for the classical Riesz potential) ∘ *If* $0 < \nu < 1$ *and* r *and* k *are integer numbers such that* $r > -\frac{\nu+1}{2}$, $k > -\frac{\nu+3}{2}$, *then for* $-1 < x < 1$ *the following hold*

$$\int_{-1}^{1}\frac{Q_m^{\frac{\nu-1}{2}+r,\frac{\nu+1}{2}+k}(t)}{|x-t|^{\nu}}dt=\frac{\pi(-1)^{r}2^{r+k+1}\Gamma(m+\nu)}{m!\Gamma(\nu)\cos\frac{\nu\pi}{2}}P_{m+r+k+1}^{\frac{\nu-1}{2}-r,\frac{\nu-3}{2}-k}(x) \tag{6.114}$$

$$m+r+k+1>0. \quad \bullet$$

Proof. This is a particular case of Theorem 6.4 for $\gamma=\frac{1}{2}$.

In the general case, the Jacobi polynomials in the right-hand side of equations (6.113) and (6.114) are the Jacobi polynomials orthogonal with non-integrable weight function [192].

In order to consider classical Jacobi polynomials, we must have, in addition to the conditions of Theorem 6.5,

$$\frac{\nu-1}{2}-r>-1, \qquad \frac{\nu-3}{2}-k>-1. \tag{6.115}$$

There is only one pair of values for r and k which simultaneously satisfies the conditions of Theorem 6.5 and (6.115): $r=0$, $k=-1$. In this case, equation (6.114) takes the following simple form:

THEOREM 6.6 ∘ *If* $0 < \nu < 1$, *then for* $-1 < x < 1$ *the following hold*

$$\int_{-1}^{1} \frac{Q_m^{\frac{\nu-1}{2},\frac{\nu-1}{2}}(t)}{|x-t|^\nu}dt = \frac{\pi\Gamma(m+\nu)}{m!\Gamma(\nu)\cos\frac{\nu\pi}{2}} P_m^{\frac{\nu-1}{2},\frac{\nu-1}{2}}(x), \tag{6.116}$$

$$m = 0, 1, 2, \ldots \quad \bullet$$

THEOREM 6.7 ∘ *If* $0 < \nu < 1$ *and* r *and* k *are integer numbers such that* $r > -1 - \frac{\nu}{2}$, $k > -1 - \frac{\nu}{2}$, *then for* $-1 < x < 1$ *the following hold*

$$\int_{-1}^{1} \frac{\operatorname{sign}(x-t)}{|x-t|^\nu} Q_m^{\frac{\nu}{2}+r,\frac{\nu}{2}+k}(t)dt = \frac{\pi(-1)^r\Gamma(m+\nu)}{2^{-r-k-1}m!\,\Gamma(\nu)\sin\frac{\pi\nu}{2}} P_{m+r+k+1}^{\frac{\nu}{2}-r-1,\frac{\nu}{2}-k-1}(x), \tag{6.117}$$

$$m + r + k + 1 \geq 0. \quad \bullet$$

Proof. This is a particular case of Theorem 6.4 for $\gamma = 0$.

There are four particular cases of Theorem 6.7 for classical Jacobi polynomials, namely

(i) $r = k = -1$;
(ii) $r = -1$, $k = 0$;
(iii) $r = 0$, $k = -1$;
(iv) $r = k = 0$,

but only three of them are different, because (ii) and (iii) lead to the same formula. We have:

THEOREM 6.8 ∘ *If* $0 < \nu < 1$, *then for* $-1 < x < 1$ *the following formulas are valid:*

$$\int_{-1}^{1} \frac{\operatorname{sign}(x-t)}{|x-t|^\nu} Q_m^{\frac{\nu}{2}-1,\frac{\nu}{2}-1}(t)dt = -\frac{\pi\Gamma(m+\nu)}{2\,m!\,\Gamma(\nu)\sin\frac{\pi\nu}{2}} P_{m-1}^{\frac{\nu}{2},\frac{\nu}{2}}(x), \tag{6.118}$$

$$m = 1, 2, 3, \ldots;$$

$$\int_{-1}^{1} \frac{\operatorname{sign}(x-t)}{|x-t|^\nu} Q_m^{\frac{\nu}{2},\frac{\nu}{2}}(t)dt = \frac{2\pi\Gamma(m+\nu)}{m!\,\Gamma(\nu)\sin\frac{\pi\nu}{2}} P_{m+1}^{\frac{\nu}{2}-1,\frac{\nu}{2}-1}(x), \tag{6.119}$$

$$m = 0, 1, 2, \ldots;$$

$$\int_{-1}^{1} \frac{\operatorname{sign}(x-t)}{|x-t|^\nu} Q_m^{\frac{\nu}{2}-1,\frac{\nu}{2}}(t)dt = -\frac{\pi\Gamma(m+\nu)}{m!\,\Gamma(\nu)\sin\frac{\pi\nu}{2}} P_m^{\frac{\nu}{2},\frac{\nu}{2}-1}(x), \tag{6.120}$$

$$m = 0, 1, 2, \ldots \quad \bullet$$

Taking in (6.118)–(6.119) $\nu \to 1$, we obtain the well-known formulas of the Hilbert transform of weighted Chebyshev polynomials of the first and the second kinds [63, formulas 10.12(47) and 10.12.(48)] and weighted Jacobi polynomials $P_m^{-\frac{1}{2},\frac{1}{2}}(x)$ [213, formula A-12.4, p.304].

6.4.4 Left Riemann–Liouville Fractional Integrals and Derivatives

THEOREM 6.9 ∘ *If* $0 < \nu < 1$, *and* r *and* k *are integer numbers such that* $r > \nu - 1$, $k > -1 - \nu$, *then for* $-1 < x < 1$ *the following hold*

$$\int_{-1}^{x} \frac{Q_m^{r,\nu+k}(t)}{(x-t)^\nu} dt = \frac{\pi(-1)^{r+1} 2^{r+k+1} \Gamma(m+\nu)}{m! \Gamma(\nu) \sin(\nu\pi)} P_{m+r+k+1}^{\nu-r-1,-k-1}(x), \quad (6.121)$$

$$m = 0, 1, 2, \ldots \qquad \bullet$$

Proof. This is a particular case of Theorem 6.4 for $\gamma = \nu/2$.

The relationship (6.121) can be written also using the symbol of fractional differentiation:

$$_{-1}D_t^{1-\nu}\left\{Q_n^{r,\nu+k}(t)\right\} = \frac{\pi(-1)^{r+1} 2^{r+k+1} \Gamma(n+\nu)}{n! \Gamma(\nu) \sin(\nu\pi)} P_{n+r+k+1}^{\nu-r-1,-k-1}(t), \quad (6.122)$$

$$\left(|t| < 1, \quad 0 < \nu < 1, \quad r > \nu - 1, \quad k > -1 - \nu, \quad r, k = \overline{0, \infty}\right).$$

In the general case, the Jacobi polynomials in the right-hand side of equation (6.121) are the Jacobi polynomials orthogonal with non-integrable weight function [192].

To consider classical Jacobi polynomials, we must have, in addition to the conditions of Theorem 6.9,

$$\nu - r - 1 > -1, \qquad -k - 1 > -1. \quad (6.123)$$

There is only one pair of values for r and k which simultaneously satisfies the conditions of Theorem 6.9 and (6.123): $r = 0$, $k = -1$. In this case, equation (6.121) becomes

THEOREM 6.10 ∘ *If* $0 < \nu < 1$, *then for* $-1 < x < 1$ *the following hold*

$$\int_{-1}^{x} \frac{Q_m^{0,\nu-1}(t)}{(x-t)^\nu} dt = -\frac{\pi \Gamma(m+\nu)}{m! \Gamma(\nu) \sin(\nu\pi)} P_m^{\nu-1,0}(x), \quad (6.124)$$

$$m = 0, 1, 2, \ldots \qquad \bullet$$

Putting $\nu = 1 - \lambda$ $(0 < \lambda < 1)$ and performing obvious substitutions of variables, we obtain

$$\int_0^y \frac{Q_m^{0,-\lambda}(2\tau - 1)}{(y-\tau)^{1-\lambda}} d\tau = \frac{\pi \Gamma(m - \lambda + 1)}{2^\lambda\, m!\, \Gamma(1-\lambda) \sin(\lambda\pi)} P_m^{-\lambda,0}(2y - 1), \quad (6.125)$$

$$0 < y < 1; \qquad m = 0, 1, 2, \ldots$$

The integral in the left-hand side of (6.125) is a multiple of the Riemann–Liouville fractional integral of order λ, which is defined by (e.g., [179, 232])

$$_aD_y^{-\lambda} f(y) = \frac{1}{\Gamma(\lambda)} \int_a^y (y-\tau)^{\lambda-1} f(\tau) d\tau, \qquad (y > a). \qquad (6.126)$$

Using this notation, we can write equation (6.125) as

THEOREM 6.11 ∘ *If* $0 < \lambda < 1$, *then for* $0 < t < 1$ *the following hold*

$$_0D_t^{-\lambda} Q_m^{0,-\lambda}(2t - 1) = \frac{\Gamma(m - \lambda + 1)}{2^\lambda\, m!} P_m^{-\lambda,0}(2t - 1), \qquad (6.127)$$

$$m = 0, 1, 2, \ldots \qquad \bullet$$

THEOREM 6.12 ∘ *If* $0 < \lambda < 1$, *then for* $0 < t < 1$ *the following holds*

$$_0D_t^{\lambda} P_m^{-\lambda,0}(2t - 1) = \frac{2^\lambda\, m!}{\Gamma(m - \lambda + 1)} Q_m^{0,-\lambda}(2t - 1), \qquad (6.128)$$

$$m = 0, 1, 2, \ldots \qquad \bullet$$

Proof. Applying the operator of Riemann–Liouville fractional differentiation to both sides of equation (6.127) and using the well-known property (e.g, [179, 232])

$$_0D_t^{\lambda} \left({}_0D_t^{-\lambda} f(t) \right) = f(t),$$

we obtain equation (6.128).

THEOREM 6.13 ∘ *If* $0 \le n - 1 < p < n$, *n is integer, then for* $0 < t < 1$ *the following holds*

$$_0D_t^p\, Q_m^{0,p-n}(2t - 1) = \begin{cases} \dfrac{\Gamma(m + p + 1)}{2^{n-p}\, m!} P_{m-n}^{p,n}(2t - 1), & (m \ge n) \\ 0, & (m < n). \end{cases} \qquad \bullet \qquad (6.129)$$

Proof. Differentiating both sides of equation (6.127) n times with respect to t and using [63, formula 10.8(17)], we obtain

$$\frac{d^n}{dt^n}\, {}_0D_t^{-\lambda} Q_m^{0,-\lambda}(2t-1) = \begin{cases} \dfrac{\Gamma(m+n-\lambda+1)}{2^\lambda\, m!} P_{m-n}^{n-\lambda,n}(2t-1), & (m \geq n) \\ 0, & (m < n). \end{cases} \tag{6.130}$$

In the left-hand side of this relationship we recognize the Riemann–Liouville fractional derivative (e.g., [179, 232]) of order $p = n - \lambda$. This allows us to write equation (6.130) as (6.129).

THEOREM 6.14 ∘ *If* $\alpha > -1$, $\beta > -1$ *and* $\lambda > 0$, *then for* $0 < t < 1$ *the following holds*

$${}_0D_t^{-\lambda} Q_m^{\alpha,\beta}(2t-1) = \frac{(-1)^m \Gamma(\lambda)\Gamma(m+\beta+1) 2^{\alpha+\beta}}{\Gamma(\beta+\lambda+2)\, m!}$$

$$\times\; (1-t)^{\alpha+\lambda}\, t^{\beta+\lambda}\; {}_2F_1\left(\left. \begin{matrix} -m+\lambda, & \alpha+\beta+m+\lambda+1 \\ & \beta+\lambda+1 \end{matrix} \right| t \right) \tag{6.131}$$

$$m = 0, 1, 2, \ldots \qquad \bullet$$

Proof. Putting in (6.105) $\gamma = \frac{\nu}{2}$ we obtain

$$\int_{-1}^{x} (x-t)^{-\nu} Q_m^{\alpha,\beta}(t)\,dt = \frac{\sin \pi(\nu-\beta)}{\Gamma(\nu)\sin(\pi\nu)} \Phi_2(x). \tag{6.132}$$

Substituting now (6.106) into (6.132), setting $\nu = 1 - \lambda$ and using [63, formula 2.9(2)] we obtain (6.131).

6.4.5 Other Spectral Relationships For the Left Riemann–Liouville Fractional Integrals

The following spectral relationships for the Chebyshev and Legendre polynomials were obtained by I. Podlubny with the help of the properties of the finite Fourier transforms [191]; the formula (6.135) can also be found in [2, formula 22.13.10]

$${}_{-1}D_t^{-1/2} \left\{ \frac{T_n(t)}{\sqrt{1+t}} \right\} = \frac{\sqrt{\pi}}{2} \Big(P_n(t) - P_{n-1}(t) \Big), \tag{6.133}$$

$$(P_{-1} \equiv 0, \quad |t| < 1, \quad n = \overline{0, \infty})$$

$$
{}_{-1}D_t^{-1/2}\left\{U_n(t)\sqrt{1+t}\right\} = \frac{\sqrt{\pi}}{2}\Big(P_n(t) + P_{n+1}(t)\Big), \tag{6.134}
$$

$$
(|t| < 1, \quad n = \overline{0, \infty}).
$$

$$
{}_{-1}D_t^{-1/2}\left\{P_n(t)\right\} = \frac{2\sqrt{\pi}}{2n+1} \cdot \frac{T_n(t) + T_{n+1}(t)}{\sqrt{1+t}}, \tag{6.135}
$$

$$
(|t| < 1, \quad n = \overline{0, \infty})
$$

There is a spectral relationship relating the Gegenbauer polynomials and the Jacobi polynomials (see [216, formula 2.21.2(9)], [91]):

$$
{}_{-1}D_t^{-\alpha}\left\{(1+t)^{\beta-1/2} C_n^{(\beta)}(t)\right\} = \sigma_n^{\alpha,\beta}(1+t)^{\alpha+\beta-1/2} P_n^{\beta-\alpha-1/2,\, \beta+\alpha-1/2}(t), \tag{6.136}
$$

$$
\sigma_n^{\alpha,\beta} = \frac{\Gamma(2\beta + n)\,\Gamma(\beta + \frac{1}{2})}{\Gamma(2\beta)\,\Gamma(\alpha + \beta + n + \frac{1}{2})},
$$

$$
(|t| < 1, \quad \alpha > 0, \quad \beta > \frac{1}{2}, \quad n = \overline{0, \infty}).
$$

R. Askey obtained another spectral relationship for the Jacobi polynomials [8]:

$$
{}_{-1}D_t^{-\mu}\left\{(1+t)^{\beta} P_n^{\alpha,\beta}(t)\right\} = \frac{\Gamma(n+\beta+1)}{\Gamma(n+\beta+\mu+1)} P_n^{\alpha-\mu,\beta+\mu}(t), \tag{6.137}
$$

$$
(|t| < 1, \quad \alpha > -1, \quad \beta > -1, \quad \mu > 0).
$$

The following formula for the Laguerre polynomials was probably first obtained by E. Kogbetliantz (see [247, Task 20, p. 383 of the Russian edition], where a reference to the Kogbetliantz's paper is given):

$$
{}_0D_t^{-\alpha}\left\{t^{\beta} L_n^{(\beta)}(t)\right\} = \frac{\Gamma(\beta+n+1)}{\Gamma(\alpha+\beta+n+1)} t^{\alpha+\beta} L_n^{\alpha+\beta}(t), \tag{6.138}
$$

$$
(t > 0, \quad \beta > -1, \quad \alpha > 0, \quad n = \overline{0, \infty}).
$$

6.4.6 Spectral Relationships For the Right Riemann–Liouville Fractional Integrals

The following spectral relationsips for the Chebyshev and Legendre polynomials were obtained by I. Podlubny with the help of the properties of the finite Fourier transforms [191]; the formula (6.141) can also be found in [2, formula 22.13.11]

$$ {}_tD_1^{-1/2}\left\{\frac{T_n(t)}{\sqrt{1-t}}\right\} = \frac{\sqrt{\pi}}{2}\Big(P_n(t)+P_{n-1}(t)\Big), \tag{6.139}$$

$$(P_{-1}\equiv 0, \quad |t|<1, \quad n=\overline{0,\infty})$$

$$ {}_tD_1^{-1/2}\left\{U_n(t)\sqrt{1-t}\right\} = \frac{\sqrt{\pi}}{2}\Big(P_n(t)-P_{n+1}(t)\Big), \tag{6.140}$$

$$(|t|<1, \quad n=\overline{0,\infty})$$

$$ {}_tD_1^{-1/2}\Big\{P_n(t)\Big\} = \frac{2\sqrt{\pi}}{2n+1}\cdot\frac{T_n(t)-T_{n+1}(t)}{\sqrt{1-t}}, \tag{6.141}$$

$$(|t|<1, \quad n=\overline{0,\infty}).$$

R. Gorenflo and Vu Kim Tuan obtained the spectral relationship for the Gegenbauer polynomials [91]:

$$ {}_tD_1^{-k/2}\Big\{Y_n^{(k/2)}(t)\Big\} = \frac{n!\,\Gamma(k/2)}{2^{(2-k)/2}\Gamma(n+k)}(1-t)^{(k-1)/2}C_n^{(k/2)}(t), \tag{6.142}$$

$$(|t|<1, \quad n=\overline{0,\infty}, \quad k=\overline{0,\infty}),$$

where the functions

$$Y_n^{(k/2)}(t) = \frac{\cos\Big((n+\frac{k}{2})\arccos t\Big)}{\sqrt{1-t^2}}$$

are orthogonal on $(-1,-1)$ with the weight $w(t)=\sqrt{1-t^2}$:

$$\int_{-1}^{1} Y_n^{(k/2)}(t)Y_m^{(k/2)}(t)\sqrt{1-t^2}dt = \frac{\pi}{2}\delta_{nm}$$

and can be expressed in terms of the Chebyshev polynomials ($U_{-1}(t)\equiv 0$):

$$Y_n^{(k/2)}(t) = (1-t^2)^{-1/2}T_{n+\frac{k}{2}}(t) \quad \text{for even } k\ , \tag{6.143}$$

$$Y_n^{(k/2)}(t)=\frac{1}{\sqrt{2}}\Big((1-t^2)^{-1/2}T_{n+\frac{k-1}{2}}(t)-(1-t^2)^{1/2}U_{n+\frac{k-3}{2}}(t)\Big), \quad \text{for odd } k. \tag{6.144}$$

The following spectral relationship for the Jacobi polynomials was obtained by R. Askey [8]:

$$ {}_tD_1^{-\mu}\Big\{(1-t)^\alpha P_n^{\alpha,\beta}(t)\Big\} = \frac{\Gamma(n+\alpha+1)}{\Gamma(n+\alpha+\mu+1)} P_n^{\alpha+\mu,\beta-\mu}(t), \qquad (6.145) $$

$$ (|t|<1, \quad \alpha>-1, \quad \beta>-1, \quad \mu>0). $$

The spectral relationship for the Laguerre polynomials was given by G. Ya. Popov [213, formula B-7.2, p.307]:

$$ {}_tD_\infty^{-\alpha}\Big\{e^{-t}L_n^{(\beta)}(t)\Big\} = e^{-t}L_n^{(\beta-\alpha)}(t), \qquad (6.146) $$

$$ (\alpha>0, \quad \beta>\alpha-1, \quad n=\overline{0,\infty}). $$

6.4.7 Solution of Arutyunyan's Equation in Creep Theory

As the first simple example of the use of the method of orthogonal polynomials we will consider the equation deduced by N.Kh.Arutyunyan [7] for the plane contact problem of linear creep theory, which can be reduced to the solution of the equation with the Riesz kernel:

$$ \int_{-1}^{1} \frac{p(\tau)d\tau}{|t-\tau|^\alpha} = f(t), \quad (|t|<1). \qquad (6.147) $$

Let us obtain the solution to equation (6.147) using the method of orthogonal polynomials.

The spectral relationship (6.116) suggests the following form of the solution:

$$ p(\tau) = \sum_{n=0}^{\infty} p_n Q_n^{\frac{\alpha-1}{2},\frac{\alpha-1}{2}}(\tau). \qquad (6.148) $$

Substitution of (6.148) into equation (6.147) and the subsequent use of the spectral relationship (6.116) gives:

$$ \sum_{n=0}^{\infty} p_n \frac{\pi\Gamma(n+\alpha)}{n!\Gamma(\alpha)\cos\frac{\alpha\pi}{2}} P_n^{\frac{\alpha-1}{2},\frac{\alpha-1}{2}}(t) = f(t), \quad (|t|<1), \qquad (6.149) $$

and using the orthogonality of the Jacobi polynomials we find the coefficients p_n $(n=\overline{0,\infty})$:

$$ p_n = f_n \frac{(2n+\alpha)\Gamma^2(n+\alpha)\Gamma(\alpha)\cos\frac{\alpha\pi}{2}}{\pi 2^\alpha \Gamma^2(n+\frac{\alpha+1}{2})}, \qquad (6.150) $$

$$f_n = \int_{-1}^{1} f(t)(1-t^2)^{(\alpha-1)/2} P_n^{\frac{\alpha-1}{2},\frac{\alpha-1}{2}}(t)dt, \quad (n = \overline{0,\infty}). \tag{6.151}$$

The formulas (6.148), (6.150), and (6.151) give the explicit solution of the equation (6.147). For the numerical computation of the solution G. Ya. Popov's quadrature formulas [213, pp. 37–39] can be used. Those quadrature formulas do not require knowledge of the roots of the Jacobi polynomials and take into account the oscillations of the integrated function.

6.4.8 Solution of Abel's Equation

As the second example let us consider the classical integral equation (6.72) of H.N.Abel.

Let us suppose that the right-hand side $f(t)$ is bounded at $t = 0$. In this case, it is known that $y(t) \sim \text{const} \cdot t^{-\alpha}$, $(t \to 0)$. Therefore, we can use the spectral relationship (6.127) and look for the solution in the form

$$y(\tau) = t^{-\alpha} \sum_{n=0}^{\infty} y_n P_n^{0,-\alpha}(2\tau - 1). \tag{6.152}$$

The usual procedure of the determination of y_n leads to the following result:

$$y_n = \frac{(2n-\alpha+1)\Gamma(n+1)}{4\Gamma(n-\alpha+1)} f_n, \quad f_n = \int_0^1 f(t) P_n^{-\alpha,0}(2t-1)(1-t)^{-\alpha}dt, \tag{6.153}$$

$$(n = \overline{0,\infty}).$$

The formulas (6.152) and (6.153) give the explicit solution of the equation (6.72). For the numerical computation of the solution the above-mentioned Popov quadrature formulas can be used.

6.4.9 Finite-part Integrals

In the above sections we give some spectral relationships for the Jacobi polynomials orthogonal with non-integrable weight function. The development of the theory of such generalized polynomials has just started recently, so in this section and in the subsequent one we give only very basic information, which is necessary for applications.

Because of the non-integrability of the weight function, the main instrument is the notion of the finite part of a divergent integral.

The definition of the finite part of a divergent integral was given by Hadamard [99], when he was considering the integral

$$I(\lambda) = \int_a^b f(x)(x-a)^\lambda dx, \qquad \lambda < -1. \tag{6.154}$$

He obtained and used the first regularization formulas for divergent integrals with non-integrable weight function like (6.154).

In the paper [192] a class of Jacobi polynomials which are orthogonal with non-integrable weight is studied. Among other results, the following regularization formula for the finite-part integrals was obtained: if $f(x)$ is continuously differentiable in $[-1,1]$, then

$$\int_{-1}^{1} \frac{f(x)dx}{(1-x)^{\alpha+1}(1+x)^{\beta+1}} = \frac{1}{4\alpha\beta}\int_{-1}^{1} \frac{\alpha-\beta-(\alpha+\beta)x}{(1-x)^\alpha(1+x)^\beta} f'(x)dx$$

$$+ \frac{(\alpha+\beta)(\alpha+\beta-1)}{4\alpha\beta}\int_{-1}^{1} \frac{f(x)dx}{(1-x)^\alpha(1+x)^\beta}, \tag{6.155}$$

where α and β must satisfy the following conditions:

$$\alpha < 1, \quad \beta < 1, \quad \alpha \neq 0, \quad \beta \neq 0, \quad \alpha+\beta \neq 0; 1.$$

If both integrals in the right-hand side of (6.155) exist in the usual sense, then the value of the right-hand side gives the finite value of the integral in the left-hand side. Mathematically, we use here analytical continuation with respect to α and β.

One of many particular cases of the regularization formula (6.155), which will be used below, is for $\alpha = \beta = \frac{1}{2}$:

$$\int_{-1}^{1} \frac{f(x)dx}{(1-x^2)^{3/2}} = -\int_{-1}^{1} \frac{xf'(x)dx}{(1-x^2)^{1/2}}. \tag{6.156}$$

We can also mention the following two particular cases of the formula (6.155):

$$\int_{-1}^{1} \frac{f(x)dx}{(1-x)^{\alpha+1}(1+x)^{1-\alpha}} = -\frac{1}{2\alpha}\int_{-1}^{1} \frac{f'(x)dx}{(1-x)^\alpha(1+x)^{-\alpha}}, \quad (0<\alpha<1), \tag{6.157}$$

$$\int_{-1}^{1} \frac{f(x)dx}{(1-x)^{\alpha+1}(1+x)^{2-\alpha}} = \frac{1}{4\alpha(1-\alpha)} \int_{-1}^{1} \frac{(2\alpha-1-x)f'(x)dx}{(1-x)^{\alpha}(1+x)^{1-\alpha}}, \quad (6.158)$$

$$(0 < \alpha < 1).$$

Using (6.156), we can easily evaluate, for instance, the following finite-part integrals:

$$I_0 = \int_{-1}^{1} \frac{dx}{(1-x^2)^{3/2}} = 0 \quad (6.159)$$

$$I_2 = \int_{-1}^{1} \frac{x^2 dx}{(1-x^2)^{3/2}} = -\int_{-1}^{1} \frac{2x^2 dx}{(1-x^2)^{1/2}} = -\pi \quad (6.160)$$

$$I_4 = \int_{-1}^{1} \frac{x^4 dx}{(1-x^2)^{3/2}} = -\int_{-1}^{1} \frac{4x^4 dx}{(1-x^2)^{1/2}} = -\frac{3\pi}{2} \quad (6.161)$$

$$\int_{-1}^{1} \frac{x^{2n+1} dx}{(1-x^2)^{3/2}} = -\int_{-1}^{1} \frac{(2n+1)x^{2n+1} dx}{(1-x^2)^{1/2}} = 0, \quad (n \geq 0). \quad (6.162)$$

In the case of *equalities* the rules applicable to integrals in the classical sense can be used also for the manipulation with finite-part integrals. For example, it holds that

$$\int_a^b = \int_a^c + \int_c^b,$$

and so on. However, manipulation with *inequalities* requires some care. For example, knowing that $f(t)$ is positive in $[a, b]$, we can say nothing about the sign of the finite-part integral of the type (6.154) or (6.156). Indeed, it may also be zero (6.159) or negative (6.160).

Estimates for finite-part integrals can be obtained using the regularization formulas.

It is interesting to note that the Riemann–Liouville fractional derivative can be written in the form of a finite-part integral

$${}_a\mathbf{D}_t^{\alpha} f(t) = \frac{1}{\Gamma(-\alpha)} \int_a^t f(\tau)(t-\tau)^{-\alpha-1} d\tau, \quad (6.163)$$

which can also be considered as a convolution of two generalized functions: $\Phi_{-\alpha}(t) = t^{-\alpha-1}/\Gamma(-\alpha)$ and $f(t)$. Therefore, the finite-part integral form of the fractional derivative is equivalent to the generalized functions approach described in Section 2.4.2. However, it seems to us that the finite-part integral approach can be in a certain sense more transparent.

6.4.10 Jacobi Polynomials Orthogonal with Non-integrable Weight Function

The authors of the Bateman Manuscript Project [64] noticed that the majority of the relationships for the classical Jacobi polynomials $P_b^{\alpha,\beta}(t)$ can be (formally) used even if $\alpha < -1$, or $\beta < -1$, or both $\alpha < -1$ and $\beta < -1$. That remark opened a way to the generalization of the Jacobi polynomials considering unrestricted values of the parameters α and β.

The first real application for such generalized Jacobi polynomials was found by G. Ya. Popov and O. V. Onishchuk [181]. They reduced the problem for a plate with a rigid inclusion to an integral equation with a so-called smooth kernel, for which the Jacobi polynomials $P_n^{\alpha,\beta}(t)$, as they proved, are the eigenfunctions. This allowed them to obtain the solution of the integral equation in the form of a Fourier–Jacobi series.

In this section we use some parts of the theory presented in [192].

For the application of Jacobi polynomials orthogonal with non-integrable weight function to the solution of fractional integral and differential equations we first of all need a tool for the evaluation of the Fourier–Jacobi coefficients of the function $f(t)$. The following formula provides such a tool:

$$\int_{-1}^{1} \frac{f(t)P_n^{-\alpha-1,-\beta-1}(t)dt}{(1-t)^{\alpha+1}(1+t)^{\beta+1}} = \frac{1}{2n}\int_{-1}^{1} \frac{f'(t)P_{n-1}^{-\alpha,-\beta}(t)dt}{(1-t)^{\alpha}(1+t)^{\beta}}, \qquad (n=\overline{1,\infty}), \tag{6.164}$$

under the assumption that $f(t)$ is continuously differentiable in the closed interval $[-1, 1]$, $0 < \alpha$, $\beta < 1$, $\alpha \neq 0$, $\beta \neq 0$, and $\alpha + \beta \neq 0; 1$.

The formula (6.164) allows easy evaluation of the squared norms of the considered Jacobi polynomials for $n \geq 1$:

$$||P_n^{-\alpha-1,-\beta-1}||^2 = \int_{-1}^{1} \left(P_n^{-\alpha-1,-\beta-1}(t)\right)^2 (1-t)^{-\alpha-1}(1+t)^{-\beta-1}dt$$

$$= \frac{1}{2n}\int_{-1}^{1} \frac{\left(P_n^{-\alpha-1,-\beta-1}(t)\right)' P_{n-1}^{-\alpha,-\beta}(t)dt}{(1-t)^{\alpha}\,(1+t)^{\beta}}$$

$$= \frac{n-\alpha-\beta-1}{4n} ||P_n^{-\alpha,-\beta}||^2 \tag{6.165}$$

$$= \frac{2^{-\alpha-\beta-1}\,\Gamma(n-\alpha)\,\Gamma(n-\beta)}{(2n-\alpha-\beta-1)\,n!\,\Gamma(n-\alpha-\beta-1)}. \tag{6.166}$$

From the formal point of view, the expression (6.166) is the same as the formula 10.8(4) from [64]. However, in our case the expression (6.166) represents the regularized value of the finite-part integral.

It follows from (6.166) that for $n \geq 1$ we have

$$||P_n^{-\alpha-1,-\beta-1}||^2 \begin{cases} > 0, & \text{for} \quad n > \alpha+\beta+1 \\ = 0, & \text{for} \quad n = \alpha+\beta+1 \\ < 0, & \text{for} \quad n < \alpha+\beta+1, \end{cases} \tag{6.167}$$

which means that squared norms of the considered Jacobi polynomials are not always positive, but can also be negative or zero, depending on a combination α, β and n. Such a norm is called an *indefinite* norm [10]. The convergence of Fourier series in such orthogonal polynomials should be investigated in indefinite metric spaces. Moreover, since we have only a finite number of polynomials with negative squared norm, those indefinite metric spaces are Pontryagin spaces (definitions can be found in [10]). We give below only a very brief overview of some results which can be useful for proper application of the considered generalized Jacobi polynomials.

Q-metrics and Q-orthogonality

Let F be a linear space of continuously differentiable functions in the closed interval $[-1, 1]$. Let us consider the real linear form

$$\{f, g\} = \int_{-1}^{1} f(t)g(t)(1-t)^{-\alpha-1}(1+t)^{-\beta-1}dt.$$

The following properties of this form are obvious:

1. $\{f, g\} = \{g, f\}$
2. $\{\lambda_1 f_1 + \lambda_2 f_2, g\} = \lambda_1\{f_1, g\} + \lambda_2\{f_2, g\}$
3. $\{f, f\}$ can be positive, negative or zero.

Indeed, it follows from (6.159), (6.160), and (6.161) that for $\alpha = 0.5$ and $\beta = 0.5$ we have

$$\{1,\,1\} = 0,$$

$$\{t,\,t\} = -\pi,$$

$$\{1-t^2,\,1-t^2\} = \pi/2.$$

The linear form $\{f,\,g\}$ is an indefinite metric (Q-metric) in F [10]. It is said that $f(t)$ is *Q-positive*, *Q-negative*, or *Q-neutral*, if $\{f,\,f\} > 0$, $\{f,\,f\} < 0$, or $\{f,\,f\} = 0$, respectively.

If $\{f,\,g\}=0$, then the functions $f(t)$ and $g(t)$ are called *Q-orthogonal*, which we denote as $f\{\perp\}g$.

For example, $f(t) = t$ is Q-negative, $g(t) = 1$ is Q-neutral, $h(t) = 1-t^2$ is Q-positive, and $f\{\perp\}g$, because $\{t,\,1\} = 0$.

Similarly to the classical Jacobi polynomials we have:

THEOREM 6.15 ∘ *$P_n^{-\alpha-1,-\beta-1}(t)$ is Q-orthogonal to all polynomials of lower order $\Pi_m(t)$:*

$$\{P_n^{-\alpha-1,-\beta-1},\,\Pi_m\} = 0, \quad (m < n). \quad \bullet$$

We also have the Buniakowski inequality:

THEOREM 6.16 ∘ *If $f(t) \in F$ and $g(t) \in F$ are not Q-negative functions, then*

$$\left(\int_{-1}^{1} \frac{f(t)g(t)dt}{(1-t)^{\alpha+1}(1+t)^{\beta+1}}\right)^2$$

$$\leq \int_{-1}^{1} \frac{f^2(t)dt}{(1-t)^{\alpha+1}(1+t)^{\beta+1}} \int_{-1}^{1} \frac{g^2(t)dt}{(1-t)^{\alpha+1}(1+t)^{\beta+1}}. \quad \bullet$$

A system of functions $S = \{s_i(t)\}_{i\in I}$, where I is an arbitrary set of indices, is called a *Q-orthogonal system*, if

$$\{s_i,\,s_j\} = \sigma_i\delta_{ij}, \qquad \sigma_i \neq 0, \quad i \in I, \quad j \in I$$

(δ_{ij} is the Kronecker delta).

THEOREM 6.17 ∘ *The system $S_{\alpha,\beta} = \left\{P_n^{-\alpha-1,-\beta-1}\right\}$ is a Q-orthogonal system.* •

A Q-orthogonal system $S \subset F$ is called a *Q-closed system*, if there is no function $h(t) \in F$ such that $h(t) \notin S$ and $h(t) \not\equiv 0$ and $h\{\perp\}S$.

THEOREM 6.18 ∘ *$S_{\alpha,\beta}$ is a Q-closed system.* •

Let us divide S in two subsystems, s^+ and S^-, consisting of the Q-positive and Q-negative functions respectively. The number of functions in a smaller subsystem is called the *range of indefiniteness* of the system S, and denoted $r(S)$.

THEOREM 6.19 ∘ *For the system $S_{\alpha,\beta}$ we have*

$$r(S_{\alpha,\beta}) = \begin{cases} [\alpha+\beta+1], & \textit{for} \quad \alpha+\beta+1 > 0, \\ 0, & \textit{for} \quad \alpha+\beta+1 < 0. \end{cases} \quad \bullet$$

The range of indefiniteness of the system of the classical Jacobi polynomials is equal to 0, and the Q-orthogonality becomes the usual orthogonality.

The Fourier–Jacobi Series

Let us suppose, as above, that $0 < \alpha < 1$, $0 < \beta < 1$, $\alpha + \beta \neq 0; 1$, and recall that F is a linear space of continuously differentiable functions in the closed interval $[-1, 1]$.

The considered generalized Jacobi polynomials, which are orthogonal with non-integrable weight function, allow formal development of functions in Fourier–Jacobi series, and such series have the uniqueness property:

THEOREM 6.20 ∘ *If for $f_1(t) \in F$ and $f_2(t) \in F$ their Fourier series in Jacobi polynomials $P_n^{-\alpha-1,-\beta-1}(t)$ are identical, then $f_1(t) = f_2(t)$.* •

And these series converge for functions from F:

THEOREM 6.21 ∘ *For the function $f(t)$ its Fourier series in Jacobi polynomials $P_n^{-\alpha-1,-\beta-1}(t)$ uniformly converges to $f(t)$ in the closed interval $[-1+\epsilon, 1-\epsilon]$, where ϵ is an arbitrary constant between 0 and 1.* •

For the evaluation of the Fourier–Jacobi coefficients of such series the formulas (6.164) and (6.166) must be used.

Chapter 7

Numerical Evaluation of Fractional Derivatives

In this chapter we describe a simple but effective method for the evaluation of fractional-order derivatives. This approach is based on the fact that for a wide class of functions, which appear in real physical and engineering applications, two definitions – Riemann–Liouville and Grünwald–Letnikov – are equivalent. This allows us to use an approximation arising from the Grünwald–Letnikov definition for the evaluation of fractional derivatives of both types.

We also formulate the principle of "short memory", which reduces the amount of computation, and give two examples of its intermediate usage: computation of heat fluxes in a blast furnace wall and numerical evaluation of finite-part integrals.

7.1 Riemann–Liouville and Grünwald–Letnikov Definitions of the Fractional-order Derivative

The Riemann–Liouville Definition

Recalling Section 2.3, the Riemann–Liouville definition of the fractional-order derivative is

$$ {}_aD_t^\alpha f(t) = \frac{1}{\Gamma(n-a)} \left(\frac{d}{dx}\right)^n \int\limits_a^t \frac{f(\tau)d\tau}{(t-\tau)^{\alpha-n+1}}, \quad (n-1 < \alpha < n). \quad (7.1)$$

The Grünwald–Letnikov Definition

Let us also recall the Grünwald–Letnikov definition (see Section 2.2):

$$ {}_aD_t^\alpha f(t) = \lim_{h \to 0} \frac{{}_a\Delta_h^\alpha f(t)}{h^\alpha}, \qquad {}_a\Delta_h^\alpha f(t) = \sum_{j=0}^{\left[\frac{t-a}{h}\right]} (-1)^j \binom{\alpha}{j} f(t - jh), \tag{7.2} $$

where $[x]$ means the integer part of x.

For a wide class of functions, important for applications, both definitions are equivalent (see Section 2.3.7). This allows one to use the Riemann–Liouville definition during problem formulation, and then turn to the Grünwald–Letnikov definition for obtaining the numerical solution.

7.2 Approximation of Fractional Derivatives

7.2.1 Fractional Difference Approach

We use the following approximation, arising from the Grünwald–Letnikov definition:

$$ {}_aD_t^\alpha f(t) \approx {}_a\Delta_h^\alpha f(t). \tag{7.3} $$

In Figs 7.1–7.4 (see page 201) fractional derivatives of order α ($0 \leq \alpha \leq 1$) of the Heaviside function, sine, cosine and logarithmic function are given. Computations were performed using approximation (7.3).

We see the transition from $\alpha = 0$ to $\alpha = 1$, for which we obtained conventional first-order derivatives. Derivatives of the Heaviside function and the cosine function are unbounded at $t = 0$. This is in agreement with the well-known asymptotics of the Riemann–Liouville fractional derivative of a function which is non-zero (but bounded) at the initial point $t = 0$ [153, 179, 232].

Since $\log(t)$ and its derivatives are infinite at $t = 0$, values for a small neighbourhood of $t = 0$ are not depicted in Fig. 7.4.

7.2.2 The Use of Quadrature Formulas

Another type of approximation can be obtained from the Riemann–Liouville definition by n-times integration by parts and subsequent approximation of the integral containing $f^{(n)}(\tau)$ (see also [179]). In this work we prefer to systematically use approximation (7.3).

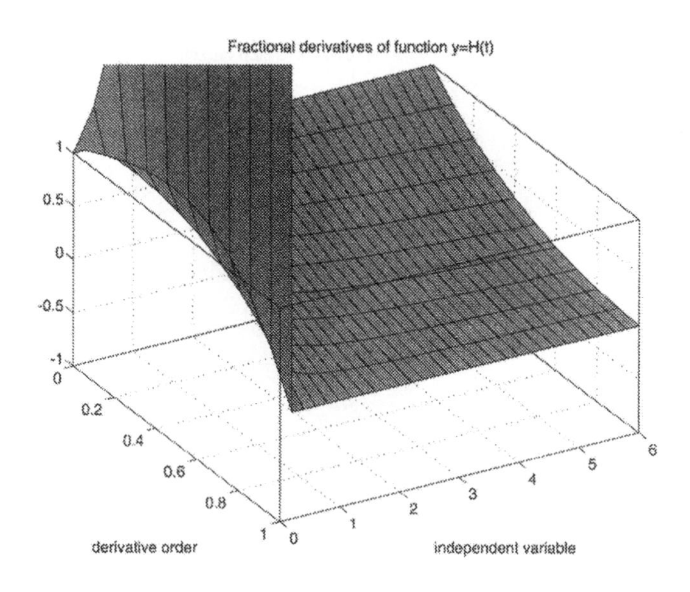

Figure 7.1: *Fractional derivatives of order* $0 \leq \alpha \leq 1$ *of the Heaviside function*

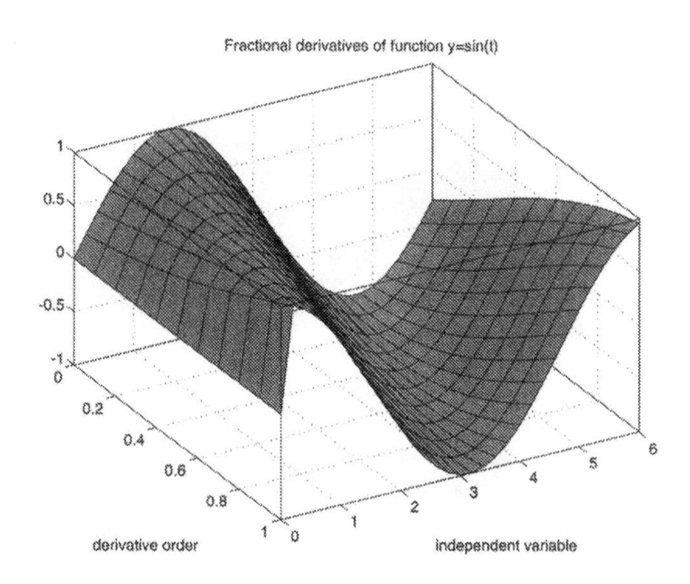

Figure 7.2: *Fractional derivatives of order* $0 \leq \alpha \leq 1$ *of* $\sin(t)$.

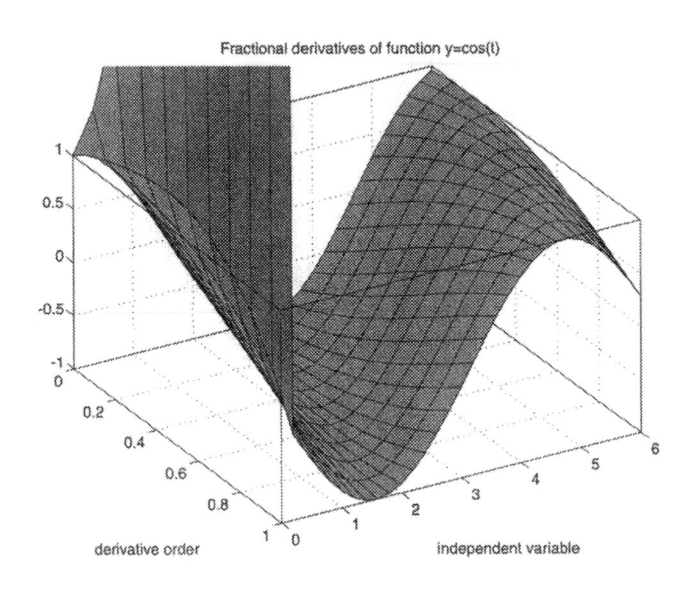

Figure 7.3: *Fractional derivatives of order* $0 \leq \alpha \leq 1$ *of* $\cos(t)$.

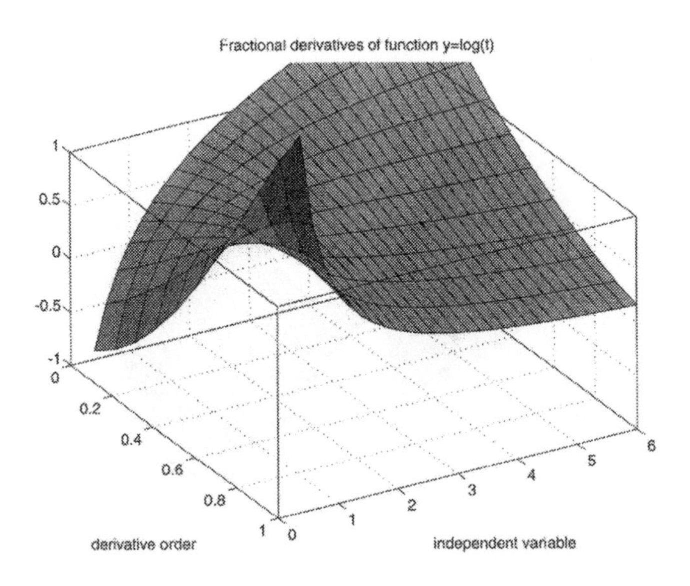

Figure 7.4: *Fractional derivatives of order* $0 \leq \alpha \leq 1$ *of* $\log(t)$.

7.3 The "Short-Memory" Principle

For $t \gg a$ the number of addends in the fractional-derivative approximation (7.3) and, therefore, in formulas (8.4), (8.25) and (8.55) (see Chapter 8) becomes enormously large. However, it follows from the expression for the coefficients in the Grünwald–Letnikov definition (7.2) that for large t the role of the "history" of the behaviour of the function $f(t)$ near the lower terminal (the "starting point") $t = a$ can be neglected under certain assumptions. Those observations lead us to the formulation of the "short-memory" principle, which means taking into account the behaviour of $f(t)$ only in the "recent past", i.e. in the interval $[t - L, t]$, where L is the "memory length":

$$ {}_aD_t^{\alpha} f(t) \approx {}_{t-L}D_t^{\alpha} f(t), \qquad (t > a + L). \tag{7.4} $$

In other words, according to the short-memory principle (7.4), the fractional derivative with the lower limit a is approximated by the fractional derivative with moving lower limit $t - L$. Due to this approximation, the number of addends in approximation (7.3) is alway no greater than $[L/h]$.

Of course, for this simplification we pay a penalty in the form of some inaccuracy. If $f(t) \leq M$ for $a \leq t \leq b$, which usually takes place in applications, then, using (7.64), we easily establish the following estimate for the error introduced by the short-memory principle:

$$ \Delta(t) = |\, {}_aD_t^{\alpha} f(t) - {}_{t-L}D_t^{\alpha} f(t)\,| \leq \frac{ML^{-\alpha}}{|\Gamma(1-\alpha)|}, \qquad .(a + L \leq t \leq b) \tag{7.5} $$

This inequality can be used for determining the "memory length" L providing the required accuracy ϵ:

$$ \Delta(t) \leq \epsilon, \quad (a + L \leq t \leq b), \qquad \text{if} \quad L \geq \left(\frac{M}{\epsilon\,|\Gamma(1-\alpha)|} \right)^{1/\alpha}. \tag{7.6} $$

To finish this section, we would like to mention that the formulated "short-memory" principle (7.4) along with the error estimation (7.5) completes, in a certain sense, the answer to Love's question formulated in [182] (the relationship for the fractional integrals with different lower limits was established in [232]).

From the historical point of view, the appearance of a similar idea ("limited after-effect" assumption) in Volterra's work [252, chapter IV], must be mentioned.

We use the short-memory principle below for computing changes in the thermal loading of blast furnace walls by means of the fractional derivative.

7.4 Order of Approximation

Let us first recall some basic facts on the approximation of integer-order derivatives.

It is well known that backward finite differences can be used for approximating integer-order derivatives. For example, for a fixed t and a small step h we can approximate the first-order derivative by the two-point backward difference:

$$y'(t) \approx \widetilde{y'(t)} = \frac{y(t) - y(t-h)}{h}, \tag{7.7}$$

which is obtained from the classical definition of the first-order derivative by omitting the operation $\lim\limits_{h \to 0}$. Due to this, there is an inaccuracy in the relationship (7.7), which depends on h, and which can be estimated under the assumption that we have the exact values of $y(t)$ and $y(t-h)$. Writing $y(t-h)$ in the form of the Taylor series, we have:

$$\widetilde{y'(t)} = \frac{y(t) - y(t-h)}{h} = y'(t) - \frac{y''(t)}{2}h + \ldots = y'(t) + O(h),$$

which means that

$$y(t) - \widetilde{y'(t)} = O(h); \tag{7.8}$$

in other words, the two-point backward difference formula gives the first-order approximation of $y'(t)$.

Let us show that formula (7.3) gives the first-order approximation for the α-th derivative. For simplicity, it is convenient to assume that $a = 0$, and that the discretization step h and the number of nodes n are related by $t = nh$, where t is the point at which the derivative is evaluated. In this case, we can write the approximation of the α-th derivative as

$$\begin{aligned} {}_0\widetilde{D_t^\alpha} f(t) &= h^{-\alpha} \sum_{j=0}^{n} (-1)^j \binom{\alpha}{j} f(t - jh) && (7.9) \\ &= h^{-\alpha} \sum_{j=0}^{n} \binom{j-\alpha-1}{j} f(t-jh). && (7.10) \end{aligned}$$

To introduce the idea of the considerations which will follow, let us take the simplest function $f_0(t) \equiv 1$ $(t \geq 0)$. We already know that its exact α-th derivative is

$$ {}_0D_t^\alpha f_0(t) = \frac{t^{-\alpha}}{\Gamma(1-\alpha)}. $$

On the other hand, the approximation (7.10) gives the approximate value

$$ {}_0\widetilde{D_t^\alpha} f_0(t) = h^{-\alpha} \sum_{j=0}^{n} \binom{j-\alpha-1}{j}. $$

Using the known summation formula for the binomial coefficients

$$ \sum_{j=0}^{n} \binom{j-\alpha-1}{j} = \binom{n-\alpha}{n}, \tag{7.11} $$

and the asymptotic formula [63, formula 1.18(4)]

$$ z^{b-a} \frac{\Gamma(z+a)}{\Gamma(z+b)} = 1 + O(z^{-1}), \tag{7.12} $$

we have for fixed t

$$ \begin{aligned} {}_0\widetilde{D_t^\alpha} f_0(t) &= h^{-\alpha} \binom{n-\alpha}{n} \\ &= \frac{t^{-\alpha}}{\Gamma(1-\alpha)} \, \frac{n^\alpha \Gamma(n-\alpha+1)}{\Gamma(n+1)} \\ &= \frac{t^{-\alpha}}{\Gamma(1-\alpha)} \Big(1 + O(h)\Big), \end{aligned} \tag{7.13} $$

and therefore for $f_0(t) \equiv 1$ $(t \geq 0)$

$$ {}_0D_t^\alpha f_0(t) - {}_0\widetilde{D_t^\alpha} f_0(t) = O(h), $$

which is similar to the relationship (7.8).

Let us now consider $f_m(t) = t^m$, $m = 1, 2, \ldots$. In this case, the exact α-th derivative is

$$ {}_0D_t^\alpha f_m(t) = \frac{\Gamma(1+m)}{\Gamma(1+m-\alpha)} \, t^{m-\alpha}, $$

and the approximation (7.10) of the exact derivative becomes

$$
{}_0\widetilde{D_t^\alpha} f_m(t) = t^{m-\alpha} n^\alpha \sum_{j=0}^{n} \binom{j-\alpha-1}{j} \left(1 - \frac{j}{n}\right)^m, \tag{7.14}
$$

or, after expanding the binomial,

$$
{}_0\widetilde{D_t^\alpha} f_m(t) = t^{m-\alpha} \sum_{r=0}^{m} (-1)^r \binom{m}{r} n^{\alpha-r} \sum_{j=0}^{n} \binom{j-\alpha-1}{j} j^r. \tag{7.15}
$$

The sum

$$
S = \sum_{j=0}^{n} \binom{j-\alpha-1}{j} j^r \tag{7.16}
$$

can be transformed to a more convenient form involving the Stirling numbers of the second kind $\sigma_n^{(m)}$, which are defined as coefficients of the expansion of x^n in a sum of factorial polynomials $x^{[i]}$ [2, Chapter 24]:

$$
x^n = \sum_{i=0}^{n} \sigma_n^{(i)} x^{[i]}, \tag{7.17}
$$

$$
x^{[i]} = x(x-1)(x-2)\ldots(x-i+1) = \frac{\Gamma(x+1)}{\Gamma(x-i+1)}. \tag{7.18}
$$

Using (7.17) and (7.18), and substituting $x = j$, we obtain:

$$
j^r = \sum_{i=1}^{r} \sigma_r^{(i)} \frac{\Gamma(j+1)}{\Gamma(j-i+1)}, \tag{7.19}
$$

and therefore

$$
\begin{aligned}
S &= \sum_{j=0}^{n} \binom{j-\alpha-1}{j} \sum_{i=1}^{r} \sigma_r^{(i)} \frac{\Gamma(j+1)}{\Gamma(j-i+1)} \\
&= \sum_{i=1}^{r} \sigma_r^{(i)} \sum_{j=i}^{n} \frac{\Gamma(j-\alpha)}{\Gamma(-\alpha)\Gamma(j-i+1)} \\
&= \sum_{i=1}^{r} \sigma_r^{(i)} \sum_{k=0}^{n-i} \frac{\Gamma(k+i-\alpha)}{\Gamma(-\alpha)\Gamma(k+1)} \\
&= \sum_{i=1}^{r} \sigma_r^{(i)} \frac{\Gamma(i-\alpha)}{\Gamma(-\alpha)} \sum_{k=0}^{n-i} \binom{k+i-\alpha-1}{k}.
\end{aligned}
$$

Now using the formula (7.11) we obtain

$$S = \sum_{i=1}^{r} \sigma_r^{(i)} \frac{\Gamma(i-\alpha)}{\Gamma(-\alpha)} \binom{n-\alpha}{n-i},$$

or finally

$$S = \sum_{j=0}^{n} \binom{j-\alpha-1}{j} j^r = \sum_{i=1}^{r} \sigma_r^{(i)} \frac{\Gamma(n-\alpha+1)}{(i-\alpha)\Gamma(-\alpha)\Gamma(n-i+1)}. \qquad (7.20)$$

Substituting (7.20) into (7.15) gives

$${}_0\widetilde{D_t^\alpha} f_m(t) = \frac{t^{m-\alpha}}{\Gamma(-\alpha)} \sum_{r=0}^{m} (-1)^r \binom{m}{r} \sum_{i=1}^{r} \sigma_r^{(i)} \frac{n^{\alpha-r}\,\Gamma(n-\alpha+1)}{(i-\alpha)\Gamma(n-i+1)}. \qquad (7.21)$$

Using the asymptotics of the gamma function (7.12) we can write

$$\frac{n^{\alpha-r}\,\Gamma(n-\alpha+1)}{\Gamma(n-i+1)} = n^{i-r}\left(n^{\alpha-i}\frac{\Gamma(n-\alpha+1)}{\Gamma(n-i+1)}\right) = n^{i-r}\Big(1+O(n^{-1})\Big).$$

Then

$$\begin{aligned} {}_0\widetilde{D_t^\alpha} f_m(t) &= \frac{t^{m-\alpha}}{\Gamma(-\alpha)} \sum_{r=0}^{m} (-1)^r \binom{m}{r} \sum_{i=1}^{r} \sigma_r^{(i)} \frac{1}{(i-\alpha)} n^{i-r}\Big(1+O(n^{-1})\Big) \\ &= \frac{t^{m-\alpha}}{\Gamma(-\alpha)} \sum_{r=0}^{m} (-1)^r \binom{m}{r} \sigma_r^{(r)} \frac{1}{(r-\alpha)}\Big(1+O(n^{-1})\Big). \end{aligned}$$

Taking into account that $\sigma_r^{(r)} = 1$ for all r, and using the summation formula ([215, formula 4.2.2(43)])

$$\sum_{k=0}^{n} \frac{(-1)^m k^m}{a+k} \binom{n}{k} = (-1)^m a^{m-1} \binom{n+a}{n}^{-1}, \qquad (m \le n),$$

we easily obtain

$$\sum_{r=0}^{m} (-1)^r \binom{m}{r} \frac{\sigma_r^{(r)}}{(r-\alpha)} = B(-\alpha, m+1),$$

and therefore, since for a fixed t we have $O(n^{-1}) = O(h)$,

$${}_0\widetilde{D_t^\alpha} f_m(t) = \frac{\Gamma(1+m)}{\Gamma(1+m-\alpha)} t^{m-\alpha} + O(h),$$

and

$${}_0D_t^\alpha f_m(t) - {}_0\widetilde{D_t^\alpha} f_m(t) = O(h).$$

This means that if a function $f(t)$ can be written in the form of a power series

$$f(t) = \sum_{m=0}^{\infty} a_m t^m,$$

then the fractional difference approximation (7.3) gives the first-order approximation for the fractional derivative of order α at any point of the convergence region of the power series.

The conditions on $f(t)$ can also be weakened.

7.5 Computation of coefficients

For implementing the fractional difference method of the computation of fractional derivatives it is necessary to compute the coefficients

$$w_k^{(\alpha)} = (-1)^k \binom{\alpha}{k}, \qquad k = 0, 1, 2, \ldots, \tag{7.22}$$

where α is the order of fractional differentiation.

One of the possible approaches is to use the recurrence relationships

$$w_0^{(\alpha)} = 1; \quad w_k^{(\alpha)} = \left(1 - \frac{\alpha+1}{k}\right) w_{k-1}^{(\alpha)}, \quad k = 1, 2, 3, \ldots \tag{7.23}$$

This approach is suitable for a fixed value of α. It allows the creation of an array of coefficients which can be used for fractional differentiation of various functions, and other similar repeated operations.

However, in some problems (e.g., in system identification) the most appropriate value of α must be found; this means that various values of α are considered, and for each particular value of α the coefficients $w_k^{(\alpha)}$ must be computed separately. In such a case, the recurrence relationships (7.23) are not very suitable. Instead, the fast Fourier transform method [105] can be used.

The coefficients $w_k^{(\alpha)}$ can be considered as the coefficients of the power series for the function $(1-z)^\alpha$:

$$(1-z)^\alpha = \sum_{k=0}^{\infty} (-1)^k \binom{\alpha}{k} z^k = \sum_{k=0}^{\infty} w_k^{(\alpha)} z^k. \tag{7.24}$$

Substituting $z = e^{-\imath\varphi}$ we have

$$(1 - e^{-\imath\varphi})^\alpha = \sum_{k=0}^{\infty} w_k^{(\alpha)} e^{-\imath k\varphi}, \tag{7.25}$$

and the coefficients $w_k^{(\alpha)}$ are expressed in terms of the Fourier transform:

$$w_k^{(\alpha)} = \frac{1}{2\pi i}\int_0^{2\pi} f_\alpha(\varphi) e^{ik\varphi} d\varphi, \qquad f_\alpha(\varphi) = (1 - e^{-\imath\varphi})^\alpha. \tag{7.26}$$

Technically, the coefficients $w_k^{(\alpha)}$ can be computed using any implementation of the fast Fourier transform. Since in this case we always obtain only a finite number of the coefficients $w_k^{(\alpha)}$, the fast Fourier transform method should always be combined with the "short-memory" principle (see Section 7.3).

7.6 Higher-order approximations

We saw that the first-order fractional difference approximation (7.10) of the α-th derivative can be written in the form

$${}_0\widetilde{D_t^\alpha} f(t) = h^{-\alpha} \sum_{k=0}^{[t/h]} w_k^{(\alpha)} f(t - kh), \tag{7.27}$$

where the weights $w_k^{(\alpha)}$ ($k = 0, 1, 2, \dots, n$, $n = [\frac{t}{h}]$), assigned to the values $f(t - kh)$, are the first $n + 1$ coefficients of the Taylor series expansion of the function

$$\omega_1^{(\alpha)}(z) = (1 - z)^\alpha = \big(\omega_1(z)\big)^\alpha. \tag{7.28}$$

The coefficients 1 and -1 in the function $\omega_1(z) = 1 - z$ are at the same time the coefficients in the two-point backward difference approximation of the first-order derivative (7.7).

We have already seen that the function $\omega_1(z)$ generates the coefficients for the first-order approximation of the first-order derivative, and its α-th power, the function $\omega_1^{(\alpha)}(z) = \big(\omega_1(z)\big)^\alpha$, generates the coefficients of the first-order approximation of the α-th order derivative.

So, we may ask: The $(p+1)$-point backward difference gives the p-th order approximation of the first-order derivative; will the α-th power of the $(p+1)$-point backward difference approximation of the first-order derivative give the p-th order approximation of the α-th derivative?

The answer to this question has been given by Ch. Lubich [127], who obtained approximations of order 2, 3, 4, 5, and 6 in the form of (7.27), where the coefficients $w_k^{(\alpha)}$ are the coefficients of the Taylor series expansions of the corresponding "generating" functions

$$\omega_2^{(\alpha)}(z) = (\frac{3}{2} - 2z + \frac{1}{2}z^2)^\alpha,$$

$$\omega_3^{(\alpha)}(z) = (\frac{11}{6} - 3z + \frac{3}{2}z^2 - \frac{1}{3}z^2)^\alpha,$$

$$\omega_4^{(\alpha)}(z) = (\frac{25}{12} - 4z + 3z^2 - \frac{4}{3}z^3 + \frac{1}{4}z^4)^\alpha,$$

$$\omega_5^{(\alpha)}(z) = (\frac{137}{60} - 5z + 5z^2 - \frac{10}{3}z^3 + \frac{5}{4}z^4 - \frac{1}{5}z^5)^\alpha,$$

$$\omega_6^{(\alpha)}(z) = (\frac{147}{60} - 6z + \frac{15}{2}z^2 - \frac{20}{3}z^3 + \frac{15}{4}z^4 - \frac{6}{5}z^5 + \frac{1}{6})^\alpha.$$

In each case, the coefficients in parentheses in the right-hand side of the expression for $\omega_p^{(\alpha)}(z)$ are the coefficients of the p-th order $(p+1)$-point backward difference approximation of the first-order derivative.

The easiest and the most efficient method of computation of the coefficients $w_k^{(\alpha)}$ for the higher-order approximations of the form (7.27) is the fast Fourier transform method, and the procedure is the same as described in Section 7.5.

7.7 Calculation of Heat Load Intensity Change in Blast Furnace Walls

In this section the fractional-order derivative is used for the calculation of changes in the heat flux intensity in a blast furnace wall. In contrast to standard approaches which rely on temperature measurements at two different points of the wall, the proposed method needs temperature measurement at one point only. Results are given and analysed for the described method and for a conventional finite-difference method. The possibility for an extensive use of the described method in the solution of similar tasks for materials with high thermal resistance (e.g., fireclay) follows from the comparison.

7.7.1 Introduction to the Problem

From the point of view of operation and technology, one of the important monitored parameters of a blast furnace is the intensity of the thermal flux in its walls. The standard methods solve the task of computing the heat flux by measuring two temperatures in the wall, subsequent simulation of the thermal field, and the computation of the heat flux at some point from the temperature difference in the vicinity of this point. These methods require the use of two temperature monitors at two points with different depths within the furnace wall. It is necessary to take into account their relatively high malfunction rate as a consequnce of higher operation temperatures and the possibilities of mechanical damage after wearing of the wall from within. The replacement of inlaid thermocouples is complicated and sometimes there is no other way of doing this than to temporarily shut the furnace down, which leads to losses. Therefore, the method based on the measurement of two temperatures in the wall often becomes unusable. We present here an attempt to solve the above problems.

We give a description of two methods which were implemented and compared. The first one (denoted in the following as method A) is unconventional. It is based on the use of fractional-order derivatives and makes it possible to efficiently use the temperature measurement at one point of the furnace wall only. It should be noted that the possibility of the use of fractional-order derivatives for the computation of the heat flux behaviour, based on the known behaviour of temperatures, was first pointed out in [179]. A second method (henceforth denoted as method B) is standard. It is based on the measurement of the furnace wall temperature at two points, and on a numerical solution of the heat conduction equation. It was used for testing the first method. A mutual qualitative and quantitative comparison of these two methods is given.

7.7.2 Fractional-order Differentiation and Integration

Let us for convenience recall the Riemann–Liouville definition of a fractional derivative and a fractional integral:

$$ {}_0D_t^{\alpha} f(t) = \frac{1}{\Gamma(n-\alpha)} \frac{d^n}{dt^n} \int_a^t \frac{f(\tau)d\tau}{(t-\tau)^{\alpha-n+1}}, \qquad 0 \le n-1 < \alpha < n \quad (7.29) $$

where n is an integer, $\Gamma(z)$ is the Gauss gamma-function, and $t > a$.

Closely related to fractional-order differentiation is fractional-order integration:

$$ {}_0D_t^{-\alpha} f(t) = \frac{1}{\Gamma(\alpha)} \int_a^t \frac{f(\tau) d\tau}{(t-\tau)^{1-\alpha}}, \qquad \alpha > 0. \tag{7.30} $$

It is necessary to keep in mind that

$$ {}_0D_t^{\alpha}({}_0D_t^{-\alpha} f(t)) = f(t), \qquad \alpha > 0. \tag{7.31} $$

which generalizes an analogous property of integer derivatives and integrals.

Let us also recall (see Chapter 4) that the Laplace transform of fractional derivatives and integrals is given by

$$ \mathcal{L}({}_0D_t^{\alpha} f(t), s) = s^{\alpha} F(s) - \sum_{k=0}^{n-1} s^k \; {}_0D_t^{\alpha-1-k} f(t) \Big|_{t=0}, \tag{7.32} $$

for arbitrary real α ($F(s)$ is the Laplace transform of the function $f(t)$). In the case of an integral of fractional order ($\alpha < 0$), the sum in the right-hand side will vanish. In the case of a fractional-order derivative n is the same as in (7.29).

7.7.3 Calculation of the Heat Flux by Fractional Order Derivatives – Method A

Derivation of the Basic Relation

Let us consider the following spatially one-dimensional heat conduction problem for a semi-infinite body (Fig. 7.5):

$$ \hat{c}\hat{\rho} \frac{\partial T}{\partial t} = \hat{\lambda} \frac{\partial^2 T}{\partial x^2}, \qquad (t > 0, \quad -\infty < x < 0) \tag{7.33} $$

$$ T(0, x) = T_0 \tag{7.34} $$

$$ T(t, 0) = T_{\text{surf}}(t) \tag{7.35} $$

$$ \left| \lim_{x \to -\infty} T(t, x) \right| < \infty \tag{7.36} $$

where

t is time [s],

x is the spatial coordinate in the direction of heat conduction [m],

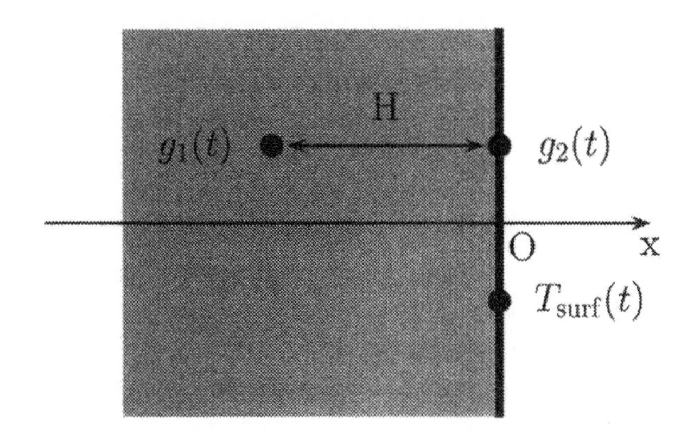

Figure 7.5: *Blast furnace wall.*

$\hat{c}$ is the heat capacity [J kg^{-1} K^{-1}],
$\hat{\rho}$ is the mass density [kg m^{-3}],
$T(t,x)$ - is the temperature [K],
$\hat{\lambda}$ is the coefficient of heat conductivity [W m^{-1} K^{-1}].
We introduce an auxiliary function

$$u(t,x) = T(t,x) - T_0, \tag{7.37}$$

which is a solution of the problem

$$\hat{c}\hat{\rho}\frac{\partial u}{\partial t} = \hat{\lambda}\frac{\partial^2 u}{\partial x^2}, \qquad (t > 0, \quad -\infty < x < 0) \tag{7.38}$$

$$u(0,x) = 0 \tag{7.39}$$

$$u(t,0) = T_{\text{surf}}(t) - T_0 \tag{7.40}$$

$$\left|\lim_{x\to-\infty} u(t,x)\right| < \infty. \tag{7.41}$$

The Laplace transform of equation (7.38) yields

$$\hat{c}\hat{\rho}U(s,x) = \hat{\lambda}\frac{d^2U(s,x)}{dx^2}. \tag{7.42}$$

The solution of equation (7.42), bounded for $x \to -\infty$, is

$$U(s,x) = U(s,0)\exp\left(x\sqrt{\frac{\hat{c}\hat{\rho}s}{\hat{\lambda}}}\right) \tag{7.43}$$

whence we find

$$\frac{dU}{dx}(s,x) = U(s,0)\sqrt{\frac{\hat{c}\hat{\rho}s}{\hat{\lambda}}}\,\exp\left(x\sqrt{\frac{\hat{c}\hat{\rho}s}{\hat{\lambda}}}\right). \tag{7.44}$$

From the relations (7.43) and (7.44) we easily find

$$\frac{1}{\sqrt{s}}\frac{dU}{dx}(s,0) = \sqrt{\frac{\hat{c}\hat{\rho}}{\hat{\lambda}}}\,U(s,0) \tag{7.45}$$

from which, after using the inverse Laplace transform, we obtain in view of (7.32)

$${}_0D_t^{-1/2}\frac{\partial u}{\partial x}(t,0) = \sqrt{\frac{\hat{c}\hat{\rho}}{\hat{\lambda}}}\,u(t,0) \tag{7.46}$$

and after using the property (7.31) and the linearity of the fractional-order differentiation operator we arrive at

$$\frac{\partial u}{\partial x}(t,0) = \sqrt{\frac{\hat{c}\hat{\rho}}{\hat{\lambda}}}\,{}_0D_t^{1/2}u(t,0). \tag{7.47}$$

We can now return to the function $T(t,x)$ with the aid of relation (7.37). Taking into account the condition (7.35), we obtain the basic analytical relation for the calculation of the heat flux at the point $x = 0$:

$$q_A(t) = \sqrt{\hat{c}\hat{\rho}\hat{\lambda}}\,{}_0D_t^{1/2}g(t), \quad g(t) = T_{\mathrm{surf}}(t) - T_0 \tag{7.48}$$

where $q_A(t) = \hat{\lambda}\frac{\partial T}{\partial x}(t,0)$ is the resulting heat flux.

The Numerical Method

The problem of determining the heat flux is now reduced to the calculation of the derivative of order $\alpha = 1/2$ in the derived formula (7.48). Since we are interested in simulations for large time intervals, the known relations (see [179] and [203]) for the calculation of fractional-order derivatives are not suitable because of an enormous number of summands in these relations and because of the accumulation of the effects of round-off errors. To reduce the computation cost and to eliminate, in a certain sense, the round-off error accumulation we apply the principle of "short memory", formulated in [203] (in this work, see section 7.3). That is, we put approximately

$$q_A(t) \approx \tilde{q}_A(t) = \sqrt{\hat{c}\hat{\rho}\hat{\lambda}}\,{}_{(t-L)}D_t^{1/2}g(t)\,, \tag{7.49}$$

where L is the "memory length". It follows from the estimations derived in [203] (in this work, see section 7.3) that in our case the normed error of this approximation is

$$\delta_0 = \frac{|q_A(t) - \tilde{q}_A(t)|}{M} = \frac{1}{\sqrt{L}\,\Gamma(\frac{1}{2})}, \quad M = \max_{[0,\,\infty]} |g(t)| \tag{7.50}$$

whence we have the following constraint for the choice of the "memory length" L:

$$L \geq \frac{1}{\pi\delta_0^2} \tag{7.51}$$

where δ_0 is the maximum admissible normalized error.

For an approximate calculation of the derivative ${}_{(t-L)}D_t^{1/2} g(t)$ we therefore use the relation (see section 7.3)

$$ {}_{(t-L)}D_t^{1/2} g(t) = \tau^{-\alpha} \sum_{i=0}^{N(t)} c_i g(t - i\tau) , \tag{7.52}$$

$$N(t) = \min\left\{ \left[\frac{t}{\tau}\right], \left[\frac{L}{\tau}\right] \right\}, \quad c_i = (-1)^i \binom{1/2}{i}$$

($[z]$ is the integer part z).

For the calculation of the coefficients c_i it is advantageous to use the recursion

$$c_0 = 1, \quad c_i = \left(1 - \frac{3}{2i}\right) c_{i-1} \tag{7.53}$$

which follows from the properties of the binomial coefficients.

7.7.4 Calculation of the Heat Flux Based on the Simulation of the Thermal Field of the Furnace Wall – Method B

For an experimental verification of the proposed method we have also implemented the standard approach. With the goal of finding the true limits of applicability of the method desribed above, we have assumed in this part that the thermophysical properties of the material of the wall depend on the temperature.

On the basis of an analysis of the thermal situation and dimensions of the furnace walls from the viewpoint of the thermal load of the furnace walls it is possible to make the simplifying assumption of one-dimensional heat conduction. Non-stationary heat conduction is described by the Fourier equation:

$$\frac{\partial}{\partial t}\left(c(T)\rho(T)T(t,x)\right) = \frac{\partial}{\partial x}\left(\lambda(T)\frac{\partial T(t,x)}{\partial x}\right) \tag{7.54}$$

where $-H < x < 0$, with the initial condition

$$T(0,x) = f(x) \tag{7.55}$$

and with boundary conditions of the first kind

$$\begin{aligned} T(t,-H) &= g_1(t) \\ T(t,0) &= g_2(t) \end{aligned} \tag{7.56}$$

where H is the thickness of the wall determined by both thermocouples (see Fig. 7.5).

For a numerical calculation of the heat distribution $T(t,x)$ we first make a spatial and temporal discretization.

In the space interval $-H \leq x \leq 0$ we choose n points

$$-H = x_1 < x_2 < \cdots < x_{n-2} < x_{n-1} < x_n = 0\,,$$

in which the temperatures will be determined at discrete time intervals of length τ. The distance between the interior points x_i and x_{i+1} ($i = 2, 3, \ldots n-2$) is

$$h = H/(n-2) \tag{7.57}$$

and the distance between the exterior point x_1 (resp. x_n) and the interior point nearest to it, x_2 (resp. x_{n-1}), is $h/2$.

After the discretization of equation (7.54) with an implicit method for the whole inner region $-H < x < 0$, we obtain the following non-linear system of algebraic equations:

$$-\left(\frac{2\lambda_{1,2}^{(k+1)}}{h} + \frac{\lambda_{2,3}^{(k+1)}}{h} + \frac{K\,c(T_2^{(k+1)})}{\tau}\right)T_2^{(k+1)} + \frac{\lambda_{2,3}^{(k+1)}}{h}T_3^{(k+1)}$$

$$= -\frac{K\,c(T_2^{(k)})}{\tau}T_2^{(k)} - \frac{2\lambda_{1,2}^{(k+1)}}{h}g_1^{(k+1)}$$

$$\frac{\lambda_{i-1,i}^{(k+1)}}{h}T_{i-1}^{(k+1)} - \left(\frac{\lambda_{i-1,i}^{(k+1)}}{h} + \frac{\lambda_{i,i+1}^{(k+1)}}{h} + \frac{Kc(T_i^{(k+1)})}{\tau}\right)T_i^{(k+1)} + \frac{\lambda_{i,i+1}^{(k+1)}}{h}T_{i+1}^{(k+1)}$$

$$= -\frac{K\, c(T_i^{(k)})}{\tau} T_i^{(k)} \quad (7.58)$$

$$\frac{\lambda_{n-2,n-1}^{(k+1)}}{h} T_{n-2}^{(k+1)} - \left(\frac{\lambda_{n-2,n-1}^{(k+1)}}{h} + \frac{2\lambda_{n-1,n}^{(k+1)}}{h} + \frac{K\, c(T_{n-1}^{(k+1)})}{\tau} \right) T_{n-1}^{(k+1)}$$

$$= -\frac{K\, c(T_{n-1}^{(k)})}{\tau} T_{n-1}^{(k)} - \frac{2\lambda_{n-1,n}^{(k+1)}}{h} g_2^{(k+1)}$$

where

$$T_i^{(k)} = T(k\tau, x_i),$$
$$g_i^{(k)} = g_i(k\tau), \; (i = 1, 2)$$
$$\lambda_{i,i+1}^{(k+1)} = \lambda \left(\frac{T_i^{(k+1)} + T_{i+1}^{(k+1)}}{2} \right),$$
$$K = h\rho$$

while from the initial condition (7.55) it follows that:

$$T_i^{(0)} = f(x_i), \qquad \text{pre} \quad i = 2, 3, \ldots, n-1. \quad (7.59)$$

The system (7.58) allows the calculation of the temperatures at the points of a chosen spatial grid for the next time step based on known temperatures at the same points of the preceding time step and known boundary conditions (7.56).

For the solution of the non-linear system (7.58) we have used the iteration method, while the arising linear algebraic systems were solved via the Gauss elimination method.

After determining the temperatures for the time step $(k+1)$ we calculate the change of the heat flux intensity for the same time at the spot of the interior measurement point (i.e. at $x = 0$) according to the relation

$$q_B^{(k+1)} \equiv \Delta q^{(k+1)} = q^{(k+1)} - q^{(0)}\,, \quad (7.60)$$

$$q^{(k+1)} = 2\, \lambda_{n-1,n}^{(k+1)} \frac{T_n^{(k+1)} - T_{n-1}^{(k+1)}}{h}\,,$$

where $q^{(0)}$ is the heat flux at the point $x = 0$ at time $t = 0$.

7.7.5 Comparison of the Methods

Fireclay SK-1, from which the wall of furnace no. 2 in VSZ Kosice, Inc. is made, is a material with a very high heat resistance (low heat conductivity). The thermophysical properties of this kind of fireclay are the following [103]:

$$\rho(T) = 1750\,, \quad \lambda(T) = 0.75 + \hat{T}\cdot 0.35\cdot 10^{-3}\,, \quad c(T) = 870 + 0.14\cdot\hat{T} \tag{7.61}$$

where the temperature $\hat{T} = T - 273$. These relations were used in the computer implementation of method B (i.e. of the test method). The distance between the two points of temperature measurement (i.e. two thermocouples) is $H = 0.15$ m.

For the numerical realization of the tested method A we chose

$$\hat{c}\hat{\rho}\hat{\lambda} = c(T_m)\rho(T_m)\lambda(T_m), \tag{7.62}$$

where $T_m = 450$ °C is the average technological temperature of the furnace wall material. The allowable normed error was $\delta_0 = 0.01$. To ensure this precision we must have $L \geq 3184$ (i.e. the minimum "memory length" cannot be shorter than 3184 seconds). $L = 3600$ has been used for the calculations. The step τ in the formula (7.52) (which at the same time is a time step for method B) was chosen to be $\tau = 60$, which corresponds to a real one-minute time interval between two measurements of the temperature.

The comparison of the results of calculations of the intensity change of the thermal flux in the fireclay wall by means of method A and method B for the boundary conditions of the form

$$g_i(t) = T_i + 20\sin\left(2\pi k\tau/120\right)\,, \qquad (i = 1, 2), \quad (k = \overline{0, \infty}) \tag{7.63}$$

(i.e. for different functions of measured temperatures $g_1(t)$ and $g_2(t) = T_{\text{surf}}(t)$) was done [204].

It was observed that for the temperature difference 300 °C the results of the computations according to method A match well with those according to method B. The maximum relative error in this case is around 15%, which is still acceptable from the viewpoint of many engineering applications.

The comparison leads to the conclusion that the method based on the use of fractional-order derivatives and of the temperature measurement in only one point can be used successfully for materials with low heat conductivity, since it correctly reflects the process both qualitatively and quantitatively.

The reason for this success is that for these materials the approximation of the wall's width by a semi-infinite body ensures very satisfactory adequacy of the model.

Numerical experiments show that in the case of materials with large heat conductivity, the proposed method is less successful; however, it can still be used for a rough estimate of the changes of the heat flux intensity, if for some reason (for instance, malfunction of one of the thermocouples) only the temperature at one point of the furnace's wall is known.

Not to be neglected is the fact that, compared with the classical methods (finite-difference method, finite-element method) the fractional derivative based method requires fewer calculations and in contrast to them allows the calculation of the heat flux at a given point without calculating the distribution of temperatures along the entire width of the wall.

From a general point of view, this chapter demonstrates that even in classical problems such as heat conduction problem, fractional-order derivatives make it possible to find new, effective non-conventional solutions to important technological problems.

7.8 Finite-part Integrals and Fractional Derivatives

Instead of the classical form of the Riemann–Liouville definition (7.1), we may use the equivalent form of that definition (6.163), which leads to integrals, which are divergent in the classical sense :

$$ {}_aD_t^{\alpha} f(t) = \frac{1}{\Gamma(-\alpha)} \int_a^t \frac{f(\tau)d\tau}{(t-\tau)^{\alpha+1}}, \qquad (\alpha \neq 0, 1, 2, ...). \tag{7.64} $$

Namely, for $\alpha > 0$ the integral in (7.64) is a divergent integral. However, it is possible to define a so-called finite-value of a divergent integral, which has real physical meaning [213].

Table 7.1: *Approximate values of the finite-part integrals I_k.*

$I_k = \int_{-1}^{1} \frac{x^k dx}{(1-x^2)^{3/2}}$	I_0	I_2	I_4
h=0.01	0.0075	-3,1366	-4.6807
h=0.001	0.0008	-3.1411	-4.7092
h=0.0001	0.0001	-3.1415	-4.7120
h=0.00005	0.0000	-3.1416	-4.7123
Exact value	0.0000	−3.1416	−4.7124

7.8.1 Evaluation of Finite-part Integrals Using Fractional Derivatives

The finite-part integral (6.155) with the non-integrable Jacobi weight can be expressed in terms of fractional derivatives:

$$\int_{-1}^{1} \frac{f(t)dt}{(1-t)^{\alpha+1}(1+t)^{\beta+1}} = \Gamma(-\beta)\ {}_0D_t^{\beta} f_1(t)\Big|_{t=1} + \Gamma(-\alpha)\,{}_0D_t^{\alpha} f_2(t)\Big|_{t=1} \tag{7.65}$$

$$f_1(t) = f(-t)(1+t)^{-\alpha-1}, \quad f_2(t) = f(t)(1+t)^{-\beta-1},$$

$$(\alpha < 1, \quad \beta < 1).$$

When approximating fractional derivatives, we obtain formulas for the numerical evaluation of the finite-part integral (7.65). When we apply relationship (7.65) to the numerical evaluation of the integrals (6.159)–(6.161) with the help of the first-order approximation (7.3), the results were in agreement (Table 7.1).

7.8.2 Evaluation of Fractional Derivatives Using Finite-part Integrals

Not only can fractional derivatives be used for the evaluation of the regularized values of finite-part integrals, another side of the relationship between these two objects is that if a numerical method for evaluation of finite-part integrals is available, then it can be immediately used for the numerical evaluation of fractional-order derivatives.

K. Diethelm suggested [40] using for the numerical evaluation of fractional derivatives quadrature formulas for finite-part integrals [41].

Let us consider the interval [0, 1], to which an arbitrary interval can be easily transformed. For a given integer m, in which the value of the fractional derivative (7.64) must be evaluated, an equidistant grid with nodes $t_j = j/m$ is introduced.

The discretization of the finite-part integral in (7.64) with this grid gives

$$ {}_0D_{t_j}^{\alpha} f(t) = \frac{1}{\Gamma(-\alpha)} \int\limits_0^{t_j} \frac{f(\tau)d\tau}{(t_j-\tau)^{\alpha+1}} = \frac{t_j^{-\alpha}}{\Gamma(-\alpha)} \int\limits_0^{1} \frac{f(t_j - t_j\xi)\, d\xi}{\xi^{\alpha+1}}, \qquad (7.66) $$

and the use of Diethelm's first-degree compound quadrature formula for finite-part integrals [41] with equidistant nodes $0, 1/j, 2/j, \ldots, 1$, leads to the following approximate formula for the evaluation of a fractional-order derivative:

$$ {}_0D_{t_j}^{\alpha} f(t) \approx \frac{t_j^{-\alpha}}{\Gamma(-\alpha)} \sum_{k=0}^{j} \omega_{kj} f\left(\frac{k}{j}\right), \qquad (j = 1, 2, \ldots, m) \qquad (7.67) $$

where the weights ω_{kj} (for $j \geq 1$) are given by the following expressions:

$$ \omega_{kj} = \frac{j^{\alpha}}{\alpha(1-\alpha)} \begin{cases} -1, & \text{for} \quad k = 0, \\ 2k^{1-\alpha} - (k-1)^{1-\alpha} - (k+1)^{1-\alpha}, & \text{for} \quad k = 1, 2, \ldots, j-1, \\ (\alpha - 1)k^{-\alpha} - (k-1)^{1-\alpha} + k^{1-\alpha}, & \text{for} \quad k = j. \end{cases} \qquad (7.68) $$

Chapter 8

Numerical Solution of Fractional Differential Equations

The numerical solution of differential equations of integer order has for a long time been a standard topic in numerical and computational mathematics. However, in spite of a large number of recently formulated applied problems, the state of the art is far less advanced for fractional-order differential equations.

In this chapter we describe a method which was experimentally verified on a number of test problems.

8.1 Initial Conditions: Which Problem to Solve?

We consider here the initial value problems only for homogeneous initial conditions which correspond to the equilibrium state at the beginning of a dynamical process:

$$f^{(k)}(0) = 0, \quad k = 0, 1, 2, \ldots, n-1 \tag{8.1}$$

where $n - 1 < \alpha < n$, and α is the order of the differential equation.

There are two main reasons for considering homogeneous initial conditions. First, this provides the equivalence of solutions of initial value problems for so-called sequential fractional-order differential equations [153] and for corresponding standard fractional-order differential equations, even if the number of initial conditions is different (see Chapter 4;

also [201]). Second, to this author's knowledge, a satisfactory approximation of the fractional derivative at its lower limit is not known.

8.2 Numerical Solution

In this section we concentrate on describing the method without studying the convergence of the method from the theoretical point of view.

The proposed numerical scheme is explicit. It was experimentally verified on a number of examples, some of which are given below, by comparing it with analytical solutions. As the introduced examples show, the proposed method works for different important cases, such as equations with constant coefficients, equations with non-constant coefficients and non-linear equations with different numbers of initial conditions. This speaks favourably of its wide applicability.

It follows from [127] that the order of approximation of equations in all examples is $O(h)$.

8.3 Examples of Numerical Solutions

In this section we give some examples of numerical solution of fractional-order differential equations of various type. We provide a comparison with some known explicit or asymptotic solutions, which demonstrates the useability of the proposed numerical approach.

8.3.1 Relaxation–oscillation Equation

Let us consider an initial value problem for one of the simplest fractional-order differential equations appearing in applied problems (e.g., [184]):

$$ {}_0D_t^{\alpha} y(t) + Ay(t) = f(t), \qquad (t > 0), \tag{8.2}$$

$$ y^{(k)}(0) = 0, \qquad (k = 0, 1, \ldots, n-1) $$

where $n - 1 < \alpha \leq n$. For $0 < \alpha \leq 2$ this equation is called the relaxation–oscillation equation.

The first-order approximation of problem (8.2) is

$$ h^{-\alpha} \sum_{j=0}^{m} w_j^{(\alpha)} y_{m-j} + Ay_m = f_m, \;\; (m = 1, 2, \ldots); \quad y_0 = 0, \tag{8.3}$$

$$t_m = mh, \; y_m = y(t_m), \; f_m = f(t_m), \quad (m = 0, 1, 2, \ldots);$$

$$w_j^{(\alpha)} = (-1)^j \binom{\alpha}{j}, \quad (j = 0, 1, 2, \ldots).$$

Using approximation (8.3), we derive the following algorithm for obtaining the numerical solution:

$$y_k = 0, (k = 1, 2, \ldots, n-1)$$

$$y_m = -Ah^\alpha y_{m-1} - \sum_{j=1}^{m} w_j^{(\alpha)} y_{m-j} + h^\alpha f_m, \quad (m = n, n+1, \ldots). \quad (8.4)$$

The results of our computations for different values of α $(1 \le \alpha \le 2)$ and $f(t) \equiv H(t)$, where $H(t)$ is the Heaviside function, are shown in Fig. 8.1. They are in perfect agreement with the analytical solutions, obtained with the help of fractional Green's function for a two-term fractional differential equation with constant coefficients (see Section 5.3). The analytical solution of the initial-value problem (8.2) is

$$y(t) = \int_0^t G_2(t-\tau) f(\tau) d\tau, \qquad G_2(t) = t^{\alpha-1} E_{\alpha,\alpha}(-At^\alpha). \qquad (8.5)$$

8.3.2 Equation with Constant Coefficients: Motion of an Immersed Plate

In this section we consider the initial value problem for the fractional differential equation which was originally formulated by R. L. Bagley and P. J. Torvik [16].

Mathematical model of the motion of a large thin plate in a Newtonian fluid

First, a basic relationship in terms of fractional derivatives for a Newtonian viscous fluid will be obtained.

Let us consider the motion of a half-space Newtonian viscous fluid induced by a prescribed transverse motion of a rigid plate on the surface (Fig. 8.2). Our aim is to show that the resulting shear stress at any point

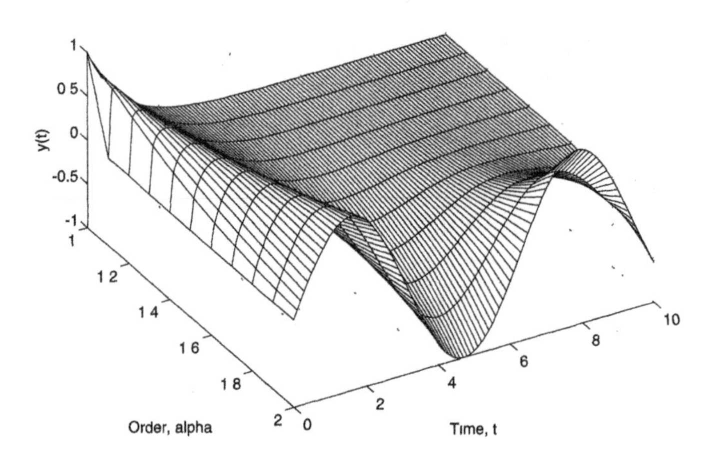

Figure 8.1: *Solutions of relaxation–oscillation equation for* $1 \leq \alpha \leq 2$.

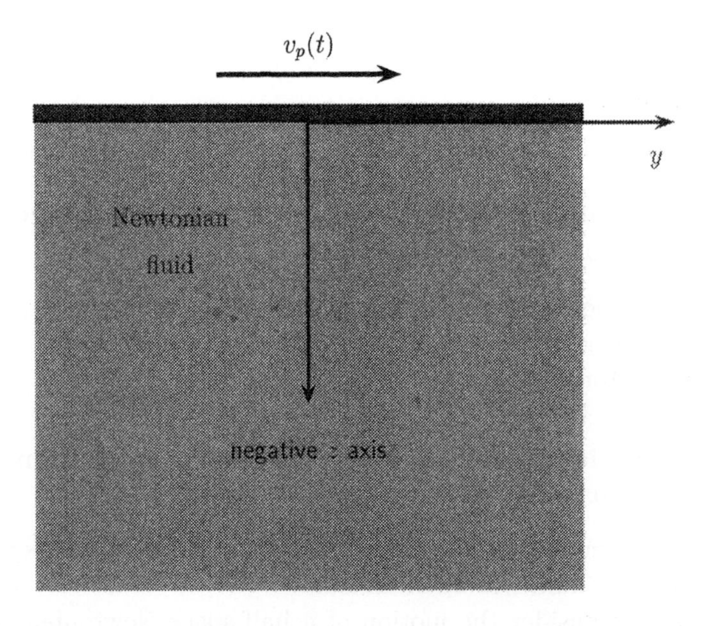

Figure 8.2: *A rigid plate in a Newtonian fluid.*

in the fluid can be expressed directly in terms of a fractional-order time derivative of the fluid velocity profile.

The equation of motion of the fluid is

$$\rho\frac{\partial v}{\partial t} = \mu\frac{\partial^2 v}{\partial z^2}, \qquad (0 < t < \infty, \quad -\infty < z < 0) \tag{8.6}$$

where ρ is the fluid density, μ is the viscosity and $v(t,z)$ is the transverse velocity, which is a function of time t and the distance z from the fluid–plate contact boundary.

We assume that initially the fluid is in equilibrium, i.e

$$v(0,z) = 0, \qquad (-\infty < z < 0) \tag{8.7}$$

and that the influence of the plate's motion vanishes for $z \to \infty$:

$$v(t,-\infty) = 0, \qquad (0 < t < \infty). \tag{8.8}$$

The fluid's velocity at $z = 0$ is equal to the given velocity of the plate:

$$v(t,0) = v_p(t) \tag{8.9}$$

Applying the Laplace transform we obtain the following boundary-value problem for an ordinary differential equation

$$\rho s V(s,z) = \mu\frac{d^2V(s,z)}{dz^2} \tag{8.10}$$

$$V(s,0) = V_p(s), \tag{8.11}$$

$$V(s,-\infty) = 0, \tag{8.12}$$

where s is the Laplace transform parameter, $V_p(s)$ is the Laplace transform of the plate's velocity and $V(s,z)$ is the fluid's velocity transform.

The solution of problem (8.10)–(8.12) can be ealily found to be

$$V(s,z) = V_p(s)\exp\left(z\sqrt{\frac{\rho s}{\mu}}\right). \tag{8.13}$$

By differentiation of (8.13) we find that

$$\frac{dV(s,z)}{dz} = \sqrt{\frac{\rho s}{\mu}}V_p(s)\exp\left(z\sqrt{\frac{\rho s}{\mu}}\right) = \sqrt{\frac{\rho s}{\mu}}V(s,z). \tag{8.14}$$

Knowing the velocity profile $v(t,z)$ in the fluid, one can obtain the shear stress $\sigma(t,z)$ by

$$\sigma(t,z) = \mu\frac{\partial v(t,z)}{\partial z}. \tag{8.15}$$

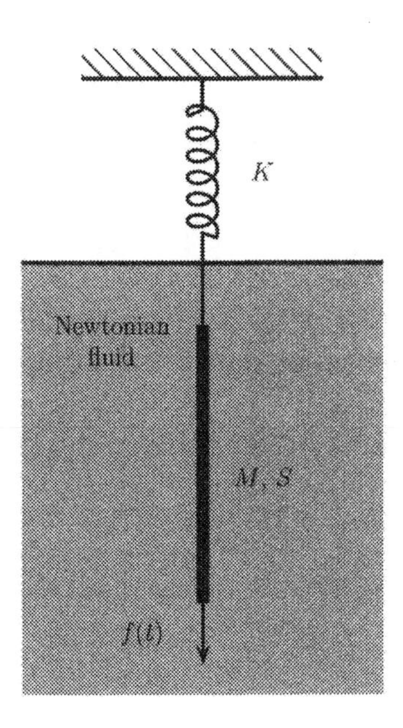

Figure 8.3: *An immersed plate in a Newtonian fluid.*

In terms of the Laplace transform, relationship (8.15) takes the form:

$$\overline{\sigma}(s, z) = \mu \frac{dV(s, z)}{dz} = \sqrt{\mu \rho s}\, V(s, z), \tag{8.16}$$

where $\overline{\sigma}(s, z)$ denotes the Laplace transform of $\sigma(t, z)$.

Comparing (8.16) and (1.80), we recognize the Laplace transform of the fractional derivative ${}_0D_t^{1/2} v(s, z)$ multiplied by $\sqrt{\mu\rho}$ in the right-hand side of (8.16). Therefore, after returning to the time domain, relationship (8.16) gives

$$\sigma(t, z) = \sqrt{\mu\rho}\, {}_0D_t^{1/2} v(s, z). \tag{8.17}$$

It must be mentioned that equation (8.17) is not a constitutive relationship for a Newtonian fluid; the constitutive relationship is (8.15). However, equation (8.17) describes the relationship between the stress and velocity for the considered particular geometry (a semi-infinite fluid domain) and loading (prescribed velocity at the boundary surface). It is important in this case that the fractional derivative is used to describe a real physical system, which was formulated in a conventional manner.

The physical interpretation of relationship (8.17) is that stress at a given point at any time is dependent on the time history of the velocity profile at that point.

Let us now consider a thin rigid plate of mass M and area S immersed in a Newtonian fluid of infinite extent and connected by a massless spring of stiffness K to a fixed point (Fig. 8.3). A force $f(t)$ is applied to the plate. We assume that the spring does not disturb the fluid and that the area of the plate is sufficiently large to produce in the fluid adjacent to the plate the velocity field and stresses related by (8.17). Moreover, to allow application of relationship (8.17), the plate–fluid system must be initially in an equilibrium state — displacements and velocities must be initially zero.

Summing forces on the plate we find that the displacement y of the plate is described by

$$My''(t) = f(t) - Ky(t) - 2S\sigma(t, 0). \tag{8.18}$$

Substituting the stress given by relationship (8.17) and taking into account that

$$v_p(t, 0) = y'(t),$$

we arrive at the following fractional-order differential equation:

$$Ay''(t) + B\, {}_0D_t^{3/2} y(t) + Cy(t) = f(t) \qquad (t > 0), \tag{8.19}$$

$$A = M, \quad B = 2S\sqrt{\mu\rho}, \quad C = K,$$

to which the intial conditions describing the equilibrium initial state of the system must be attached:

$$y(0) = 0, \qquad y'(0) = 0. \tag{8.20}$$

Numerical solution of the Bagley–Torvik equation

Let us consider the following initial value problem for the inhomogeneous Bagley–Torvik equation [16]:

$$Ay''(t) + B\, {}_0D_t^{3/2} y(t) + Cy(t) = f(t), \quad (t > 0); \tag{8.21}$$

$$y(0) = 0, \qquad y'(0) = 0. \tag{8.22}$$

Let us take the time step h. The first-order approximation of the problem (8.21)–(8.22) is

$$Ah^{-2}(y_m - 2y_{m-1} + y_{m-2}) + Bh^{-3/2} \sum_{j=0}^{m} w_j^{(3/2)} y_{m-j} + Cy_m = f_m, \tag{8.23}$$

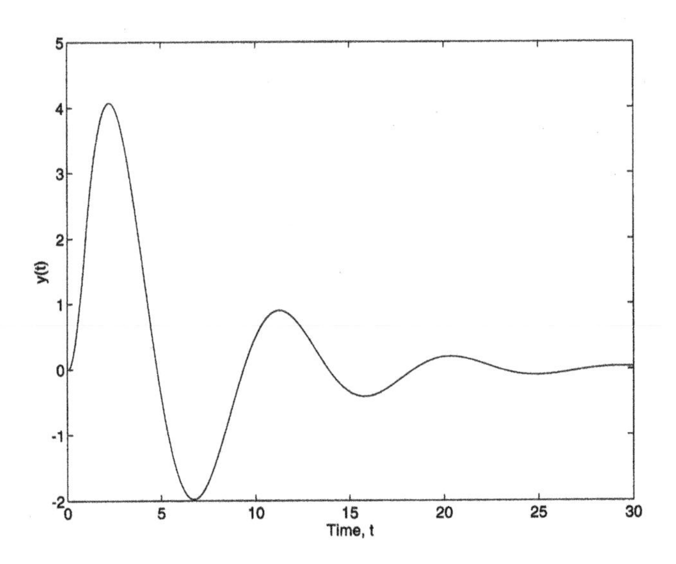

Figure 8.4: *Numerical solution of the Bagley–Torvik equation.*

$$(m = 2, 3, \ldots),$$

$$y_0 = 0, \qquad \frac{y_1 - y_0}{h} = 0, \tag{8.24}$$

where $y_m = y(mh), \quad f_m = f(mh), \quad (m = 0, 1, 2, \ldots)$.

Using approximation (8.23)–(8.24), we derive the following algorithm for obtaining the numerical solution:

$$y_0 = 0, \qquad y_1 = 0,$$

$$y_m = \frac{h^2(f_m - Cy_{m-1}) + A(2y_{m-1} - y_{m-2}) - B\sqrt{h}\sum_{j=1}^{m} w_j^{(3/2)} y_{m-j}}{A + B\sqrt{h}}, \tag{8.25}$$

$$(m = 2, 3, \ldots).$$

The results of our computations according to algorithm (8.25) are in agreement with the analytical solution, obtained with the help of the fractional Green's function for a three-term fractional differential equation with constant coefficients (see Section 5.4). The analytical solution

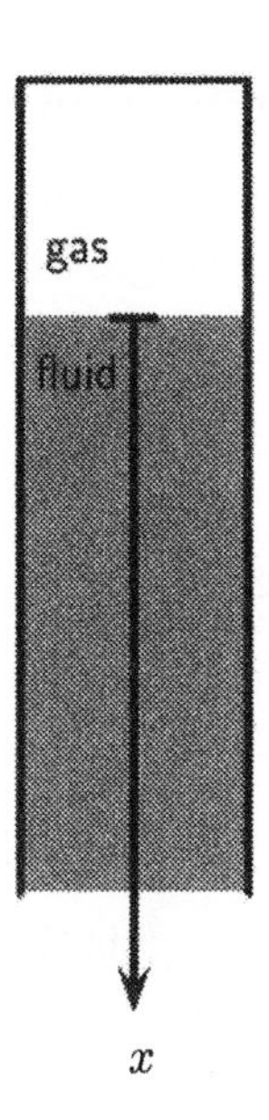

Figure 8.5: *Solution of a gas in a fluid: problem formulation*

of the initial-value problem (8.21)–(8.22) is

$$y(t) = \int_0^t G_3(t-\tau)f(\tau)d\tau, \tag{8.26}$$

$$G_3(t) = \frac{1}{A}\sum_{k=0}^{\infty} \frac{(-1)^k}{k!}\left(\frac{C}{A}\right)^k t^{2k+1} E^{(k)}_{\frac{1}{2},2+\frac{3k}{2}}(-\frac{B}{A}\sqrt{t}), \tag{8.27}$$

$$E^{(k)}_{\lambda,\mu}(y) \equiv \frac{d^k}{dy^k}E_{\lambda,\mu}(y) = \sum_{j=0}^{\infty} \frac{(j+k)!\; y^j}{j!\;\Gamma(\lambda j + \lambda k + \mu)}, \qquad (k = 0, 1, 2, ...).$$

In Fig. 8.4 the results of computations are given for

$$f(t) = f_*(t) = \begin{cases} 8, & (0 \le t \le 1) \\ 0, & (t > 1), \end{cases} \qquad A = 1, \quad B = 0.5, \quad C = 0.5.$$

8.3.3 Equation with Non-constant Coefficients: Solution of a Gas in a Fluid

The following example illustrates the use of the proposed method for fractional-order differential equations with non-constant coefficients.

Mathematical model of solution of a gas in a fluid

Yu. I. Babenko [11] gives the following mathematical model of the process of a solution of a gas in a fluid (Fig.8.5):

$$\frac{\partial}{\partial \tau}\left(V_0 f(\tau/\theta)\cdot P(\tau,0)\frac{M}{RT}\right) = F\mathcal{D}\left.\frac{\partial C}{dx}\right|_{x=0} \tag{8.28}$$

$$-\sqrt{\mathcal{D}}\left.\frac{\partial C}{dx}\right|_{x=0} = {}_0D_\tau^{1/2}(C_s(\tau) - C_0), \qquad (0 < \tau < \theta) \tag{8.29}$$

$$C(0,x) = C_0, \quad P(0,x) = P_0 = C_0/\kappa, \quad (0 < x < \infty); \tag{8.30}$$

where V_0 is the initial gas volume; θ is the time of the gas compression to zero volume; $f(t/\theta)$ is a function describing a change of the gas volume, such as $f(0) = 1$ and $f(1) = 0$; M is the gas molar weight; R is the universal gas constant; $\mathcal{D}$ is the coefficient of diffusion of the gas in the fluid; F is the contact surface between the gas and the fluid; $C(t,x)$ is the gas concentration; and $P(t,x)$ is the unknown gas pressure. The gas pressure near the contact surface $P(t,0)$ is to be found. The Ox axis goes down from the contact surface, for which $x = 0$. The gas temperature T is assumed to be constant. In other words, the gas compression is slow enough. The depth of the fluid is infinite.

Equation (8.28) describes the change of the mass of the gas volume due to diffusion through the contact surface. The mass change depends on the change of the gas concentration near the contact surface, which is given by equation (8.29). This makes consideration of the mass transfer for $x > 0$ unnecessary.

The problem (8.28)–(8.29) can be written in dimensionless form as

$$\frac{\partial}{\partial t}(c(t,0)f(t)) = \lambda\left.\frac{\partial c}{\partial \xi}\right|_{\xi=0}, \qquad (0 < t < 1) \tag{8.31}$$

$$-\left.\frac{\partial c}{\partial \xi}\right|_{\xi=0} = {}_0D_t^{1/2}(c(t,0) - 1)) \tag{8.32}$$

$$c(0,\xi) = 1 \tag{8.33}$$

where

$$p = c = \frac{C}{C_0} = \frac{P}{P_0}, \quad \xi = x\sqrt{\mathcal{D}\theta}; \quad t = \tau/\theta; \quad \lambda = \kappa RT\sqrt{\mathcal{D}\theta}/(MV_0).$$

Inserting (8.32) into (8.31) we obtain the following initial-value problem for determining the dimensionless gas pressure $p(t) \equiv p(t,0)$ near the contact surface:

$$\frac{d}{dt}(f(t)p(t)) + \lambda_0 D_t^{1/2}(p(t)-1) = 0, \quad (0 < t < 1) \tag{8.34}$$

$$p(0) = 1. \tag{8.35}$$

It is convenient to introduce the function

$$y(t) = p(t) - 1,$$

which allows consideration of problem (8.34)–(8.35) in the form

$$\frac{d}{dt}\Big(f(t)(y(t)+1)\Big) + \lambda_0 D_t^{1/2} y(t) = 0, \qquad (0 < t < 1) \tag{8.36}$$

$$y(0) = 0. \tag{8.37}$$

We arrived at the inhomogeneous (due to the presense of $f(t)$) linear fractional differential equation with zero initial condition. This allows us to develop a procedure of a numerical solution similar to the previous example. However, this problem allows us to obtain analytical solutions for some particular cases.

Analytical solutions for some particular cases

If the change of gas volume is described by the function expandable in a fractional power series

$$f(t) = \sum_{n=0}^{\infty} b_n t^{n/2}, \quad b_0 = 1, \tag{8.38}$$

then the solution of problem (8.36)–(8.37) can also be found in the form of a fractional power series (see Section 6.2.2):

$$y(t) = \sum_{n=1}^{\infty} a_n t^{n/2} \tag{8.39}$$

where the coefficients a_n satisfy the following recurrence relationships:

$$a_1 = -b_1, \qquad \sum_{k=0}^{n} a_{n+1-k} b_k + \lambda a_n \frac{\Gamma\left(\frac{n}{2}+1\right)}{\Gamma\left(\frac{n+3}{2}\right)} = -b_{n+1}. \tag{8.40}$$

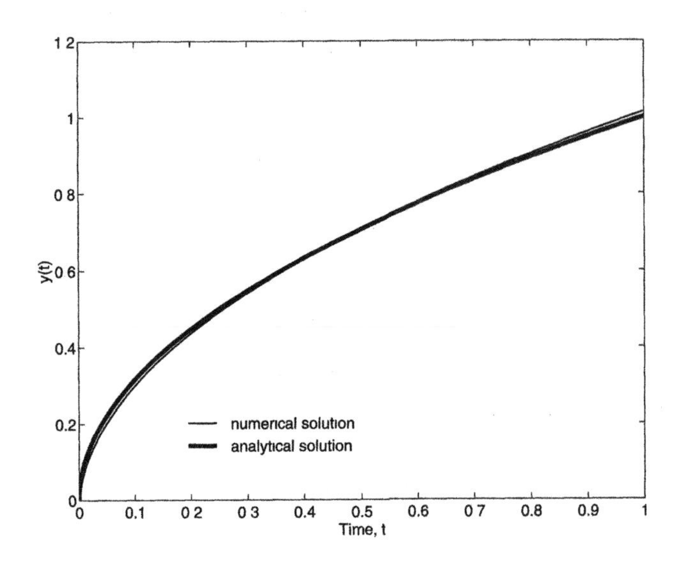

Figure 8.6: *Solution of a gas in a fluid: numerical example.*

Because of the construction of the solution (8.39), the initial condition (8.37) is satisfied automatically.

If we take, for example,

$$f(t) = 1 - \sqrt{t}, \tag{8.41}$$

and

$$\lambda = \frac{\Gamma\left(\frac{n+3}{2}\right)}{\Gamma\left(\frac{n+2}{2}\right)}$$

then the solution $y(t)$ is given by a finite sum. For example, [1]

$$\lambda = \frac{\Gamma(2)}{\Gamma(3/2)} = \frac{2}{\sqrt{\pi}}, \qquad y(t) = \sqrt{t} \tag{8.42}$$

$$\lambda = \frac{\Gamma(5/2)}{\Gamma(2)} = \frac{3\sqrt{\pi}}{4}, \qquad y(t) = \sqrt{t} + \left(1 - \frac{3\pi}{8}\right) t. \tag{8.43}$$

[1]There is a mistake in Babenko's book [11] on p. 107, where $p(\tau)$ corresponding to $\lambda = 3\sqrt{\pi}/4$ is given: instead of the expression $p(\tau) = 1 + \sqrt{\tau} + (1 - \sqrt{\pi}/2)\tau$, it should be $p(\tau) = 1 + \sqrt{\tau} + (1 - 3\pi/8)\tau$.

Numerical solution

Let us consider the initial value problem (8.36)–(8.37).

To construct a numerical algorithm, we write the problem in the form

$$F(t)y'(t) + G(t)\,{}_0D_t^{1/2}y(t) + y(t) = -1, \qquad (0 < t < 1); \qquad (8.44)$$

$$y(0) = 0,$$

where $F(t) = f(t)/f'(t)$, $G(t) = \lambda/f'(t)$. The first-order approximation of problem (8.44) is

$$(y_m - y_{m-1})F_m h^{-1} + G_m h^{-1/2}\sum_{j=0}^{m} w_j^{(1/2)} y_{m-j} + y_m = -1, \quad (m = 1, 2, \ldots) \qquad (8.45)$$

$$y_0 = 0.$$

Using approximation (8.45), we derive the following algorithm for the numerical solution of problem (8.44):

$$y_m = -F_m^{-1}\left(G_m\sqrt{h} + h\right) y_{m-1} - F_m^{-1}h + y_{m-1} - F_m^{-1}G_m\sqrt{h}\sum_{j=1}^{m} w_j^{(1/2)} y_{m-j}, \qquad (8.46)$$

$$(m = 1, 2, \ldots); \qquad y_0 = 0.$$

The results of our computations are in agreement with the analytical solutions obtained in the previous section.

For instance, if $f(t) = 1 - \sqrt{t}$ and $\lambda = \dfrac{2}{\sqrt{\pi}}$, then the analytical solution to problem (8.44) is $y(t) = \sqrt{t}$. Comparison of this analytical solution and the numerical solution obtained by (8.46) for $h = 0.001$ is given in Fig. 8.6.

8.3.4 Non-Linear Problem: Cooling of a Semi-infinite Body by Radiation

In this section we demonstrate the applicability of the proposed numerical method to non-linear fractional differential equations. The obtained numerical solution is compared with asymptotic solutions for small and large values of the independent variable t.

Problem formulation

Let us consider the following initial-boundary value problem describing the process of cooling of a semi-infinite body by radiation:

$$\frac{\partial u}{\partial t} = \frac{\partial^2 u}{\partial x^2}, \qquad (0 < x < \infty; \quad 0 < t < \infty) \tag{8.47}$$

$$\frac{\partial u}{\partial x}(t, 0) = \alpha\, u^4(t) \tag{8.48}$$

$$u(t, \infty) = u(0, x) = u_0. \tag{8.49}$$

We are interested in finding the surface temperature $u(0, t)$ for $t > 0$. In Section 7.7 we have obtained for $\frac{\partial u}{\partial x}(t, 0)$ a representation via fractional derivative of $u(t, x)$ with respect to time t (see formula (7.47)), which is valid if $u(t, x)$ satisfies equation (8.47) and conditions (8.49).

For this problem, we have

$$\frac{\partial u}{\partial x}(t, 0) = {}_0D_t^{1/2}(u_0 - u(t, 0)),$$

and after the substitution of this relationship into boundary condition (8.48) we obtain the following one-dimensional initial-value problem for the non-linear fractional differential equation:

$${}_0D_t^{1/2}y(t) - \alpha(u_0 - y(t))^4 = 0, \qquad (t > 0) \tag{8.50}$$

$$y(0) = 0 \tag{8.51}$$

where $y(t) = u_0 - u(0, t)$, and $u(0, t)$ is the surface temperature which must be found. Therefore, we need to find $y(t)$.

We need this substitution of the unknown function to obtain zero initial conditions for the construction of the numerical algorithm.

Asymptotic solution

Using the power series method, we obtained the following asymptotic representations for $y(t)$, which are in agreement with solutions given in [11]:

$$y(t) \approx \frac{2\,\alpha\, u_0^4\, t^{1/2}}{\sqrt{\pi}}, \qquad (t \ll 1) \tag{8.52}$$

$$y(t) \approx u_0\,(1 - \frac{1}{\alpha\sqrt{\pi}})\, t^{-1/8}, \qquad (t \gg 1). \tag{8.53}$$

We will use solutions (8.52) and (8.53) below for comparison with the numerical solution.

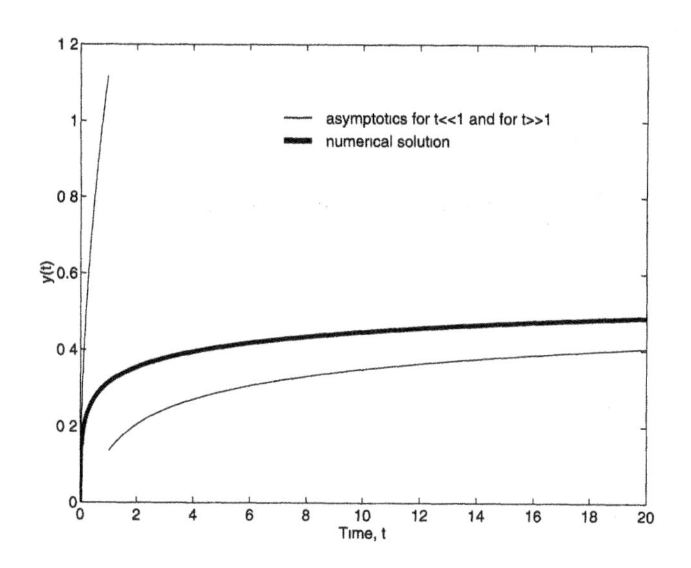

Figure 8.7: *Cooling of a semi-infinite body by radiation ($\alpha = 1$, $u_0 = 0$, $h = 0.02$).*

Numerical solution

Let us take the time step h and denote, as usual, $t_m = mh$, $y_m = y(t_m)$, $(m = 0, 1, 2, \ldots)$. Approximating the fractional-order derivative in (8.50) by (7.3), we obtain the following approximation for the problem (8.50)–(8.51):

$$y_0 = 0; \qquad h^{-1/2} \sum_{j=0}^{m} c_j y_{m-j} - \alpha \left(u_0 - y_m\right)^4 = 0, \quad (m = 1, 2, 3, \ldots) \tag{8.54}$$

where $c_j = (-1)^j \binom{1/2}{j}$.

Approximation (8.54) leads to the numerical solution algorithm described by

$$y_0 = 0; \quad y_m = h^{1/2} \alpha \left(u_0 - y_{m-1}\right)^4 - \sum_{j=1}^{m} c_j y_{m-j}, \quad (m = 1, 2, 3, \ldots). \tag{8.55}$$

The algorithm (8.55) allows step-by-step calculation of the values $y_m = y(mh)$. The results of computations for $\alpha = 1$, $u_0 = 1$, $h = 0.02$ are shown in Fig. 8.7.

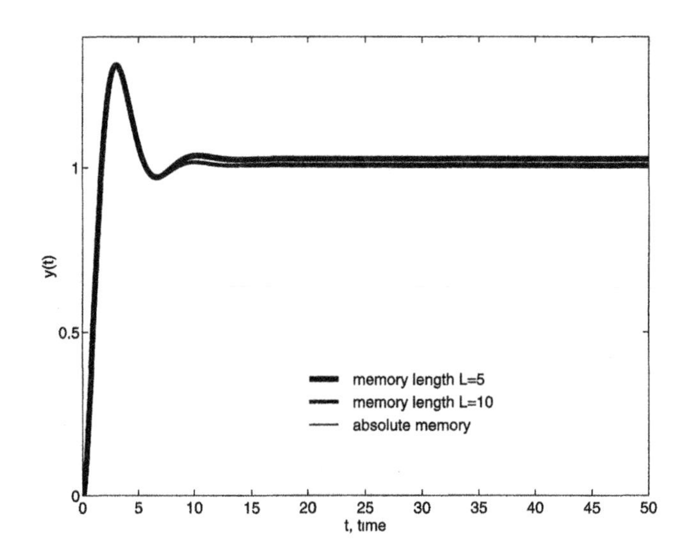

Figure 8.8: *Solution of the problem (8.56) for* $f(t) \equiv 1$.

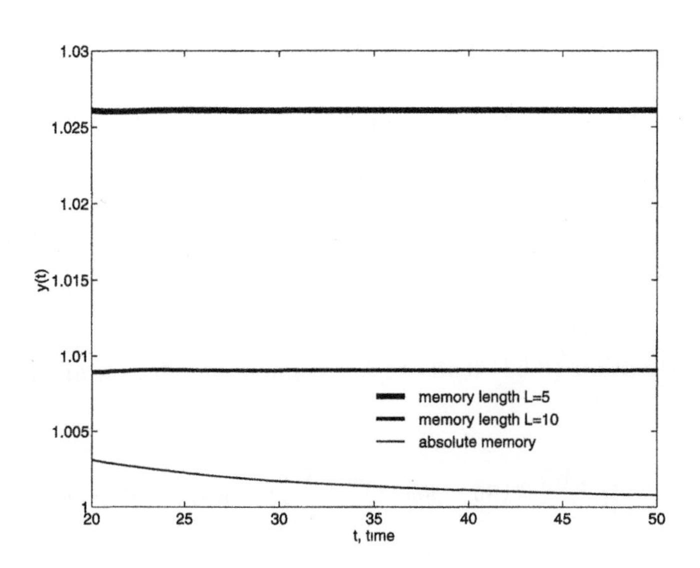

Figure 8.9: *Solution of the problem (8.56) for* $f(t) \equiv 1$ *(zoom).*

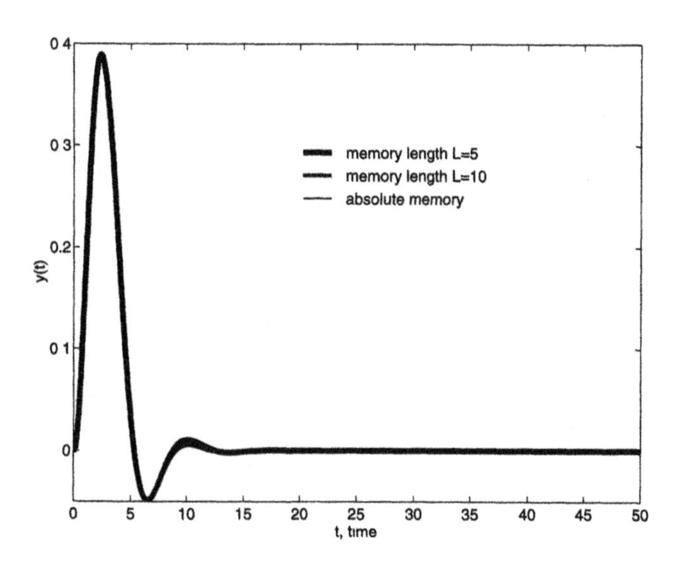

Figure 8.10: *Solution of the problem (8.56) for* $f(t) = te^{-t}$.

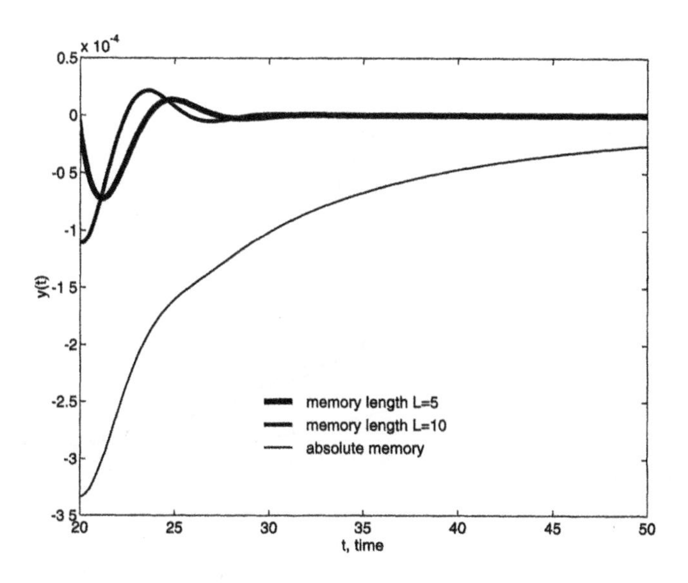

Figure 8.11: *Solution of the problem (8.56) for* $f(t) = te^{-t}$ *(zoom)*.

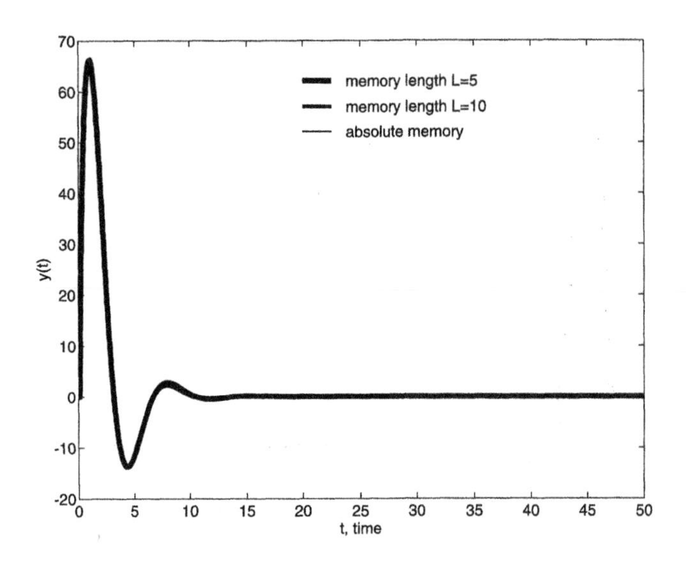

Figure 8.12: *Solution of the problem (8.56) for* $f(t) = t^{-1}e^{-1/t}$.

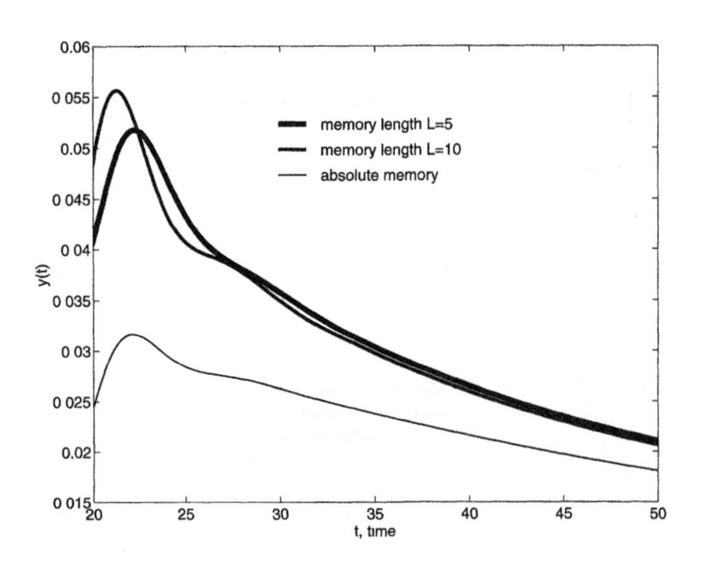

Figure 8.13: *Solution of the problem (8.56) for* $f(t) = t^{-1}e^{-1/t}$ *(zoom).*

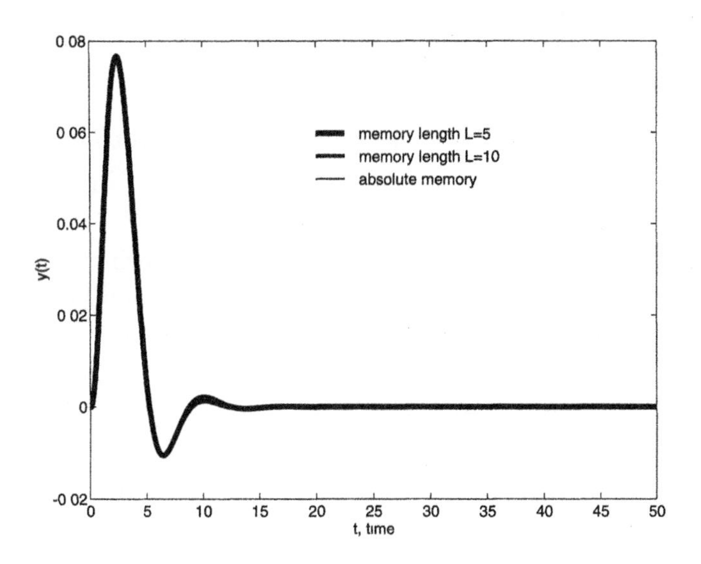

Figure 8.14: *Solution of the problem (8.56) for* $f(t) = e^{-t}\sin(0.2t)$.

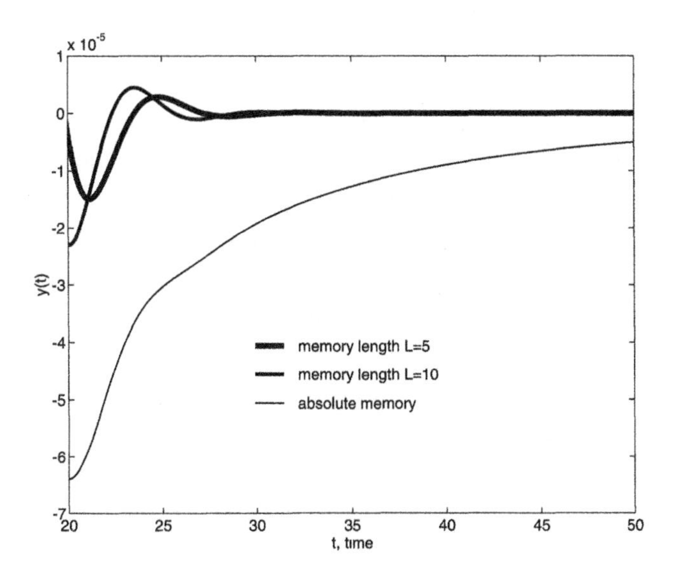

Figure 8.15: *Solution of (8.56) for* $f(t) = e^{-t}\sin(0.2t)$ *(zoom)*.

8.4 The "Short-Memory" Principle in Initial Value Problems for Fractional Differential Equations

In all the examples given above, the use of the short-memory principle leads to the simple replacement of $\sum_{j=1}^{m}$ by $\sum_{j=1}^{M}$, where $M = \min\left\{m, \left[\frac{L}{h}\right]\right\}$ and L is the memory length.

To illustrate the usefulness of the short-memory principle for the numerical solution of initial-value problems for fractional differential equations, we give on Figs (8.8)–(8.15) numerical solutions of the problem

$$ {}_0D_t^{3/2}y(t) + y(t) = f(t), \qquad (t > 0) \tag{8.56} $$

$$ y(0) = y'(0) = 0 $$

for the following particular cases of the right-hand side $f(t)$:

1. $f(t) \equiv 1$ (Figs 8.8 and 8.9);
2. $f(t) = te^{-t}$ (Figs 8.10 and 8.11);
3. $f(t) = t^{-1}e^{-1/t}$ (Figs 8.12 and 8.13);
4. $f(t) = e^{-t}\sin(0.2t)$ (Figs 8.14 and 8.15).

Numerical solutions were computed using the time step $h = 0.1$ for the interval $0 \le t \le 50$. One can see that even taking the memory length $L = 5$ gives satisfactory accuracy.

We also found that using the short-memory principle leads in many cases to suppression of the influence of accumulating rounding error during long-time simulations — due to a smaller number of addends.

Chapter 9

Fractional-order Systems and Controllers

At present, a growing number of works by many authors from various fields of science and engineering deal with dynamical systems described by fractional-order equations which means equations involving derivatives and integrals of non-integer order.

These new models are more adequate than the previously used integer-order models. This was demonstrated, for instance, in [24, 170, 70]. Important fundamental physical considerations in favour of the use of fractional-derivative based models were given in [30, 254]. Fractional-order derivatives and integrals provide a powerful instrument for the description of memory and hereditary properties of different substances. This is the most significant advantage of the fractional-order models in comparison with integer-order models, in which, in fact, such effects are neglected.

However, because of the absense of appropriate mathematical methods, fractional-order dynamical systems were studied only marginally in the theory and practice of control systems. Some sucessful attempts were undertaken in [13, 140, 9, 110, 184], but generally the study in the time domain has been almost avoided.

In this chapter effective and easy-to-use tools for the time-domain analysis of fractional-order dynamical systems, which are described in the previous chapters, are used for solving problems of control theory. The concept of a $PI^{\lambda}D^{\mu}$-controller, involving fractional-order integrator and fractional-order differentiator, is introduced. An example is provided to demonstrate the necessity of such controllers for the more efficient

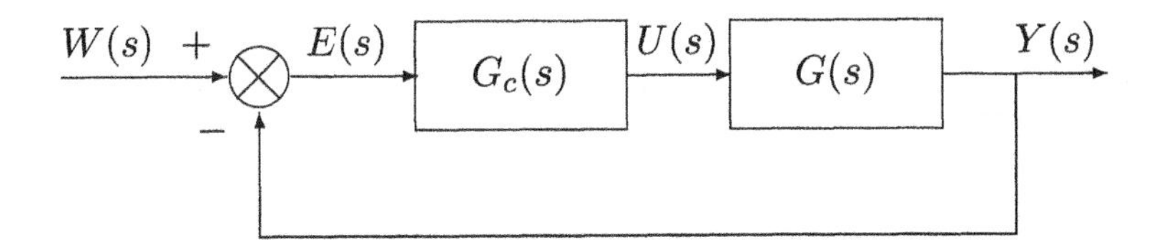

Figure 9.1: *Simple unity-feedback control system.*

control of fractional-order dynamical systems.

The idea of using fractional-order controllers for the control of dynamical systems belongs to A. Oustaloup, who developed the so-called CRONE controller (CRONE is an abbreviation of *Commande Robuste d'Ordre Non Entier*), which is described in a series of his books [183, 185, 186, 187] along with examples of applications in various fields. A. Oustaluop demonstrated the advantage of the CRONE controller in comparison with the PID-controller. The $PI^{\lambda}D^{\mu}$-controller, considered in this chapter, is a new type of fractional-order controller, which also shows better performance when used for the control of fractional-order systems than the classical PID-controller.

9.1 Fractional-order Systems and Fractional-order Controllers

This chapter is a natural continuation of Chapter 4, which we recommend to readers interested in the system response to an arbitrary input. However, here we turn from purely mathematical aspects of the fractional calculus to application of the fractional calculus in control theory.

9.1.1 Fractional-order Control System

Let us consider the simple unity-feedback control system shown in Fig. 9.1, where $G(s)$ is the transfer function of the controlled system, $G_c(S)$ is the transfer of the controller, $W(s)$ is an input, $E(s)$ is an error, $U(s)$ is the controller's output, and $Y(s)$ is the system's output.

Contrary to the traditional approach, we will consider *transfer functions of arbitrary real order*. We call such systems fractional-order systems. They include, in particular, traditional integer-order systems. It is important to realize that the words "fractional-order system" mean just "systems which are better described by fractional-order mathematical models".

9.1.2 Fractional-order Transfer Functions

Let us consider the fractional-order transfer function (FOTF) given by the following expression:

$$G_n(s) = \frac{1}{a_n s^{\beta_n} + a_{n-1} s^{\beta_{n-1}} + \ldots + a_1 s^{\beta_1} + a_0 s^{\beta_0}}, \tag{9.1}$$

where β_k, $(k = 0, 1, \ldots, n)$ is an arbitrary real number,
$\beta_n > \beta_{n-1} > \ldots > \beta_1 > \beta_0 > 0$,
a_k $(k = 0, 1, \ldots, n)$ is an arbitrary constant.

In the time domain, the FOTF (9.1) corresponds to the n-term fractional-order differential equation (FDE)

$$a_n D^{\beta_n} y(t) + a_{n-1} D^{\beta_{n-1}} y(t) + \ldots + a_1 D^{\beta_1} y(t) + a_0 D^{\beta_0} y(t) = u(t) \tag{9.2}$$

where $D^\gamma \equiv {}_0D_t^\gamma$ is Caputo's fractional derivative of order γ with respect to the variable t and with the starting point at $t = 0$ [23, 24]:

$${}_0D_t^\gamma y(t) = \frac{1}{\Gamma(1-\delta)} \int_0^t \frac{y^{(m+1)}(\tau) d\tau}{(t-\tau)^\delta}, \tag{9.3}$$

$$(\gamma = m + \delta, \quad m \in Z, \quad 0 < \delta \leq 1).$$

If $\gamma < 0$, then one has a fractional integral of order $-\gamma$:

$${}_0I_t^{-\gamma} y(t) = {}_0D_t^\gamma y(t) = \frac{1}{\Gamma(-\gamma)} \int_0^t \frac{y(\tau) d\tau}{(t-\tau)^{1+\gamma}}, \quad (\gamma < 0). \tag{9.4}$$

The Laplace transform of the fractional derivative defined by (9.3) is 2.253

$$\int_0^\infty e^{-st} D^\gamma y(t) dt = s^\gamma Y(s) - \sum_{k=0}^{m} s^{\gamma-k-1} y^{(k)}(0). \tag{9.5}$$

For $\gamma < 0$ (i.e., for the case of a fractional integral) the sum on the right-hand side must be omitted.

It is worth mentioning here that from the pure mathematical point of view there are different ways to interpolate between integer-order multiple integrals and derivatives. The most widely known and precisely studied is the Riemann–Liouville definition of fractional derivatives (e.g., [179, 232, 153]). The main advantage of Caputo's definition in comparison with the Riemann–Liouville definition is that it allows consideration of easily interpreted conventional initial conditions such as $y(0) = y_0, y'(0) = y_1$, etc. Moreover, Caputo's derivative of a constant is bounded (namely, equal to 0), while the Riemann–Liouville derivative of a constant is unbounded at $t = 0$. The only exception is if one takes $t = -\infty$ as the starting point (lower limit) in the Riemann–Liouville definition. In this case, the Riemann–Liouville fractional derivative of a constant is also 0, and this was used in [174]. However, one interested in transient processes could not accept placement of the starting point at $-\infty$, and in such cases Caputo's definition seems to be the most appropriate compared to others.

Formula (9.5) is a particular case of a more general formula (2.259) given in Section 2.8.5 for the Laplace transform of the so-called sequential fractional derivative (2.170).

To find the unit-impulse and unit-step response of the fractional-order system described by FDE (9.2), we need to evaluate the inverse Laplace transform of the function $G_n(s)$.

The problem of the Laplace inversion of (9.1), however, can appear in any field of applied mathematics, physics, engineering, etc., where the Laplace transform method is used. This fact along with the absense of the necessary inversion formula in tables and handbooks on the Laplace transform motivated us to give a general solution to this problem in the following two sections.

9.1.3 New Function of the Mittag-Leffler Type

The so-called Mittag-Leffler function in two parameters $E_{\alpha,\beta}(z)$ was introduced by Agarwal [3]. His definition was later modified by the authors of [65] to be

$$E_{\alpha,\beta}(z) = \sum_{j=0}^{\infty} \frac{z^j}{\Gamma(\alpha j + \beta)}, \qquad (\alpha > 0, \quad \beta > 0). \tag{9.6}$$

Its k-th derivative is given by

$$E_{\alpha,\beta}^{(k)}(z) = \sum_{j=0}^{\infty} \frac{(j+k)!\ z^j}{j!\ \Gamma(\alpha j + \alpha k + \beta)}, \qquad (k = 0, 1, 2, \ldots). \tag{9.7}$$

We find it convenient to introduce the function

$$\mathcal{E}_k(t, y; \alpha, \beta) = t^{\alpha k+\beta-1} E^{(k)}_{\alpha,\beta}(yt^{\alpha}), \quad (k = 0, 1, 2, \ldots). \tag{9.8}$$

Its Laplace transform was (in other notation) evaluated in Chapter 4:

$$\int_0^{\infty} e^{-st}\mathcal{E}_k(t, \pm y; \alpha, \beta)dt = \frac{k!\, s^{\alpha-\beta}}{(s^{\alpha} \mp y)^{k+1}}, \qquad (Re(s) > |y|^{1/\alpha}). \tag{9.9}$$

Another convenient property of $\mathcal{E}_k(t, y; \alpha, \beta)$, which we use in this chapter, is its simple fractional differentiation (see Section 1.2.3):

$$_0D_t^{\lambda}\mathcal{E}_k(t, y; \alpha, \beta) = \mathcal{E}_k(t, y; \alpha, \beta - \lambda), \quad (\lambda < \beta). \tag{9.10}$$

Other properties of the function $\mathcal{E}_k(t, y; \alpha, \beta)$, such as special cases, its asymptotic behaviour, etc., can be obtained from (9.6)–(9.8) and the known properties [65] of the Mittag-Leffler function $E_{\alpha,\beta}(z)$.

9.1.4 General Formula

Relationship (9.9) allows us to evaluate the inverse Laplace transform of (9.1) as was done in Chapter 5.

Let $\beta_n > \beta_{n-1} > \ldots > \beta_1 > \beta_0 > 0$. Then using (9.9) gives the final expression for the inverse Laplace transform of the function $G_n(s)$:

$$\begin{aligned} g_n(t) &= \frac{1}{a_n}\sum_{m=0}^{\infty}\frac{(-1)^m}{m!}\sum_{\substack{k_0+k_1+\ldots+k_{n-2}=m\\ k_0\geq 0;\ldots,k_{n-2}\geq 0}}(m; k_0, k_1, \ldots, k_{n-2}) \\ &\times \prod_{i=0}^{n-2}\left(\frac{a_i}{a_n}\right)^{k_i}\mathcal{E}_m(t, -\frac{a_{n-1}}{a_n}; \beta_n - \beta_{n-1}, \beta_n + \sum_{j=0}^{n-2}(\beta_{n-1} - \beta_j)k_j) \end{aligned} \tag{9.11}$$

where $(m; k_0, k_1, \ldots, k_{n-2})$ are the multinomial coefficients [2, chapter 24].

Further inverse Laplace transforms can be obtained by combining (9.10) and (9.11). For instance, let us take

$$F(s) = \sum_{i=1}^{N} b_i s^{\alpha_i} G_n(s), \tag{9.12}$$

where $\alpha_i < \beta_n$, $(i = 1, 2, \ldots, N)$. Then the inverse Laplace transform of $F(s)$ is

$$f(t) = \sum_{i=1}^{N} b_i D^{\alpha_i} g_n(t), \tag{9.13}$$

where the fractional derivatives of $g_n(t)$ are evaluated with the help of (9.10).

9.1.5 The Unit-impulse and Unit-step Response

The unit-impulse response of the fractional-order system with the transfer function (9.1) is given by formula (9.11), i.e. $y_{impulse}(t) = g_n(t)$.

To find the unit-step response $y_{step}(t)$, one has to integrate (9.11) with the help of (9.10). The result is:

$$\begin{aligned} y_{step}(t) &= \frac{1}{a_n} \sum_{m=0}^{\infty} \frac{(-1)^m}{m!} \sum_{\substack{k_0+k_1+\ldots+k_{n-2}=m \\ k_0 \geq 0; \ldots, k_{n-2} \geq 0}} (m; k_0, k_1, \ldots, k_{n-2}) \\ &\times \prod_{i=0}^{n-2} \left(\frac{a_i}{a_n}\right)^{k_i} \mathcal{E}_m\Big(t, -\frac{a_{n-1}}{a_n}; \beta_n - \beta_{n-1}, \\ &\beta_n + \sum_{j=0}^{n-2} (\beta_{n-1} - \beta_j) k_j + 1\Big). \end{aligned} \tag{9.14}$$

9.1.6 Some Special Cases

For illustration, we give the following three particular cases of (9.11) and (9.14).

1)

$$G_2(s) = \frac{1}{as^{\alpha} + b}, \qquad (\alpha > 0)$$

$$\left.\begin{array}{r} y_{impulse}(t) = g_2(t) \\ y_{step}(t) = {}_0D_t^{-1}\, g_2(t) \end{array}\right\} = \frac{1}{a}\mathcal{E}_0(t, -\frac{b}{a}; \alpha, \alpha + \left\{\begin{array}{c} 0 \\ 1 \end{array}\right\}) \tag{9.15}$$

2)

$$G_3(s) = \frac{1}{as^{\beta} + bs^{\alpha} + c}, \qquad (\beta > \alpha > 0)$$

$$\left. \begin{array}{r} y_{impulse}(t) = g_3(t) \\ y_{step}(t) = {}_0D_t^{-1}\, g_3(t) \end{array} \right\}$$

$$= \frac{1}{a}\sum_{k=0}^{\infty} \frac{(-1)^k}{k!} \left(\frac{c}{a}\right)^k \mathcal{E}_k(t, -\frac{b}{a}; \beta - \alpha, \beta + \alpha k + \left\{ \begin{array}{c} 0 \\ 1 \end{array} \right\}) \qquad (9.16)$$

3)

$$G_4(s) = \frac{1}{as^\gamma + bs^\beta + cs^\alpha + d}, \qquad (\gamma > \beta > \alpha > 0)$$

$$\left. \begin{array}{r} y_{impulse}(t) = g_4(t) \\ y_{step}(t) = {}_0D_t^{-1}\, g_4(t) \end{array} \right\} = \frac{1}{a}\sum_{m=0}^{\infty} \frac{1}{m!} \left(\frac{-d}{a}\right)^m \sum_{k=0}^{m} \binom{m}{k} \left(\frac{c}{d}\right)^k$$

$$\times\, \mathcal{E}_m(t, -\frac{b}{a}; \gamma - \beta, \gamma + \beta m - \alpha k + \left\{ \begin{array}{c} 0 \\ 1 \end{array} \right\}). \qquad (9.17)$$

Integrating the unit-step response with the help of (9.10), we obtain the unit-ramp response. Double integration of the unit-step response gives the response for the parabolic input. All these standard test input signals are frequently used in control theory, and the above formulas provide explicit analytical expressions for the corresponding system responses.

9.1.7 $PI^\lambda D^\mu$-controller

As will be shown in an example below, a suitable way to the more efficient control of fractional-order systems is to use fractional-order controllers. We propose a generalization of the PID-controller, which can be called the $PI^\lambda D^\mu$-controller because it involves an integrator of order λ and differentiator of order μ. The transfer function of such a controller has the form:

$$G_c(s) = \frac{U(s)}{E(s)} = K_P + K_I s^{-\lambda} + K_D s^\mu, \qquad (\lambda,\, \mu > 0). \qquad (9.18)$$

The equation for the $PI^\lambda D^\mu$-controller's output in the time domain is:

$$u(t) = K_P e(t) + K_I D^{-\lambda} e(t) + K_D D^\mu e(t). \qquad (9.19)$$

Taking $\lambda = 1$ and $\mu = 1$, we obtain a classical PID-controller. $\lambda = 1$ and $\mu = 0$ give a PI-controller. $\lambda = 0$ and $\mu = 1$ give a PD-controller. $\lambda = 0$ and $\mu = 0$ give a gain.

All these classical types of PID-controllers are the particular cases of the fractional $PI^{\lambda}D^{\mu}$-controller (9.18). However, the $PI^{\lambda}D^{\mu}$-controller is more flexible and gives an opportunity to better adjust the dynamical properties of a fractional-order control system.

9.1.8 Open-loop System Response

Let us delete the feedback in Fig. 9.1 and consider the obtained open loop with the $PI^{\lambda}D^{\mu}$-controller (9.18) and the fractional-order controlled system with the transfer function $G_n(s)$ given by expression (9.1).

In the time domain, this open-loop system is described by the fractional-order differential equation

$$\sum_{k=0}^{n} a_k D^{\beta_k} y(t) = K_P w(t) + K_I D^{-\lambda} w(t) + K_D D^{\mu} w(t). \tag{9.20}$$

The transfer function of the considered open-loop system is

$$G_{open}(s) = \left(K_P + K_I s^{-\lambda} + K_D s^{\mu}\right) G_n(s). \tag{9.21}$$

Since (9.21) has the same structure as (9.12), the inverse Laplace transform for $G_{open}(s)$ can be found with the help of formula (9.13). Therefore, the unit-step response of the considered fractional-order open-loop system is

$$g_{open}(t) = K_P g_n(t) + K_I D^{-\lambda} g_n(t) + K_D D^{\mu} g_n(t), \tag{9.22}$$

where $g_n(t)$ is given by (9.11).

To find the unit-step response, one should integrate (9.22) using formula (9.10).

9.1.9 Closed-loop System Response

To obtain the unit-impulse and unit-step response for a closed-loop control system (Fig.9.1) with the $PI^{\lambda}D^{\mu}$-controller and the fractional-order controlled system with the transfer function $G_n(s)$ given by expression (9.1), one needs, first, to replace $w(t)$ with $e(t) = w(t) - y(t)$ in equation (9.20). This step results in

$$\begin{aligned} \sum_{k=0}^{n} a_k D^{\beta_k} y(t) &+ K_P y(t) + K_I D^{-\lambda} y(t) + K_D D^{\mu} y(t) \\ &= K_P w(t) + K_I D^{-\lambda} w(t) + K_D D^{\mu} w(t). \end{aligned} \tag{9.23}$$

From (9.23) one obtains the following expression for the transfer function of the considered closed-loop system:

$$G_{closed}(s) = \frac{K_P s^{\lambda} + K_I + k_D s^{\mu+\lambda}}{\sum\limits_{k=0}^{n} a_k s^{\beta_k+\lambda} + K_P s^{\lambda} + K_I + K_D s^{\mu+\lambda}}. \tag{9.24}$$

The unit-impulse response $g_{closed}(t)$ is then obtained by the Laplace inversion of (9.24), which could be performed by rearranging in decreasing order of differentiation the addends in the denominator of (9.24) and applying after that relationships (9.11) and (9.13). To find the unit-step response, one should integrate the obtained unit-impulse response with the help of (9.10).

9.2 Example

In this section we give an example showing the usefulness of the $PI^{\lambda}D^{\mu}$-controllers in comparison with conventional PID-controllers. We consider a fractional-order system, which plays *the role* of "reality", and its integer-order approximation, which plays the role of a "model". We emphasize that, at first glance, the model, obtained in the usual manner, fits the data obtained from "reality" well.

However, the PD-controller, designed on the basis of the model, is shown to be not so suitable for the control of "reality" as one should expect.

A good way to improve the control is to use a controller of a similar "nature" to "reality", i.e. a fractional-order PD^{μ}-controller. At this stage we assume that the fractional-order transfer function has been identified exactly.

It is important to realize that often, in fact, the structure of the model is postulated (in our example, the second order differential equation model) and then the parameters of the model (in our case, the coefficients of the differential equation) are determined to provide suitable fitting of data obtained from the real object. However, there are numerous real systems which are better described by fractional-order differential equations. For such systems classical integer-order models, even of high order, will give less adequate results than fractional-order models. From this point of view, the example demonstrates some of the possible effects arising from the difference of the nature of "reality" and the "model". It also indicates the necessity of the development of methods for identification of parameters of fractional-order models, including the most appropriate *order of the model* (not the order of the real object).

9.2.1 Fractional-order Controlled System

Let us consider a fractional-order controlled system with the transfer function

$$G(s) = \frac{1}{a_2 s^{\beta} + a_1 s^{\alpha} + a_0} \tag{9.25}$$

where we take $a_2 = 0.8$, $a_1 = 0.5$, $a_0 = 1$, $\beta = 2.2$, $\alpha = 0.9$.

The fractional-order transfer function (9.25) corresponds in the time domain to the three-term fractional-order differential equation

$$a_2 y^{(\beta)}(t) + a_1 y^{(\alpha)}(t) + a_0 y(t) = u(t) \tag{9.26}$$

with zero initial conditions $y(0) = 0$, $y'(0) = 0$, $y''(0) = 0$.

The unit-step response is found by (9.16):

$$y(t) = \frac{1}{a_2} \sum_{k=0}^{\infty} \frac{(-1)^k}{k!} \left(\frac{a_0}{a_2} \right)^k \mathcal{E}_k(t, -\frac{a_1}{a_2}; \beta - \alpha, \beta + \alpha k + 1). \tag{9.27}$$

9.2.2 Integer-order Approximation

For comparison purposes, let us approximate the considered fractional-order system by a second-order system. Noticing that $\beta = 2.2$ and $\alpha = 0.9$ are close to 2 and 1, respectively, one may expect a good approximation. Using the least-squares method for the determination of coefficients of the resulting equation, we obtained the following approximating equation corresponding to (9.26):

$$\tilde{a}_2 y''(t) + \tilde{a}_1 y'(t) + \tilde{a}_0 y(t) = u(t) \tag{9.28}$$

with $\tilde{a}_2 = 0.7414$, $\tilde{a}_1 = 0.2313$, $\tilde{a}_0 = 1$.

The comparison of the unit-step response of systems described by (9.26) (original system) and (9.28) (approximating system) is shown in Fig. 9.2. The agreement seems to be satisfactory enough to build up the control strategy on the description of the original fractional-order system by its approximation.

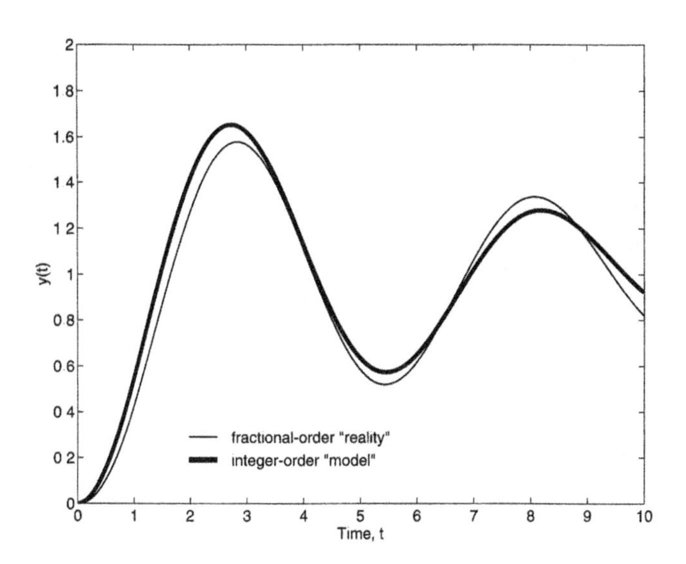

Figure 9.2: *Unit-step response of the fractional-order system (thin line) and its approximation (thick line).*

9.2.3 Integer-order PD-controller

Since the above comparison of the unit-step responses shows good agreement, one may try to control the original system (9.26) by a controller designed for its approximation (9.28). This approach is, in fact, frequently used in practice, when one controls the real object by a controller designed for the model of that object.

The PD-controller with the transfer function

$$\tilde{G}_c(s) = \tilde{K} + \tilde{T}_d s \tag{9.29}$$

was designed so that a unit step signal at the input of the closed-loop system in Fig. 9.1 will induce at the output an oscillatory unit-step response with stability measure $St = 2$ (this is equivalent to the requirement that the system must settle within 5% of the unit step at the input in 2 seconds: $T_s \leq 2$s) and damping ratio $\xi = 0.4$. In this case, the coefficients for (9.29) take on the values $\tilde{K} = 20.5$ and $\tilde{T}_d = 2.7343$.

For comparison purposes, we also computed the integral of the absolute error (IAE)

$$I(t) = \int_0^t |e(t)|dt$$

for $t = 5$ s: $I(5) = 0.8522$.

Let us now apply this controller, designed for the optimal control of the approximating integer-order system (9.28), to the control of the approximated fractional-order system (9.26).

The differential equation of the closed loop with the fractional-order system defined by (9.25) and the integer-order controller defined by (9.29) has the following form:

$$a_2 y^{(\beta)}(t) + \tilde{T}_d y'(t) + a_1 y^{(\alpha)}(t) + (a_0 + \tilde{K}) y(t) = \tilde{K} w(t) + \tilde{T}_d w'(t). \quad (9.30)$$

This is a four-term fractional differential equation, and the unit-step response of this system is found with the help of (9.17):

$$\begin{aligned} y(t) &= \frac{1}{a_2} \sum_{m=0}^{\infty} \frac{(-1)^m}{m!} \left(\frac{a_0 + \tilde{K}}{a_2} \right)^m \\ & \sum_{k=0}^{m} \binom{m}{k} \left(\frac{a_1}{a_0 + \tilde{K}} \right)^m \left\{ \tilde{K} \mathcal{E}_m(t, -\frac{\tilde{T}_d}{a_2}; \beta - 1, \beta + m - \alpha k + 1) \right. \\ & \left. + \tilde{T}_d \mathcal{E}_m(t, -\frac{\tilde{T}_d}{a_2}; \beta - 1, \beta + m - \alpha k) \right\}. \quad (9.31) \end{aligned}$$

A comparison of the unit-step response of the closed-loop integer-order (approximating) system and the closed-loop fractional-order (approximated) system with the same integer-order controller, optimally designed for the approximating system, is shown in Fig. 9.3.

One can see that the dynamical properties of the closed loop with the fractional-order controlled system and the integer-order controller, which was designed for the integer-order approximation of the fractional-order system, are considerably worse than the dynamic properties of the closed loop with the approximating integer-order system. The system stabilizes slower and has larger surplus oscillations. Computations show that, in comparison with the integer-order "model", in this case the IAE within 5 s time interval is larger by 76%. Moreover, the closed loop with the fractional-order controlled system is more sensitive to changes in controller parameters. For example, under a change of $\tilde{T}_d$ to the value 1, the closed loop with the fractional-order system ("reality") is already unstable, whereas the closed loop with the approximating integer-order system (the "model") still shows stability (Fig. 9.4).

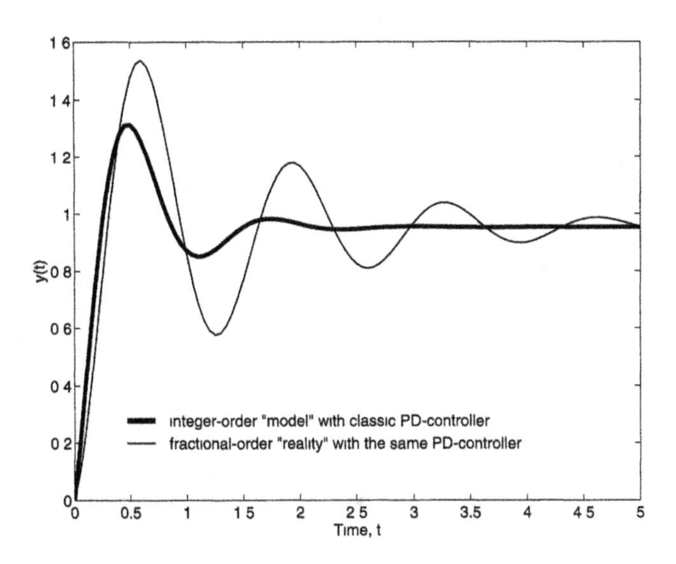

Figure 9.3: *Unit-step response of the closed-loop integer-order (thick line) and fractional-order (thin line) systems with the same integer-order controller, designed for the approximating integer-order system.*

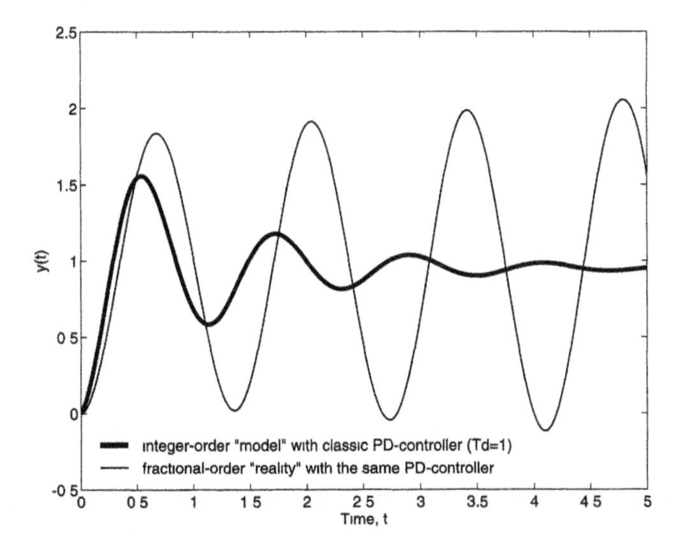

Figure 9.4: *Unit-step response of the closed-loop integer-order (thick line) and fractional-order (thin line) systems with the same integer-order controller, designed for the integer-order system, for* $\tilde{T}_d = 1$.

9.2.4 Fractional-order Controller

We see that disregarding the fractional order or the original system (9.26), replacing it with the approximating integer-order system (9.28) and application of the controller, designed for the approximating system, to the control of the original fractional-order system, is not generally adequate.

An alternative and more successful approach in our example is to use the fractional-order PD^{μ}-controller characterized by the fractional-order transfer function

$$G_c(s) = K + T_d s^{\mu}. \tag{9.32}$$

Let us take $\alpha < \mu < \beta$. The differential equation of the closed-loop control system with the fractional-order system transfer (9.25) and the fractional-order controller transfer (9.32) can be written in the form:

$$a_2 y^{(\beta)}(t) + T_d y^{(\mu)}(t) + a_1 y^{(\alpha)}(t) + (a_0 + K) y(t) = K w(t) + T_d w^{(\mu)}(t). \tag{9.33}$$

We are interested in the unit-step response of this system.

Using (9.17), (9.13) and (9.10), the following solution to equation (9.33) is obtained:

$$\begin{aligned} y(t) &= \frac{1}{a_2} \sum_{m=0}^{\infty} \frac{(-1)^m}{m!} \left(\frac{a_0 + K}{a_2} \right)^m \\ &\sum_{k=0}^{m} \binom{m}{k} \left(\frac{a_1}{a_0 + K} \right)^k \Big\{ K \mathcal{E}_m(t, -\frac{T_d}{a_2}; \beta - \mu, \beta + \mu m - \alpha k + 1) \\ &\qquad + T_d \mathcal{E}_m(t, -\frac{T_d}{a_2}; \beta - \mu, \beta + \mu m - \alpha k + 1 - \mu) \Big\} \end{aligned} \tag{9.34}$$

In Fig. 9.5, the comparison of the unit-step response of the closed loop with the fractional-order system controlled by a fractional-order PD^{μ}-controller with $K = \tilde{K}$, $T_d = 3.7343$ and $\mu = 1.15$ (the values of the parameters were found by computational experiments) and the unit-step response of the closed loop with the same system controlled by the integer-order PD-controller, designed for the approximating integer-order system, is given.

One can see that the use of the fractional-order controller leads to the improvement of the control of the fractional-order system.

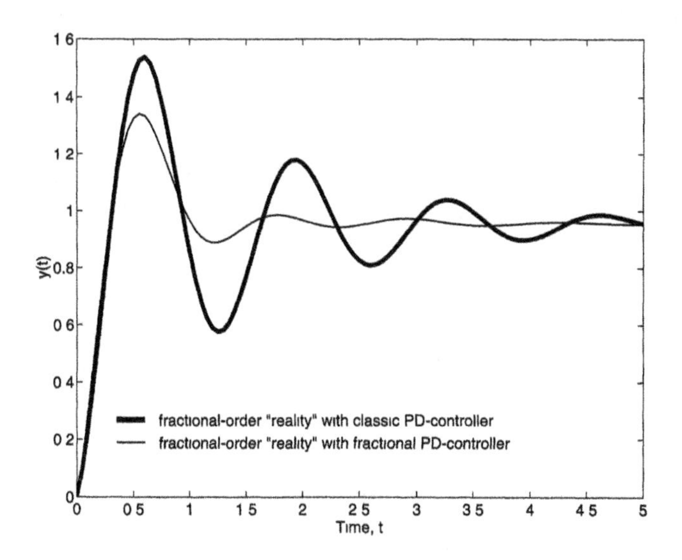

Figure 9.5: *Unit-step response of the closed-loop fractional-order system with the conventional PD-controller, designed for the approximating integer-order system (thick line), and with the PD^{μ}-controller (thin line).*

9.3 On Fractional-order System Identification

In this section we briefly discuss an approach to identification of parameters of fractional-order models of real dynamical systems. The method is illustrated on the example of identification of parameters of fractional-order models of a re-heating furnace.

A set of measured values y_i^* $(i = \overline{0, M})$ was obtained for the transfer function of a real experimental re-heating furnace. Then three models were developed for this object.

The first model was obtained using classical integer-order derivatives. Assuming that the system can be described by the second-order differential equation

$$a_2 y''(t) + a_1 y'(t) + a_0 y(t) = u(t), \qquad (9.35)$$

the following values were obtained for the coefficients of the modelling equation:

$$a_2 = 1.8675, \quad a_1 = 5.5184, \quad a_0 = 0.0063,$$

which minimize the criterion Q

$$Q = \frac{1}{M+1} \sum_{i=0}^{M} (y_i^* - y_i),$$

where y_i is the output of the model at the point of the i-th measurement. In this case, the minimal value of Q is

$$Q_1 = 1.5 \cdot 10^{-3}.$$

The second model was obtained under the assumption that the system can be described by the three-term fractional differential equation

$$b_2 y^{(\alpha)}(t) + b_1 y^{(\beta)}(t) + b_0 y(t) = u(t). \tag{9.36}$$

In this case, the following values for orders α and β and for the coefficients b_0, b_1 and b_2 were obtained:

$$\alpha = 2.5708, \quad \beta = 0.8372,$$

$$b_2 = 0.7943, \quad b_1 = 5.2385, \quad b_0 = 1.5560$$

giving for the criterion the value of

$$Q_2 = 1.3 \cdot 10^{-4}.$$

Third, the considered object was also modelled by a two-term fractional differential equation. In this case we must put $a_2 = 0$ in equation (9.36), so the term with the α-th derivative disappears. The remaining parameters of the two-term fractional model

$$b_1 y^{(\beta)}(t) + b_0 y(t) = u(t) \tag{9.37}$$

take on the values

$$\beta = 1.0315, \quad b_1 = 6.2868, \quad b_0 = 1.8508,$$

and the corresponding value of the criterion is

$$Q_3 = 4.3 \cdot 10^{-4}.$$

The result of fitting the unit-step response of the furnace using (9.37) is presented in Fig. 9.6.

The comparison of these three models leads to interesting observations.

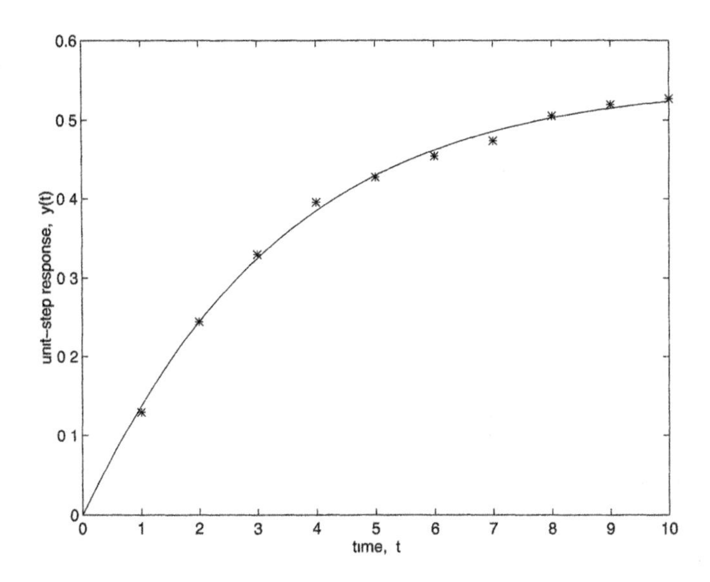

Figure 9.6: *Fractional-order model of a re-heating furnace.*

Note that the integer-order model (9.35) is just a particular case of the more general fractional-order model (9.36). If the integer-order model is the best model among the models described by three-term equations, then the identification of the parameters of the fractional-order model (9.36) should give $\alpha = 2$, $\beta = 1$, and $b_k = a_k \quad (k = 0, 1, 2)$. However, this did not happen; this indicates that the integer-order model (9.35) is less adequate than the fractional-order model (9.36).

The real explanation of this difference between the integer-order approach and the fractional-order model is *not* the larger number of parameters (we have five parameters in (9.36) against three parameters in (9.35)), but the different "nature" of the models, which allows us to use the same number of parameters for achieving higher adequacy of the resulting model. This higher level of adequacy is demonstrated by the third model (9.37), in which we also have three parameters, like in (9.35), but get a lower value of the criterion Q.

9.4 Conclusion

We have shown that the proposed concept of the fractional-order $PI^{\lambda}D^{\mu}$-controller is a good way for the adequate control of fractional-order dynamical systems.

Of course, for the physical realization of the $PI^{\lambda}D^{\mu}$-controller specific circuits are necessary: they must perform Caputo's fractional-order differentiation and integration. It should be mentioned that such fractional integrators and differentiators have already been described in [179] and [180].

All the results of computations were also verified by the numerical solution of the initial-value problems for the corresponding fractional-order differential equations by the numerical method described in Chapter 8.

The most important limitation of the method presented in this chapter is that only linear systems with constant coefficients can be treated. On the other hand, it allows consideration of a new class of dynamical systems (systems of arbitrary real order) and new types of controllers.

Finally, the example of identification of parameters of fractional-order models of real objects, considered in Section 9.3 shows that for fruitful applications of fractional-order models of dynamical systems and fractional-order controllers further development of effective methods for identification of the *structure* of a fractional-order mathematical model of a real object, as well as methods for identification of the *model parameters*, is necessary.

Chapter 10

Survey of Applications of the Fractional Calculus

In this chapter a survey of applications of the fractional calculus in various fields of science is given. It covers the widely known classical fields, such as Abel's integral equation and viscoelasticity, and also less well-known fields, including analysis of feedback amplifiers, capacitor theory, fractances, generalized voltage dividers, fractional-order Chua–Hartley systems, electrode–electrolyte interface models, fractional multipoles, electric conductance of biological systems, fractional-order models of neurons, fitting of experimental data, and others.

This survey cannot be considered as a complete one, but as a collection of sample applications, which can be used for further developments using analogies in the mathematical description of real problems arising in different fields of science. Moreover, in some cases we also use particular applications for illustrating the methods described in the previous chapters.

10.1 Abel's Integral Equation

The Abel integral equation is well studied, and there exist many sources devoted to its applications in different fields. Among many existing books on various aspects of Abel integral equations the monographs [90] and [84] must be mentioned, in which special attention is paid to applications.

Because of this, in this section we pay attention mainly to those types of integral equations which appear in applications and which can be reduced to Abel's integral equation.

10.1.1 General Remarks

The simplest and most well-known example of a fractional-order system goes back to H. N. Abel [1]. The integral equation

$$\frac{1}{\Gamma(\alpha)} \int_0^t \frac{\varphi(\tau)d\tau}{(t-\tau)^{1-\alpha}} = f(t), \quad (t > 0), \tag{10.1}$$

where $0 < \alpha < 1$, is called Abel's integral equation. Its solution is given by the well-known formula

$$\varphi(t) = \frac{1}{\Gamma(1-\alpha)} \frac{d}{dt} \int_0^t \frac{f(\tau)d\tau}{(t-\tau)^{\alpha}}, \quad (t > 0),$$

which we prefer to write in the reversed form as

$$\frac{1}{\Gamma(1-\alpha)} \frac{d}{dt} \int_0^t \frac{f(\tau)d\tau}{(t-\tau)^{\alpha}} = \varphi(t) \quad (t > 0). \tag{10.2}$$

In terms of fractional-order derivatives, equations (10.1) and (10.2) take on the form

$${}_0D_t^{-\alpha}\varphi(t) = f(t), \quad (t > 0) \tag{10.3}$$

and

$${}_0D_t^{\alpha} f(t) = \varphi(t), \quad (t > 0) \tag{10.4}$$

respectively.

Transfer functions corresponding to equations (10.3) and (10.4) are

$$g_1(s) = s^{\alpha} \tag{10.5}$$

and

$$g_2(s) = s^{-\alpha}. \tag{10.6}$$

Therefore, in the case of equation (10.1) or, which is the same, equation (10.3), we deal with the system of order $-\alpha$. If the system's behaviour is described by equation (10.2) or by its equivalent (10.4), then we have a system of order α.

10.1.2 Some Equations Reducible to Abel's Equation

Solution of many applied problems lead to integral equations, which at the first sight have nothing in common with Abel's integral equation, and due to this impression additional efforts are undertaken for the development of analytical or numerical procedure for solving these equations. However, their transformation to the form of Abel's integral equation may often be convenient for rapidly obtaining the solution; this is the reason for giving some typical examples of equations which can be reduced to Abel's equation. Many types of such equations along with solution formulas can be found in [245].

Equations with non-moving integration limits

a) Let us consider the equation

$$\int_0^\infty \frac{\varphi(\sqrt{s^2+y^2})}{\sqrt{s^2+y^2}} ds = \frac{f(y)}{2y}. \tag{10.7}$$

Denoting

$$\frac{\varphi(r)}{r} = F(r^2),$$

we can rewrite the equation (10.7) as

$$\int_0^\infty F(s^2+y^2)ds = \frac{f(y)}{2y}.$$

Substitution of variables $x = y^2$, $\xi = s^2$ gives

$$\int_0^\infty \xi^{-1/2} F(x+\xi)d\xi = \frac{f(\sqrt{x})}{\sqrt{x}}. \tag{10.8}$$

Then the further substitution $\tau = 1/(x+\xi)$ leads to

$$\int_0^{1/x} \left(\frac{1}{x} - \tau\right) t^{-3/2} F\left(\frac{1}{\tau}\right) d\tau = f(\sqrt{x}),$$

and denoting

$$t = \frac{1}{x}, \qquad \psi(\tau) = t^{-3/2} F\left(\frac{1}{\tau}\right),$$

we arrive at an equation of the type (10.1), with $\alpha = 1/2$:

$$\int_0^t (t-\tau)^{-1/2}\psi(\tau)d\tau = f\left(\frac{1}{\sqrt{t}}\right). \tag{10.9}$$

The solution of equation (10.9) can be found with the help of formula (10.4):

$$\psi(t) = \frac{1}{\Gamma(\frac{1}{2})}\, {}_0D_t^{1/2} f\left(\frac{1}{\sqrt{t}}\right), \tag{10.10}$$

and performing backward substitution we obtain the solution of the equation (10.7) in terms of fractional derivatives:

$$\varphi\left(\frac{1}{\sqrt{t}}\right) = \frac{t}{\sqrt{\pi}}\, {}_0D_t^{1/2} f\left(\frac{1}{\sqrt{t}}\right). \tag{10.11}$$

b) With the help of the same chain of substitutions the equation

$$\int_0^\infty \frac{\varphi(\sqrt{s^2+y^2})}{\sqrt{s^2+y^2}} s^2 ds = \frac{f(y)}{2y} \tag{10.12}$$

can be reduced to Abel's integral equation of the form

$$\int_0^t (t-\tau)^{1/2}\psi(\tau)d\tau = tf\left(\frac{1}{\sqrt{t}}\right). \tag{10.13}$$

In this case $\alpha = 3/2$; using the formula (10.4) we obtain

$$\psi(t) = \frac{1}{\Gamma\left(\frac{3}{2}\right)}\, {}_0D_t^{3/2}\Big(tf(\frac{1}{\sqrt{t}})\Big) = \frac{2}{\sqrt{\pi}}\, {}_0D_t^{3/2}\Big(tf(\frac{1}{\sqrt{t}})\Big), \tag{10.14}$$

and the return to $\varphi(r)$ can be done using the relationship

$$\varphi(r) = r^6\psi(r^2).$$

c) The equation

$$\int_0^\infty \tau^{1/2}\varphi(t+\tau)d\tau = f(t) \tag{10.15}$$

is similar to equation (10.8) and can be solved in the same way.

d) The equation of the type

$$\int_0^{\infty} e^{-\tau}\tau^{1/2}\varphi(t+\tau)d\tau = f(t) \tag{10.16}$$

can be reduced to the equation of the type (10.15), in which $f(t)$ must be replaced with $e^{-t}f(t)$, with the help of the obvious substitution

$$y(t) = e^{-t}\varphi(t).$$

e) Poisson's integral equation

$$\int_0^{\pi/2} \psi(r\cos\omega)\sin^{2\nu+1}\omega\, d\omega = f(r) \tag{10.17}$$

can also be reduced to Abel's equation.

After the substitution $x = r\cos\omega$ we have

$$\int_0^r \left(1 - \frac{x^2}{r^2}\right)\psi(x)dx = rf(r),$$

and denoting

$$y = \frac{1}{r^2}, \qquad \rho(y) = \frac{1}{\sqrt{y}}\, f\left(\frac{1}{\sqrt{y}}\right)$$

we obtain the equation

$$\int_0^{1/\sqrt{y}} (1 - yx^2)^{\nu}\psi(t)dt = \rho(y),$$

which can be written as

$$\int_0^{1/\sqrt{y}} \left(\frac{1}{z} - x^2\right)^{\nu}\psi(x)dx = y^{-\nu}\rho(y). \tag{10.18}$$

Performing in (10.18) the subsequent substitutions

$$\tau = x^2, \qquad t = \frac{1}{y},$$

and denoting

$$\varphi(\tau) = \frac{\psi(\sqrt{\tau})}{\sqrt{\tau}}, \qquad g(t) = 2t^{\nu}\rho(\frac{1}{t})$$

we arrive at Abel's integral equation

$$\int_0^t (t-\tau)^{\nu}\varphi(\tau)d\tau = g(t) \qquad (10.19)$$

with the solution

$$\varphi(t) = \frac{1}{\Gamma(\nu+1)}\, {}_0D_t^{\nu} g(t).$$

Equations with moving integration limits

a) In numerous applied problems an integral equation of the following type appears:

$$\int_0^y \frac{1}{(y^2-x^2)^{\beta}}\, \psi(x)dx = f(y). \qquad (10.20)$$

Performing the substitutions

$$\tau = x^2, \qquad t = y^2,$$

and denoting

$$\varphi(\tau) = \frac{\psi(\sqrt{\tau})}{\sqrt{\tau}},$$

we arrive at Abel's integral equation:

$$\int_0^t \frac{1}{(t-\tau)^{\beta}}\varphi(\tau)d\tau = 2f(\sqrt{t}), \qquad (10.21)$$

with the solution

$$\varphi(t) = \frac{2}{\Gamma(1-\beta)}\, {}_0D_t^{1-\beta} f(\sqrt{t}),$$

and therefore the solution of the equation (10.20) is given by the formula

$$\psi(\sqrt{t}) = \frac{2\sqrt{t}}{\Gamma(1-\beta)}\, {}_0D_t^{1-\beta} f(\sqrt{t}). \qquad (10.22)$$

b) In other cases there appears an equation similar to (10.20), but with moving lower integration limit:

$$\int_x^b \frac{\psi(r) r dr}{(r^2 - x^2)^\beta} = f(x). \tag{10.23}$$

Performing the substitutions

$$\tau = b^2 - r^2, \qquad t = b^2 - x^2$$

and denoting

$$\varphi(\tau) = \psi(\sqrt{b^2 - \tau}),$$

we arrive at Abel's integral equation

$$\int_0^t \frac{\varphi(\tau) d\tau}{(t - \tau)^\beta} = \frac{1}{2} f(\sqrt{b^2 - t}) \tag{10.24}$$

with the solution

$$\varphi(t) = \frac{1}{2\Gamma(1-\beta)} \, {}_0D_t^{1-\beta} f(\sqrt{b^2 - t}),$$

which means that the solution of equation (10.23) is given by the following formula:

$$\psi(\sqrt{b^2 - t}) = \frac{1}{2\Gamma(1-\beta)} \, {}_0D_t^{1-\beta} f(\sqrt{b^2 - t}). \tag{10.25}$$

c) The equation

$$\int_\theta^{\pi/2} \frac{\rho(\varphi) d\varphi}{(\cos\theta - \cos\varphi)^\beta} = F(\theta), \tag{10.26}$$

which also often appears in applications, can be reduced to Abel's equation too.

Performing the substitutions

$$\tau = \cos\varphi, \qquad t = \cos\theta,$$

and then denoting

$$y(\tau) = \frac{\rho(\arccos\tau)}{\sqrt{1 - \tau^2}}, \qquad f(t) = F(\arccos t),$$

we obtain Abel's equation

$$\int_0^t \frac{1}{(t-\tau)^\beta} y(\tau) d\tau = f(t), \tag{10.27}$$

which has the solution

$$y(t) = \frac{1}{\Gamma(1-\beta)} \, {}_0D_t^{1-\beta} f(t).$$

Therefore, the solution of equation (10.26) is given by the formula

$$\rho(\arccos t) = \frac{\sqrt{1-t^2}}{\Gamma(1-\beta)} \, {}_0D_t^{1-\beta} F(\arccos t). \tag{10.28}$$

10.2 Viscoelasticity

Viscoelasticity seems to be the field of the most extensive applications of fractional differential and integral operators, and perhaps the only one in which there have been published broad surveys (see, e.g., [138, 136, 228]). The considerations discussed below show that the use of fractional derivatives for the mathematical modelling of viscoelastic materials is quite natural. It should be mentioned that the main reasons for the theoretical development are mainly the wide use of polymers in various fields of engineering.

We will consider a range of approaches to the linear theory of viscoelasticity from integer-order models to fractional calculus models.

10.2.1 Integer-order Models

Let us recall the well-known relationships between stress and strain for solids (Hooke's law)

$$\sigma(t) = E\epsilon(t) \tag{10.29}$$

and for Newtonian fluids

$$\sigma(t) = \eta \frac{d\epsilon(t)}{dt}, \tag{10.30}$$

where E and η are constants.

Relationships (10.29) and (10.30) are not universal laws, they are only mathematical models for an ideal solid material and for an ideal fluid, neither of which exist in the real world. In fact, real materials combine properties of those two limit cases and lie somewhere between ideal solids and ideal fluids, if materials are sorted with respect to their firmness.

The development of integer-order models of linear viscoelasticity is depicted in Fig. 10.1. The Hooke elastic element is represented as a spring, while the Newton viscous element is shown as a dashpot. It is common practice in rheology to manipulate with such representations instead of corresponding equations.

At the first stage, Hooke's (elastic) and Newton's (viscous) elements were combined with the aim of combining the properties of both. There are two possible combinations: parallel and serial. The serial connection of the two basic elements gives Maxwell's model of viscoelasticity; connecting them in parallel gives Voigt's model. However, both these models have obvious disadvantages.

In the case of the Maxwell model, which is described by the relationship

$$\frac{d\epsilon}{dt} = \frac{1}{E}\frac{d\sigma}{dt} + \frac{\sigma}{\eta}, \tag{10.31}$$

we have

$$\sigma = const \quad \Rightarrow \quad \frac{d\epsilon}{dt} = const, \tag{10.32}$$

which means that if stress is constant, then strain grows infinitely; this does not correspond to experimental observations.

In the case of the Voigt model σ and ϵ are related by

$$\sigma = E\epsilon + \eta\frac{d\epsilon}{dt}, \tag{10.33}$$

from which it follows that

$$\epsilon = const \quad \Rightarrow \quad \sigma = const, \tag{10.34}$$

and we see that the Voigt model of viscoelasticity does not reflect the experimentally observed stress relaxation.

At the second stage (or level), the above disadvantages of the Maxwell and the Voigt model were subjects for enhancement.

The serial connection of the Voigt viscoelastic element and the Hooke elastic element gives Kelvin's model of viscoelasticity:

$$\frac{d\sigma}{dt} + \alpha\sigma = E_1\left(\frac{d\epsilon}{dt} + \beta\epsilon\right); \tag{10.35}$$

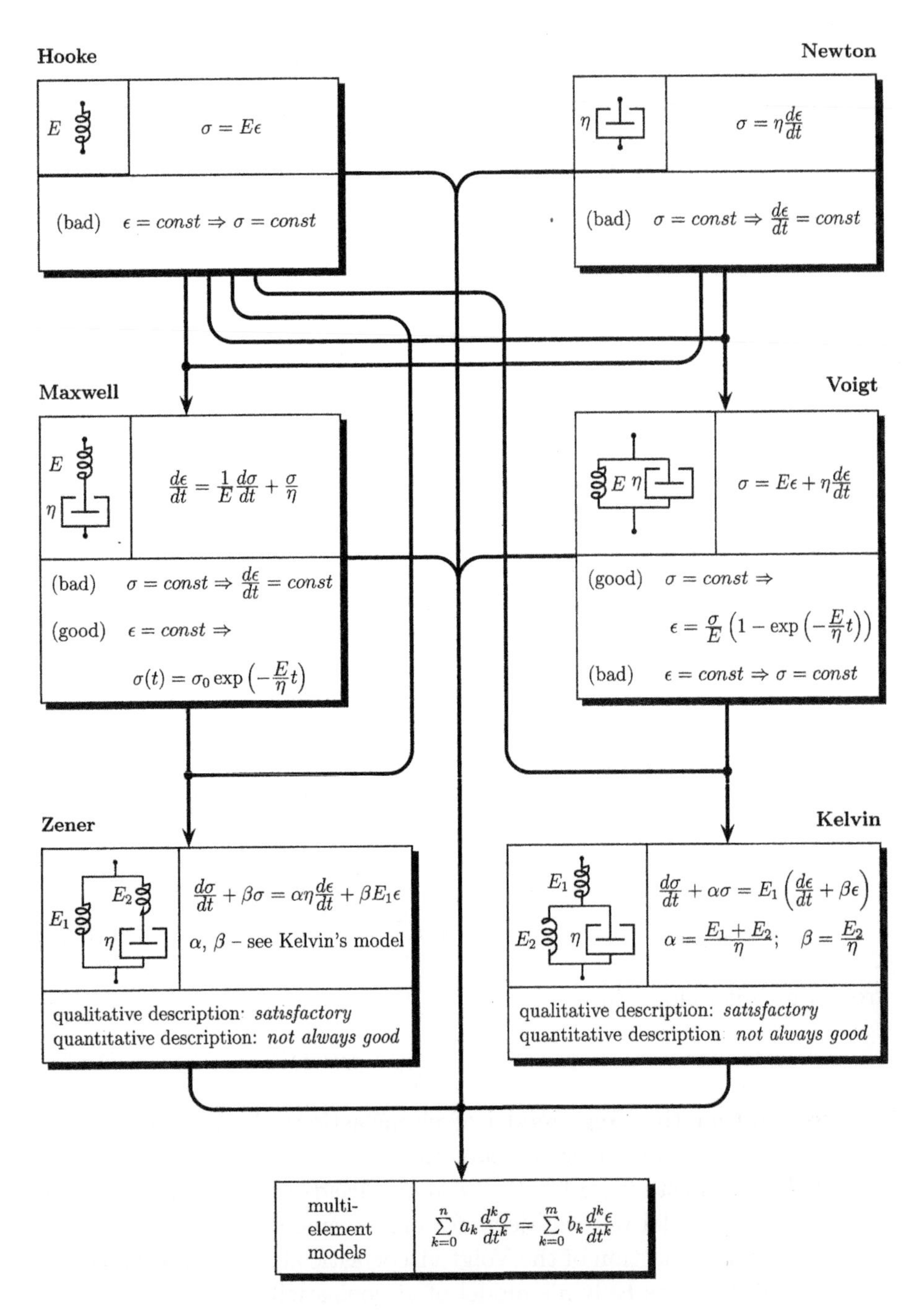

Figure 10.1: *Development of linear models of viscoelasticity.*

connecting the Maxwell viscoelastic element and the Hooke element gives Zener's model of viscoelasticity:

$$\frac{d\sigma}{dt} + \beta\sigma = \alpha\eta\frac{d\epsilon}{dt} + \beta E_1\epsilon, \tag{10.36}$$

where α and β for both models are given by

$$\alpha = \frac{E_1 + E_2}{\eta}; \quad \beta = \frac{E_2}{\eta}.$$

Both Kelvin's and Zener's models give good qualitative descriptions, but are not considered as satisfactory from the quantitative point of view [141, 246]. Because of this, there were also developed further, more complex reological models of viscoelastic materials, consisting of several Kelvin or Maxwell elements combined with Hooke's elastic element. These models result in more complex relationships relating stress and strain, in which linear combinations of derivatives of stress and strain appear (see, e.g., [246]). In the most general case in this way we arrive at the model of the form

$$\sum_{k=0}^{n} a_k \frac{d^k\sigma}{dt^k} = \sum_{k=0}^{m} b_k \frac{d^k\epsilon}{dt^k}, \tag{10.37}$$

and in each particular case the best adequacy was achieved for $n = m$ (this property starts from the Kelvin and the Zener models, for which $n = m = 1$).

Using (10.35), (10.36), or (10.37) as the basic laws of deformation of viscoelastic materials leads to complicated differential equations of high order, which causes difficulties in formulating and solving many applied problems, in spite of the fact that the resulting differential equations are linear (due to linearity of the basic laws of deformation).

However, there is a nice solution which preserves the linearity of models and at the same time provides a higher level of adequacy.

10.2.2 Fractional-order Models

Noting that stress is proportional to the zeroth derivative of strain for solids and to the first derivative of strain for fluids, it is natural to suppose, as has been done by G. W. Scott Blair [236], that for "intermediate" materials stress may be proportional to the stress derivative of "intermediate" (non-integer) order:

$$\sigma(t) = E\ {}_0D_t^{\alpha}\epsilon(t), \quad (0 < \alpha < 1), \tag{10.38}$$

where E and α are material-dependent constants. The shortest outline of how this idea appeared can be found in the appendix to G. W. Scott Blair's paper [239].

Approximately at the same time, A. N. Gerasimov [77] suggested a similar generalization of the basic law of deformation, which can be written in the form using the Caputo fractional derivative

$$\sigma(t) = \kappa \ {}_{-\infty}^{\ C}D_t^{\alpha} \epsilon(t),$$

or, since for the lower terminal at $-\infty$ the Caputo derivative coincides with the Riemann–Liouville fractional derivative,

$$\sigma(t) = \kappa \ {}_{-\infty}D_t^{\alpha} \epsilon(t), \qquad (0 < \alpha < 1), \tag{10.39}$$

where κ is a material constant (generalized viscosity). A. N. Gerasimov also considered two problems describing the movement of a viscous fluid between two moving surfaces. These problems led to the equations (in our notation):

$$\rho \frac{\partial^2 y}{\partial t^2} = \kappa \, D^{\alpha} \Big(\frac{\partial^2 y}{\partial x^2} \Big), \tag{10.40}$$

$$\rho x^3 \frac{\partial^2 y}{\partial t^2} = \kappa \frac{\partial}{\partial x} \Big(x^3 \frac{\partial}{\partial x} \left(D^{\alpha} y \right) \Big), \tag{10.41}$$

$$y = y(x, t); \qquad D^{\alpha} \equiv \ {}_{-\infty}D_t^{\alpha}.$$

It must be mentioned that A. N. Gerasimov was the first to deduce and solve fractional-order partial differential equations for particular applied problems.

Yet another formulation of a generalization of the basic laws of deformation was suggested by G. L. Slonimsky [241]:

$$\epsilon(t) = \frac{1}{\kappa} \ {}_0D_t^{-\alpha}, \qquad (\kappa = const; \quad 0 < \alpha < 1). \tag{10.42}$$

Under the condition $\epsilon(0) = 0$ Scott Blair's and Slonimsky's laws, (10.38) and (10.42), are equivalent. Also, the solutions given by Gerasimov for the equations (10.41) and (10.41) are based on the assumption that the unknown function and all given functions are equal to zero for $t < 0$; under this assumption, Gerasimov's formula (10.39) becomes

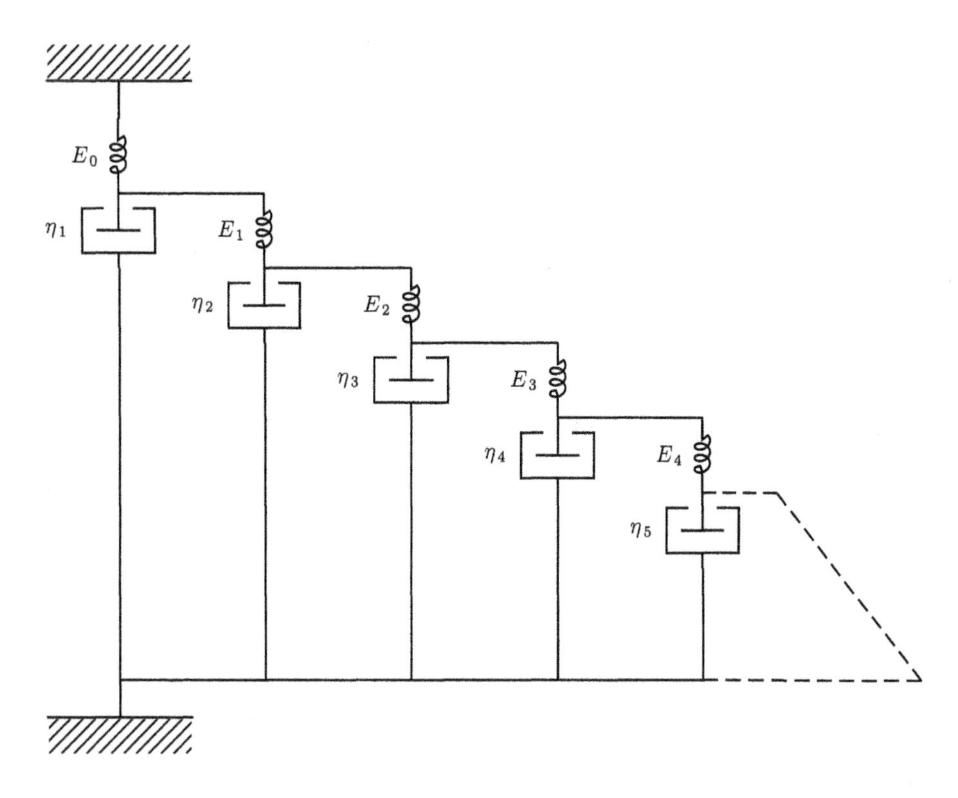

Figure 10.2: *Schiessel and Blumen's fractance-type model.*

equivalent to Scott Blair's and Slonimsky's. Therefore, instead of considering these approaches separately, we may refer to the Scott Blair law (10.38).

Since, as mentioned above, complex multi-element models, consisting of Hooke and Newton elements, were used for modelling the viscoelastic behaviour of real materials, it is natural to try to obtain multi-element models of this type also for the Scott-Blair viscoelastic element.

Such multi-element models, which consist of an infinite number of classical springs (Hooke) and dashpots (Newton) ordered hierachically in a self-similar structures (see Figs 10.2 and 10.3), were suggested by H. Schiessel and A. Blumen (see [233]) and by N. Heymans and J.-C. Bauwens [106]. In both cases the suggested models are structures of the type called a *fractance* (see Section 10.5). Adjusting the parameters of the structural parts of these models, it is possible to achieve for the whole model an equation of the form (10.38).

Now, having three basic elements (Hooke, Newton, Scott Blair) for

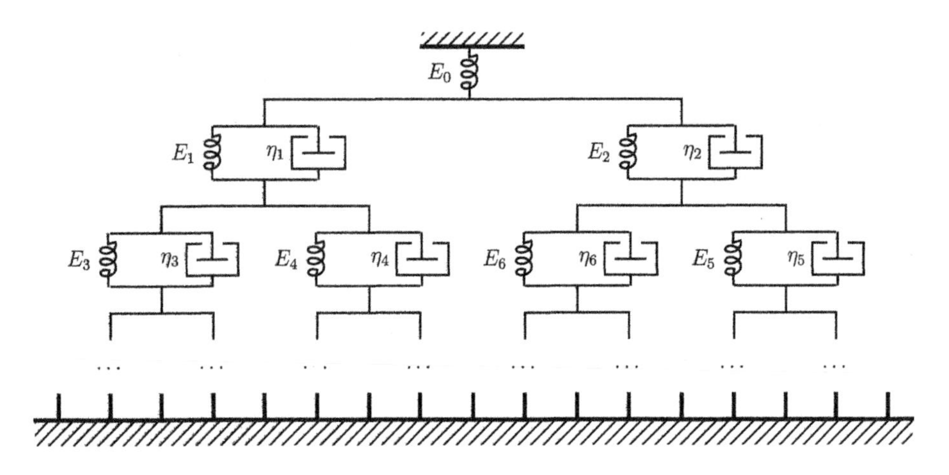

Figure 10.3: *Heymans and Bauwens fractance-type model.*

constructing rheological models, there are two options: to combine more than one of them, or to use only the Scott Blair element, since it contains two other elements as particular cases.

N. Heymans and J.-C. Bauwens suggested a generalization of the classical Maxwell model (see Fig. 10.1) by replacing both the elastic element and the viscous element by the Scott Blair element. On the other hand, they generalized the Zener model (see Fig. 10.1) by replacing only the viscous element by the Scott Blair element.

In the paper by H. Schiessel, R. Metzler, A. Blumen and T. F. Nonnenmacher [233] all four basic classical models, (the Maxwell, Voigt, Zener, and Kelvin models) were generalized by replacing *all* Hooke's and Newton's classical elements by the Scott Blair element, and then studied in detail. It must be mentioned that various types of classical models of viscoelasticity were generalized and studied by M. Caputo and F. Mainardi [29] much earlier, using the Caputo fractional derivatives and starting with the deformation law equation, which is now called the four-parameter model.

The Hooke law (10.29), which is a one-parameter model, and the Scott Blair law (10.38), which is a two-parameter model (the parameters are E and α) can also be further generalized by adding further terms on both sides, containing arbitrary-order derivatives of stress and strain. This leads to the three-parameter generalized Voigt model:

$$\sigma(t) = b_0\, \epsilon(t) + b_1\, D^{\alpha} \epsilon(t), \tag{10.43}$$

to the three-parameter generalized Maxwell model:

$$\sigma(t) + a_1 D^\alpha \sigma(t) = b_0\, \epsilon(t), \tag{10.44}$$

to the five-parameter generalized Zener model:

$$\sigma(t) + a_1 D^\alpha \sigma(t) = b_0\, \epsilon(t) + b_1\, D^\beta \epsilon(t), \tag{10.45}$$

and so on. Further, more general models have been suggested by H. Schiessel, R. Metzler, A. Blumen and T. F. Nonnenmacher [233].

However, the generalized Zener model (10.45) can be simplified, since it was observed experimentally that the modelling of most materials results in $\alpha = \beta$ [223, 18]. In addition to the experimental observations, R. L. Bagley and P. J. Torvik proved theoretically that the five-parameter model (10.45) satisfies the thermodynamic constraints if $\alpha = \beta$. This conclusion gives the four-parameter model,

$$\sigma(t) + a_1 D^\alpha \sigma(t) = b_0\, \epsilon(t) + b_1\, D^\alpha \epsilon(t), \tag{10.46}$$

which provides a satisfactory description of most real materials. It is interesting to note that among the integer-order models of viscoelasticity (see Fig. 10.1) only the Zener and the Kelvin models are those in which the highest order of derivative of stress is equal to the highest order of derivative of strain.

We see that the four-parameter model (10.46) could also be formally obtained from the integer-order Zener and Kelvin models by replacing the first-order derivatives by the fractional derivatives of the same order. Similarly, the most general linear model of viscoelasticity can also be formally obtained from (10.37) by replacing integer-order derivatives with fractional derivatives:

$$\sum_{k=0}^{n} a_k D^{\alpha_k} \sigma(t) = \sum_{k=0}^{m} b_k D^{\beta_k} \epsilon(t), \tag{10.47}$$

and it is possible that the best results may be achieved if $n = m$ and $\alpha_k = \beta_k$, $(k = 0, 1, 2, \ldots)$.

10.2.3 Approaches Related to the Fractional Calculus

Besides the pure fractional calculus approach to linear viscoelasticity, two closely related approaches must be mentioned.

The above considerations were devoted to the "transition" of linear viscoelasticity from integer-order models to fractional from the viewpoint of a mathematical description of the laws of deformation in terms of derivatives. However, fractional-order models of viscoelasticity may also be derived by starting from the so-called power-law stress relaxation in real materials, first clearly formulated by P. G. Nutting [172] in the form

$$\epsilon = at^{\alpha}\sigma^{\beta}, \tag{10.48}$$

where a, α, and β are the model parameters.

Taking $\beta = 1$ and denoting $c_0 = 1/a$, we see that for a constant strain ($\epsilon = const$) the stress relaxation is described by the power-law relationship

$$\sigma(t) = c_0 \epsilon t^{-\alpha}. \tag{10.49}$$

On the other hand, for a constant stress ($\sigma = const$) the strain is given by

$$\epsilon(t) = \frac{\sigma}{c_0} t^{\alpha}. \tag{10.50}$$

As shown by T. F. Nonnenmacher [169], it follows from equation (10.49), or respectively equation (10.50), that the functions $\sigma(t)$ and $\epsilon(t)$ satisfy the fractional differential equations:

$$D^{\alpha}\sigma(t) = \frac{\Gamma(1-\alpha)t^{-\alpha}}{\Gamma(1-2\alpha)}\sigma(t), \tag{10.51}$$

$$D^{\alpha}\epsilon(t) = \Gamma(1+\alpha)t^{-\alpha}\epsilon(t). \tag{10.52}$$

This indicates that there is a close relation between the power law representation of viscoelastic behaviour and fractional derivatives. The similarities and the differences between the power law approach and the fractional calculus approach in viscoelasticity are discussed by R. L. Bagley [12].

Besides the fractional calculus model and the power law approach, there is also another approach, involving integrals of convolution type. This approach, which in fact is a particular implementation of V. Volterra's idea [252], was developed and extensively presented mainly by Yu. N. Rabotnov ([217], see also the textbooks [218, 219]). It is essentually based on the use of the Rabotnov function $\ni_{\alpha}(\beta, t)$, which is a particular case of the Mittag-Leffler function (see equation (1.68)). This means that, in fact, Rabotnov's theory is also related to the fractional calculus approach and implicitly involves fractional integrals and derivatives.

All the above approaches to generalizations of the laws of deformation have been found useful for solving practical problems of viscoelasticity, if the results are properly interpreted. Many authors made significant contributions to the development of fractional-order models of viscoelasticity and their applications (in alphabetical order; this is not a complete list): H. Beyer and S. Kempfle [19] M. Caputo [22, 23, 24, 25, 26], M. Caputo and F. Mainardi [29, 30], M. Enelund, Å. Fenander, and P. Olsson [58], M. Enelund and B. L. Josefson [59], Å. Fenander [66], Ch. Friedrich [70, 71, 72, 73], Ch. Friedrich and H. Braun [74], L. Gaul, S. Kempfle, and P.Klein [75], A. N. Gerasimov [77], W. G. Glöckle and T. F. Nonnenmacher [80, 81, 170] B. Gross [97], N.Heymans and J.-C. Bauwens [106], R. C. Koeller [118], H. H. Lee and C.-S. Tsai [122], N. Makris and M. C. Constantinou [129, 130], F. Mainardi [136, 138, 139, 132] R. Metzler, W.Schick, H.-G. Kilian, and T. F. Nonnenmacher [151], T. F. Nonnenmacher [169], P. G. Nutting [172], T. Pritz [214], Yu. N. Rabotnov [217, 218, 219], L. Rogers [223], Yu. A. Rossikhin and M. V. Shitikova [228, 229], H. Schiessel, R. Metzler, A. Blumen and T. F. Nonnenmacher [233], G. W. Scott Blair [236, 237, 239], G. L. Slonimsky [241], A. I. Tseytlin [250], and others.

However, today's intensive development of this field and its advanced state comparing to other fields is undoubtedly due to a series of works by R. L. Bagley and co-authors [12, 13, 14, 15, 16, 17, 18], in which the advantages of the fractional calculus approach were presented with ultimate clarity using both theoretical and experimental arguments, and also due to the need for a better description of the properties of materials used in industry.

10.3 Bode's Analysis of Feedback Amplifiers

In his study on feedback amplifier design [20], first published in 1945, H. W. Bode considered a system characterized by the frequency response of the form [20, §18.2, eq. (18–5)]

$$Z(\omega) = \frac{A}{B}(i\omega)^{-n} \tag{10.53}$$

which corresponds to the transfer function

$$g_z(s) = \frac{A}{B}s^{-n} \tag{10.54}$$

where A and B are known constant, and n is the number of stages in a feedback amplifier. In his analysis Bode allowed n to be an arbitrary real number and arrived at the conclusion that the optimal number of stages in a feedback amplifier is non-integer [20, §18.9]. Therefore, he in fact performed a frequency-domain analysis of the performance of a system of non-integer order $-n$ with fractional-order transfer function (10.54).

However, after that he described how to choose a suitable integer number of stages, which is not necessarily closest to the optimal non-integer value of n.

10.4 Fractional Capacitor Theory

Fractional-order capacitor models were most probably first formally suggested and investigated by G. E. Carlson and C. A. Halijak [32, 33, 34, 31].

The fractional capacitor theory, presented recently by S. Westerlund and L. Ekstam [255], who obviously did not know about Carlson and Halijak's work, is based on the revision of a physical law. It leads to a family of fractional-order systems.

S. Westerlund starts with M. J. Curie's well-forgotten empirical law dating from 1889

$$i(t) = \frac{U_0}{h_1 t^{\nu}}, \qquad (0 < \nu < 1, \quad t > 0), \tag{10.55}$$

where h_1 is a constant related to the capacitance of the capacitor and the kind of dielectric, and ν is a constant related to the losses of the capacitor. The transfer function of the model capacitor is found to be

$$H(s) = C_\phi s^{\nu}, \qquad (0 < \nu < 1), \tag{10.56}$$

where C_ϕ is a model constant close to what is usually called the capacitance.

The capacitor's impedance is described by the transfer function

$$Z(s) = \frac{1}{C_\phi\, s^{\nu}}, \qquad (0 < \nu < 1). \tag{10.57}$$

S. Westerlund has achieved a successful fitting of experimental data by the two-term model described by the transfer function of the form

$$Z(s) = \frac{1}{s^{\nu_1} C_1 + s^{\nu_2} C_2}, \tag{10.58}$$

where $\nu_1 = 0.82$, $\nu_2 = 0.9946$, C_1 and C_2 are certain constants of the same nature as the previously mentioned C_ϕ.

Westerlund's approach to the "fractional" revision of the traditional capacitor theory can be very useful in view of the huge number of empirical laws of the type (10.55) in different fields of science and engineering, the most popular of which is represented now by the theory of fractals.

Interesting conclusions may probably be deduced from the observation that M. J. Curie's law (10.55), describing the relaxation of current in a capacitor, has the same form as a particular case of P. G. Nutting's power law in viscoelasticity, given by equation (10.49).

10.5 Electrical Circuits

There are two types of electrical circuits which are related to the fractional calculus.

Circuits of the first types are supposed to consist of capacitors and resistors, which are described by conventional (integer-order) models; however, the circuit itself may have non-integer order properties, becoming a so-called *fractance.*

Circuits of the second type may consist of resistors, capacitors (both modelled in the classical sense), and fractances.

10.5.1 Tree Fractance

The first example of an electrical circuit related to fractional calculus is the *fractance* — an electrical circuit having properties which lie between resistance and capacitance. The term *fractance* was suggested by A. Le Méhauté [121] for denoting electrical elements with non-integer order impedance.

An example of a *tree fractance* element is given in Fig. 10.4, where an infinite self-similar circuit consisting of resistors of resistance R and capacitors of capacitance C is depicted.

As has been shown by M. Nakagawa and K. Sorimachi [161], the impedance of the fractance shown in Fig. 10.4 is

$$Z(i\omega) = (R/C)^{1/2}\,\omega^{-1/2}\exp(-\pi i/4), \tag{10.59}$$

which corresponds to the fractional-order transfer function

$$Z(s) = (R/C)^{1/2}\,s^{-1/2}. \tag{10.60}$$

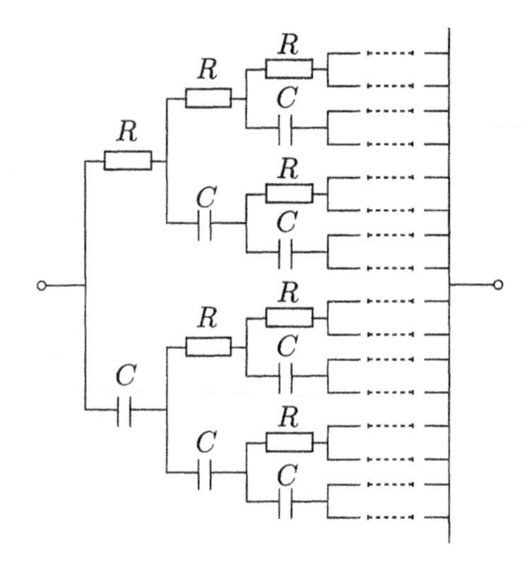

Figure 10.4: *Tree fractance.*

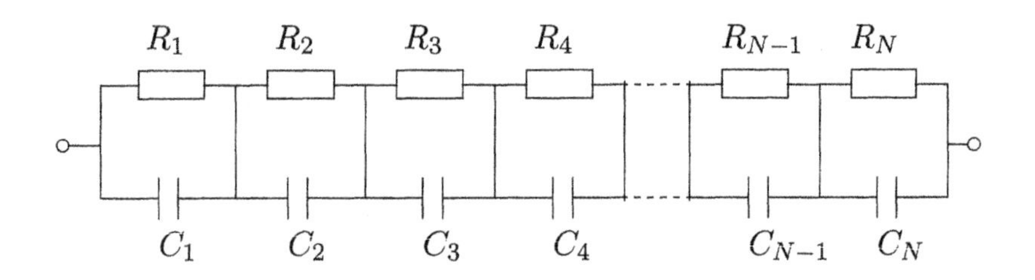

Figure 10.5: *Chain fractance.*

In practice, of course, the infinite circuit must be truncated, so the real fractance always consists of a finite number of stages (levels). However, as in the case of the domino ladder, the number of stages may be determined to achieve the required approximation.

Fractances can be used for analogue fractional differentiation and integration. They can also be used in electroengineering as a circuit element of a new type.

10.5.2 Chain Fractance

The second example of a fractance is a *chain fractance* (Fig. 10.5), suggested by G. E. Carlson and C. A. Halijak [34] and by K. B. Oldham and C. G. Zoski [179, 180].

The chain fractance consists of N resistor-capacitor pairs connected in a chain.

If $R_k = R$, $C_k = C$ $(k = \overline{1, N-1})$ and $R_N = R/2$, $C_N = 0$, then, as has been demonstrated in [179], the transfer function of the chain fractance of Fig. 10.5 is approximately equal to

$$G(s) \approx \sqrt{\frac{R}{Cs}} \tag{10.61}$$

and — over a certain time range, namely for $6RC \leq t < \frac{1}{6}N^2RC$ — this chain fractance serves as a fractional integrator of order $\frac{1}{2}$. The required accuracy and the time interval length can be achieved by an appropriate choice of R, C, and N.

This idea has been further developed in [180], where the following recipe for the design of a fractional integrator of order $1 - \nu$ $(0 < \nu < 1)$ is given.

First, one has to choose the order of fractional integration α and to compute $\nu = 1 - \alpha$. Then the lower and the upper limits t_m and t_M must be selected for the time interval in which the fractional integration is performed.

After choosing ν, the values of the capacitive and resistive geometric values G and g are calculated from the equations

$$\log G = \frac{3}{2}\nu^{2/3}, \qquad \log g = \frac{1-\nu}{\nu} \log G. \tag{10.62}$$

The number N of stages in the chain fractance, which are necessary for providing 2% accuracy, must satisfy the inequality

$$N + 1 \geq \frac{5.5 + \log(t_M/t_m) - 3\nu^{2/3}}{\log Gg}. \tag{10.63}$$

The values of the largest resistor–capacitor pair must satisfy the condition

$$R_1 C_1 = \frac{111\, t_M \exp(-3\nu^{2/3})}{Gg}, \tag{10.64}$$

and the actual values for R_1 and C_1 may be determined mainly by component availability. The remaining components are calculated as

$$R_k = \frac{R_1}{g^{k-1}}, \quad C_k = \frac{C_1}{G^{k-1}}, \qquad (k = \overline{2, N}). \tag{10.65}$$

As for the tree fractance, the chain fractance can also be used for analogue fractional differentiation and integration, and as a circuit element of a new type.

10.5.3 Electrical Analogue Model of a Porous Dyke

A. Oustaloup [184] has deduced an electrical analogue of water flow through a porous fractal [142] dyke, and that analogue has exactly the structure of the chain fractance shown in Fig. 10.5.

Resistances R_k and capacitances C_k were computed by using the formulas

$$R_1 = R; \quad R_k = R/\alpha^{k-1}, \quad (k = \overline{2, N}) \tag{10.66}$$

$$C_1 = C; \quad C_k = C/\eta^{k-1}, \quad (k = \overline{2, N}) \tag{10.67}$$

where α and η $(\alpha, \eta > 1)$ are the parameters of the recursive dynamical model of the dyke.

The impedance of the circuit in the Laplace domain has been shown to be

$$Z(s) = \left(\frac{\omega_0}{s}\right)^{\lambda}, \quad \lambda = \frac{1}{1 + \frac{\log \eta}{\log \alpha}} \tag{10.68}$$

where ω_0 is the so-called transition frequency. The transfer function (10.68) is of non-integer order $-\lambda$.

Finally, with the use of this electrical analogue, the transfer function of the water–dyke system has been obtained in the form

$$F(s) = \frac{1}{1 + (\tau s)^{1+\lambda}}, \quad \tau = \left(\frac{M}{\omega_0}\right)^{\frac{1}{1+\lambda}} \tag{10.69}$$

where M is the mass of water. We see that the transfer function (10.69) describes a dynamical system of order $\nu = 1 + \lambda$, $1 < \nu < 2$.

10.5.4 Westerlund's Generalized Voltage Divider

Both the tree fractance and the chain fractance, which are discussed above, consist of elements (resistors and capacitors) described by classical integer-order models, but demonstrate properties, which lies between resistors and capacitors They themselves can be used as elements of circuits and such circuits will then also contain elements described by fractional-order mathematical models (differential equations or transfer functions).

Moreover, such circuits can be obtained if generalized models of resistors, capacitors, and induction coils are taken.

For example, S. Westerlund suggested the following generalization of a classical voltage divider shown in Fig. 10.6.

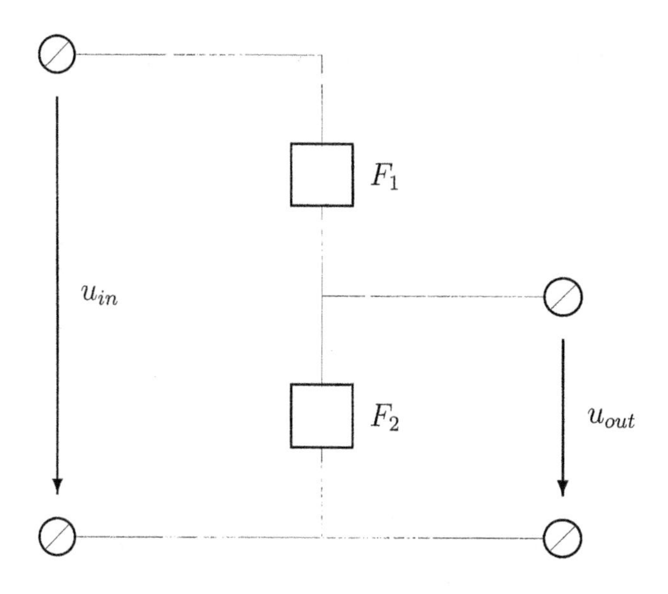

Figure 10.6: *Generalized voltage divider.*

The fractional-order impedances F_1 and F_2 may represent impedances not only of Westerlund's capacitors, classical resistors and induction coils, but also impedances of tree fractances and chain fractances.

The transfer function of a voltage divider circuit has the following general form:

$$H(s) = \frac{k}{s^{\alpha} + k}, \tag{10.70}$$

where α can have the range $-2 < \alpha < 2$ and k is a constant depending on the values of the components of the voltage divider. S. Westerlund mentions that negative α corresponds to a high-pass filter and positive α corresponds to a low-pass filter. He also lists some particular cases of the transfer function (10.70) for voltage dividers consisting of different combinations of resistors (R), capacitors (C), and induction coils (L). The impedances F_1, F_2, and the constant k, α in the Laplace domain, considered by S. Westerlund, are:

1) $F_1 = Ls, \quad F_2 = Cs^{-\nu}, \quad k = L^{-1}C^{-1}, \quad \alpha = 1 + \nu$;
2) $F_1 = R, \quad F_2 = Cs^{-\nu}, \quad k = R^{-1}C^{-1}, \quad \alpha = \nu$;
3) $F_1 = C_1 s^{-\nu_1}, \quad F_2 = C_2 s^{-\nu_2}, \quad k = C_1/C_2, \quad \alpha = \nu_1 - \nu_2$;
4) $F_1 = Cs^{-\nu}, \quad F_2 = R, \quad k = RC, \quad \alpha = -\nu$;
5) $F_1 = Cs^{-\nu}, \quad F_2 = Ls, \quad k = LC, \quad \alpha = -1 - \nu$.

If the input $u_{in}(t)$ is the unit-step signal with the Laplace transform $U_{in}(s)$, then the Laplace transform $U_{out}(s)$ of the output signal $u_{out}(t)$ is

$$U_{out}(s) = \frac{ks^{-1}}{s^{\alpha}+k}, \tag{10.71}$$

and the inversion using the Laplace transform formula (1.80) for the Mittag-Leffler function $E_{\alpha,\beta}(t)$ or the formula (9.15) for the function $\mathcal{E}_k(t,y;\alpha,\beta)$, defined by (9.8), gives

$$u_{out}(t) = k\, t^{\alpha} E_{\alpha,\,\alpha+1}(-kt^{\alpha}), \tag{10.72}$$

or

$$u_{out}(t) = k\, \mathcal{E}_0(t,-k;\alpha,\alpha+1). \tag{10.73}$$

On the other hand, some interesting properties of the solution for different values of α can be investigated by performing the evaluation of the inverse Lapace transform in the complex domain.

Let us consider $\alpha > 0$. Cutting the complex plane along the negative real half-axis and using the Cauchy theorem, we have

$$\begin{aligned} u_{out}(t) &= \frac{1}{2\pi i}\int\limits_{c-i\infty}^{c+i\infty} \frac{k\, e^{st}}{s(s^{\alpha}+k)} ds \\ &= \sum_{m}\left[\operatorname{Res}\frac{k\, e^{st}}{s(s^{\alpha}+k)}\right]_{s=s_m} - \frac{1}{2\pi i}\int\limits_{\mathrm{ABCDEF}} \frac{k\, e^{st}}{s(s^{\alpha}+k)} ds, \end{aligned} \tag{10.74}$$

where the sum is taken for those m for which $s_m = k^{1/\alpha} e^{i\pi(2m+1)/\alpha}$ are the poles of the integrated function lying inside the domain bounded by the contour ABCDEFA (Fig. 10.7).

Then we can write

$$\int\limits_{\mathrm{ABCDEF}} = \int\limits_{\mathrm{A}}^{\mathrm{B}} + \int\limits_{\mathrm{B}}^{\mathrm{C}} + \int\limits_{\mathrm{C}}^{\mathrm{D}} + \int\limits_{\mathrm{D}}^{\mathrm{E}} + \int\limits_{\mathrm{E}}^{\mathrm{F}}.$$

The integrals from A to B and from E to F tend to 0 as $R \to \infty$. Using the substitution $s = \epsilon e^{i\eta}$ it can be shown that the integral from C to D tends to $-2\pi i$ as $\epsilon \to 0$. For the two remaining integrals along the negative half-axis we have:

$$\int\limits_{\mathrm{B}}^{\mathrm{C}} + \int\limits_{\mathrm{D}}^{\mathrm{E}} = -k\int\limits_{0}^{\infty} \frac{e^{-rt} e^{i\pi} dr}{re^{i\pi}(r^{\alpha}e^{i\pi\alpha}+k)} - k\int\limits_{0}^{\infty} \frac{e^{-rt} e^{-i\pi} dr}{re^{-i\pi}(r^{\alpha}e^{-i\pi\alpha}+k)}$$

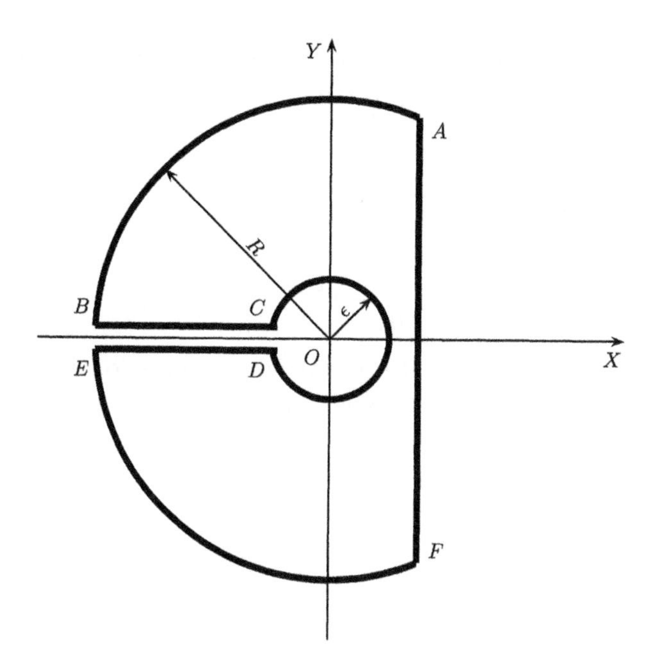

Figure 10.7: *Contour ABCDEFA.*

$$= 2ik \sin(\pi\alpha) \int_0^\infty \frac{r^{\alpha-1} e^{-rt} dt}{r^{2\alpha} + 2kr^\alpha \cos(\pi\alpha) + k^2}$$

$$= \frac{2i \sin(\pi\alpha)}{\alpha} \int_0^\infty \frac{\exp -(kx)^{1/\alpha} t \, dx}{x^2 + 2x\cos(\pi\alpha) + 1}, \tag{10.75}$$

where the substitution $r = (kx)^{1/\alpha}$ was used.

Now let us turn to residues. If $0 < \alpha < 1$, then there are no poles of the integrated function in the selected sheet of the Riemann surface, and the contribution of the sum of the residues in (10.74) is zero.

If $1 < \alpha < 2$, then we have two poles s_m, which correspond to $m = 0$ and $m = 1$. Then

$$\sum_m \left[\text{Res} \frac{k\, e^{st}}{s(s^\alpha + k)} \right]_{s=s_m} = \sum_0^1 \lim_{s \to s_m} \frac{(s - s_m)\, k\, e^{st}}{s(s^\alpha + k)} = -\frac{2}{\alpha} e^{\sigma^+ t} \cos(\omega_0^+ t), \tag{10.76}$$

$$\sigma^+ = -k^{1/\alpha} \cos\frac{\pi}{\alpha}, \qquad \omega_0^+ = k^{1/\alpha} \sin\frac{\pi}{\alpha}.$$

Since $1 < \alpha < 2$, then $\sigma > 0$.

Substituting all the intermediate results in (10.74), we obtain

$$u_{out}(t) = 1 - \frac{\sin(\pi\alpha)}{\pi\alpha} \int_0^\infty \frac{\exp -(kx)^{1/\alpha} t\, dx}{x^2 + 2x\cos(\pi\alpha) + 1} + \begin{cases} 0, & \text{if } \; 0 < \alpha < 1 \\ -\frac{2}{\alpha} e^{-\sigma^+ t} \cos(\omega_0^+ t), & \text{if } \; 1 < \alpha < 2. \end{cases} \tag{10.77}$$

In the case $-2 < \alpha < 0$ the following expression for the unit-step response of Westerlund's generalized voltage divider can be obtained in a similar way:

$$u_{out}(t) = \frac{\sin(\pi\alpha)}{\pi\alpha} \int_0^\infty \frac{\exp -(x/k)^{1/\alpha} t\, dx}{x^2 + 2x\cos(\pi\alpha) + 1} + \begin{cases} 0, & \text{if } \; -1 < \alpha < 0 \\ \frac{2}{\alpha} e^{-\sigma^- t} \cos(\omega_0^- t), & \text{if } \; -2 < \alpha < -1 \end{cases} \tag{10.78}$$

where

$$\sigma^- = -k^{-1/\alpha} \cos\frac{\pi}{\alpha}, \qquad \omega_0^- = -k^{-1/\alpha} \sin\frac{\pi}{\alpha}.$$

We see that for $1 < |\alpha| < 2$ the unit-step response of Westerlund's generalized voltage divider contains oscillation terms, in which $\sigma^{\pm}$ play the role of attenuation constants, and $\omega_0^{\pm}$ the role of the resonance frequencis.

10.5.5 Fractional-order Chua–Hartley System

The classical Chua circuit depicted in Fig. 10.8 is described by the following non-linear system of three differential equations:

$$\begin{aligned} \frac{dx(t)}{dt} &= -a_1 x + a_1 y - b_1 g(x), \\ \frac{dy(t)}{dt} &= a_2 x - a_2 y + b_2 z, \\ \frac{dz(t)}{dt} &= -a_3 y, \end{aligned} \tag{10.79}$$

where

$$a_1 = \frac{1}{RC_1}, \quad a_2 = \frac{1}{rC_2}, \quad b_1 = \frac{1}{C_1}, \quad b_2 = \frac{1}{C_2},$$

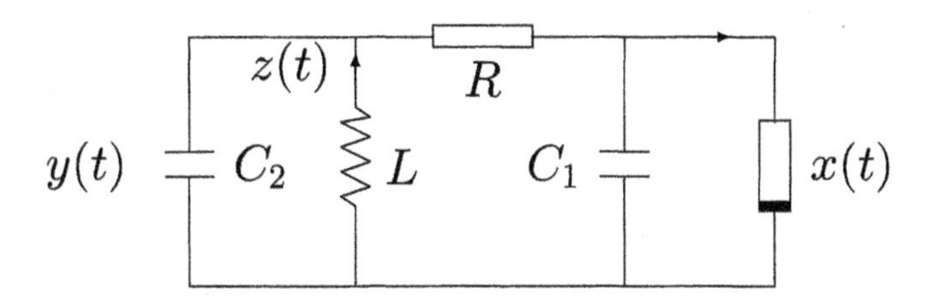

Figure 10.8: *Classical Chua circuit.*

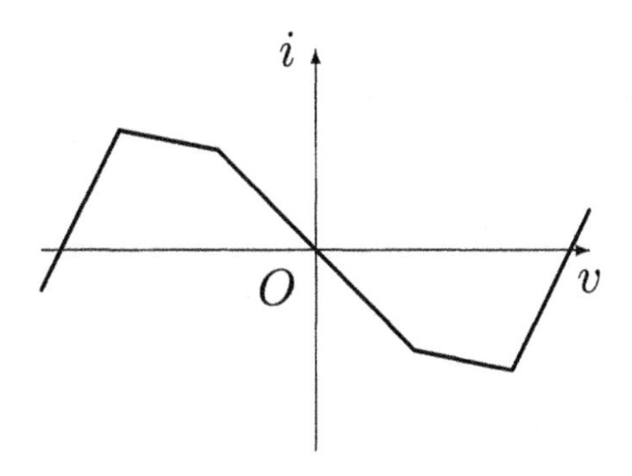

Figure 10.9: *Piecewise-linear $i - v$ characteristics of Chua's resistor.*

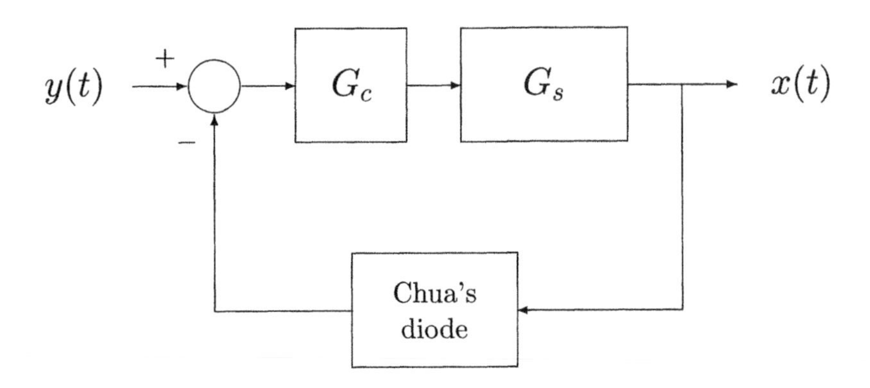

Figure 10.10: *Feedback control of Chua's circuit.*

and $g(x)$ is a piecewise-linear resistor characteristic which has the shape shown in general in Fig. 10.9.

Chua's circuit is extensively studied from the viewpoint of chaos, bifurcations, and multistable behaviour, and there are numerous papers on this, one of the most widely studied circuits today (see, for example, [189]).

Chua's circuit can also be described by the closed-loop control diagram with Chua's resistor in the feedback, as shown in Fig. 10.10, with

$$G_c(s) = \frac{1}{s}, \qquad G_s(s) = \frac{a_1(s^2 + a_2 s + a_3 b_2)}{s^2 + a_2 s + a_3 b_2 - a_1 a_2}. \tag{10.80}$$

In the paper by T. Hartley *et al.* [102] the piecewise-linear nonlinearity $g(x)$ was replaced by an appropriate cubic polynomial which yields similar behaviour, and the following particular values of the coefficients were taken:

$$a_1 = \alpha, \quad a_2 = 1, \quad a_3 = \frac{100}{7}, \quad b_1 = \frac{2\alpha}{7}, \quad b_2 = 1,$$

which gives the transfer function of the system

$$G_s(s) = \frac{\alpha(s^2 + s + 100/7)}{s\,(s^2 + s + 100/7 - \alpha)}; \tag{10.81}$$

the transfer function of the controller was, in fact, taken to be

$$G_c(s) = \frac{1}{s^{q-1}}, \tag{10.82}$$

where the exponent q is allowed to be non-integer. For $q > 1$ the controller becomes a fractional-order differentiator, and for $q < 1$ we have a fractional-order integrator.

For this particular system T. Hartley *et al.* [102] give computational results which demonstrate that, contrary to the widely accepted opinion that chaos cannot occur in continuous-time systems of order less than three, fractional-order systems of order less than three *can* display chaotic behaviour. In particular, the lowest value of q which yields chaos was $q = 0.9$.

In the time domain, this system is described by the following system of three differential equations, one of which contains two fractional derivatives: [1]

$$
\begin{aligned}
{}_0D_t^q x(t) &= \alpha\, {}_0D_t^{q-1}\Big(y(t) - x(t)\Big) - \frac{2a}{7}\Big(4x(t) - x^3(t)\Big), \\
\frac{dy(t)}{dt} &= x(t) - y(t) + z(t), \\
\frac{dz(t)}{dt} &= -\frac{100}{7}y(t).
\end{aligned}
\qquad (10.83)
$$

We will end this section with two remarks.

First, for the physical realization of the generalized Chua circuit any type of fractance can be used (tree fractance, chain fractance, Westerlund's capacitor, or a combination of these elementary fractances).

Second, the conclusion made by T. Hartley *et al.* [102] that there is a need for "a clarification of the definition of order [of a system] which can no longer be considered only by the total number of differentiations or by the highest power of the Laplace variable", is in agreement with our observation that for fractional-order differential equations the number of terms is more important than orders of derivatives appearing in such equations (see Chapter 5, in which equations are classified using the number of terms, not by orders of derivatives).

---

[1] The system

$$
\begin{aligned}
{}_0D_t^q x(t) &= \alpha\Big(y(t) + \frac{x(t) - 2x^3(t)}{7}\Big), \\
{}_0D_t^q y(t) &= x(t) - y(t) + z(t), \\
{}_0D_t^q z(t) &= -\frac{100}{7}y(t),
\end{aligned}
$$

given in [102], is not equivalent to the closed loop shown in Fig. 10.10 with the transfer functions given by (10.81) and (10.82). Instead, it gives different expressions for the transfer functions in which only non-integer powers of s appear: $G_c(s) = 1/s^q$ and $G_s(s) = \alpha(s^{2q} + s^q + 100/7)/(s^{2q} + s^q + 100/7 - \alpha)$.

10.6 Electroanalytical Chemistry

Due mainly to the works of K. B. Oldham and his co-authors [178, 175, 96, 93, 94, 176, 95, 177, 114], electrochemistry is one of those fields in which fractional-order integrals and derivatives have a strong position and bring practical results.

Although the idea of using a half-order fractional integral of current, ${}_0D_t^{-1/2}i(t)$, can be found also in the works of other authors (see, e.g., [6], [109]), it was the paper by K. B. Oldham [175] which definitely opened a new direction in the methods of electrochemistry called *semi-integral electroanalysis*, accomplished later by *semidifferential electroanalysis* suggested by M. Goto and D. Ishii [92].

One of the important subjects for study in electrochemistry is the determination of the concentration of analysed electroactive species near the electrode surface. The method suggested by K. B. Oldham and J. Spanier [178] allows, under certain conditions, replacement of a problem for the diffusion equation by a relationship on the boundary (electrode surface). Based on this idea, K. B. Oldham [175] suggested the utilization in experiments the characteristic described by the function

$$m(t) = {}_0D_t^{-1/2}i(t),$$

which is the fractional integral of the current $i(t)$, as the observed function, whose values can be obtained by measurements. Then the subject of main interest, the surface concentration $C_s(t)$ of the electroactive species, can be evaluated as

$$C_s(t) = C_0 - k\,{}_0D_t^{-1/2}i(t), \tag{10.84}$$

where k is a certain constant described below, and C_0 is the uniform concentration of the electroactive species throughout the electrolytic medium at the initial equilibrium situation characterized by a constant potential, at which no electrochemical reaction of the considered species is possible.

The relationship (10.84) was obtained by considering the following problem for a classical diffusion equation [96]:

$$\frac{\partial C(x,t)}{\partial t} = D_* \frac{\partial^2 C(x,t)}{\partial x^2}, \qquad (0 < x < \infty; \quad t > 0), \tag{10.85}$$

$$C(\infty, t) = C_0, \quad C(x, 0) = C_0,$$

$$\left[D_* \frac{\partial C(x,t)}{\partial x}\right]_{x=0} = \frac{i(t)}{nAF},$$

where D_* is the diffusion coefficient, A is the electrode area, F is Faraday's constant, and n is the number of electrons involved in the reaction (oxidation of electroactive species); the constant k in equation (10.84) is expressed as $k = 1/(nAF\sqrt{D_*})$. The solution procedure uses the Laplace tranform method and was given in [178]; it is very similar to the procedure used in Section 7.7.3.

There are several interesting features in this approach.

First, $m(t)$ is a characteristic *intermediate* between the current $i(t)$ and the passed charge $q(t)$, which is just the integral of the current:

$$q(t) = {}_0D_t^{-1} i(t).$$

Second, this approach involves no assumptions about the kinetics of the electrode process, the properties of the electrode surface, etc. In a certain sense, this is a sort of modelling "in the large": contributions of particular features of the process are "embedded" in the non-integer order of integration.

Third, instead of the classical diffusion equation (10.85), it is possible to consider the fractional-order diffusion equation

$${}_0D_t^{\alpha} C(x,t) = \tilde{D}_* \frac{\partial^2 C(x,t)}{\partial x^2},$$

with $0 < \alpha < 1$, where $\tilde{D}_*$ is the fractional diffusion coefficient. Then the surface concentration $C_s(t)$ will be related to $m_\alpha(t)$,

$$m_\alpha(t) = {}_0D_t^{\alpha/2} i(t).$$

The well-established and widely experimentally verified fractional-derivative based methods of electroanalytical chemistry can be successfully used in other fields, such as diffusion, heat conduction, mass transfer, etc., where similar basic equation appear.

10.7 Electrode–Electrolyte Interface

Another direction in the application of fractional-order models was motivated by the limitations of electric batteries, which always exhibit a limited current output due to the fact that microscopic electrochemical processes at the electrode–electrolyte interface have a finite rate and limit the current output. To circumvent this limitation, porous electrodes have been used because they have a large surface. However, has been known since the work by I. Wolfe dated 1926 (cf. [113]) that at metal–electrolyte interfaces the impedance $Z(\omega)$ does not exhibit the expected capacitive behaviour for small angular frequencies ω. Instead, for $\omega \to 0$

$$Z(\omega) \propto (i\omega)^{-\eta}, \quad (0 < \eta < 1) \tag{10.86}$$

or in the Laplace domain

$$Z(s) \propto s^{-\eta}. \tag{10.87}$$

This means that the electrode–electrolyte interface is an example of a fractional-order process.

The value η is closely related to the roughness of the interface, with η approaching unity as the surface is made infinitely smooth.

There were different models suggested for the relationship between η and the fractal dimension of the interface d_s $(2 < d_s < 3)$. It seems that no experiment has been able to confirm or to contradict the following models by different authors because it is difficult to measure the fractal dimension of real objects embedded in three-dimensional space.

A. Le Mehaute [121] proposed the relationship

$$\eta = d_s^{-1}. \tag{10.88}$$

L. Nyikos and T.Pajkossy suggested the relationship

$$\eta = (d_s - 1)^{-1}. \tag{10.89}$$

T. Kaplan *et al.* [113] have found

$$\eta = 3 - d_s. \tag{10.90}$$

The physical model proposed by the authors of [113] is presented by the self-affine Cantor block with N stages (levels), which is modelled by an N-stage electrical circuit of fractance type, i.e. similar to the one shown in Fig. 10.4.

Under certain assumptions, the impedance of the fractance circuit has been obtained in the form

$$Z(\omega) = K(i\omega)^{-\eta}, \tag{10.91}$$

where $\eta = 2 - \log(N^2)/\log a$, K and a are constant, and $a < N^2$ implies $0 < \eta < 1$.

We see that T. Kaplan *et al.* arrived at the model of a system of non-integer order $-\eta$.

10.8 Fractional Multipoles

Recently N. Engheta [60] suggested a definition for fractional-order multipoles of electric charge densities. The notion of fractional-order multipoles serves as an interpolation between the cases of integer-order point multipoles, such as point monopoles, point dipoles, point quadrupoles, etc. The approach, suggested by N. Engheta, is based on the fractional-order differentiation of the Dirac delta function (see formula (2.160)), and allows formulation of electric source distributions whose potentials are obtained by fractional differentiation or integration of potentials of integer-order point multipoles.

Since the terms *monopole, dipole, quadrupole,* etc., are related to powers of 2 (namely, 2^0, 2^1, 2^2, etc.), the fractional-order multipoles are called 2^α-poles.

In the three-dimensional case, N. Engheta found that the potential function of a point multipole with a 2^α pole along the z axis, $0 < \alpha < 1$, can be expressed in terms of the Riemann–Liouville fractional derivative with the lower terminal $t = -\infty$:

$$\Phi_{2^\alpha,z}(x,y,z) = \frac{ql^\alpha}{4\pi\epsilon} \, {}_{-\infty}D_t^\alpha \left(\frac{1}{\sqrt{x^2+y^2+z^2}} \right), \tag{10.92}$$

where q is the so-called electric monopole moment, and ϵ is a known physical constant (permittivity of homogeneous isotropic space).

The constant, which is taken in the form of l^α, where l has dimension of length, is introduced for getting the traditional dimension of the resulting volume charge density as Coulomb/m^3.

Evaluation of the fractional derivative (10.92) gives [60]

$$\Phi_{2^\alpha,z}(x,y,z) = \frac{q\,l^\alpha\,\Gamma(1+\alpha)}{4\pi\epsilon(x^2+y^2+z^2)^{(1+\alpha)/2}} P_\alpha \left(-\frac{z}{\sqrt{x^2+y^2+z^2}} \right), \tag{10.93}$$

where $P_\alpha(z)$ is the Legendre function of the first kind and of non-integer degree α [63].

It is obvious that the electrostatic potential functions for a monopole,

$$\Phi_1(x, y, z) = \frac{q}{4\pi\epsilon} \frac{1}{\sqrt{x^2 + y^2 + z^2}},$$

and for the dipole,

$$\Phi_2(x, y, z) = \frac{q}{4\pi\epsilon} \frac{\cos\theta}{x^2 + y^2 + z^2}, \qquad \cos\theta = \frac{z}{\sqrt{x^2 + y^2 + z^2}},$$

are particular cases of the function $\Phi_{2^\alpha,z}(x, y, z)$ for $\alpha = 0$ and $\alpha = 1$.

In this example of application of the fractional calculus it is interesting that a *static* object is considered, and the fractional derivative with the lower terminal $t = -\infty$ is applied with respect to the spatial variable.

In another paper [61] N. Engheta gives examples of structures containing wedges and cones, whose potentials can be described as electrostatic potentials of sets of charge distributions, which behave like fractional-order multipoles. The orders of the corresponding fractional-order multipoles depend on the wedge angle (in the two-dimensional case) and on the cone angle (in the three-dimensional case). The contour plots of the corresponding potentials are similar to the plots of stress concentration in problems of fracture mechanics in the presense of singularities of the boundary. In both cases, the known local behaviour of the solution near singular points of the boundary can be efficiently utilized during the numerical solution procedure.

10.9 Biology

10.9.1 Electric Conductance of Biological Systems

In his work on the electrical conductance of membranes of cells of biological organisms [37], published in 1933, K. S. Cole gave the following expression for the so-called membrane reactance:

$$X(\omega) = X_0\, \omega^{-\alpha}, \tag{10.94}$$

which obviously corresponds to the transfer function

$$g_x(s) = \tilde{X}_0\, s^{-\alpha}. \tag{10.95}$$

X_0 and α are constants and ω is the current frequency.

K. S. Cole also listed several values of α obtained experimentally by other authors for various types of cells: $\alpha = 0.45$ for guinea pig liver and muscle, $\alpha = 0.25$ for potato, $\alpha = 0.5$ for *Arbacia* egg, $\alpha = 0.37$ for frog muscle, and $\alpha = 0.88$ for blood.

10.9.2 Fractional-order Model of Neurons

The characteristic jerky movement of the eye which is observed at the beginning and at the end of a period of rotation of the head is called nystagmus. It is actually a reflex that provides visual fixation on stationary points while the head rotates. When rotation starts, the eyes first move slowly in the direction opposite to the direction of rotation, providing visual fixation; this is called the vestibulo-ocular reflex [234]. After reaching the limit of this movement, the eyes quickly go back to a new fixation point, and then again move slowly in the direction opposite to the direction of rotation.

These movements of the eyes are controlled by the premotor neurons and the motoneurons. Both types of neurons process the eye position signals.

In the paper [5] T. J. Anastasio pointed out the disadvantages of classical integer-order approaches to modelling the behaviour of premotor neurons in the vestibulo-ocular reflex, and suggested a fractional-order model in the form of the relationship in the Laplace domain:

$$\frac{R(s)}{V(s)} = \frac{\tau_1 (s\tau_2 + 1) s^{\alpha_d - \alpha_\imath}}{s\tau_1 + 1}, \tag{10.96}$$

where $R(s)$ is the Laplace transform of the premotor neuron discharge rate $r(t)$, $V(s)$ is the Laplace transform of the head angular velocity $v(t)$, τ_1 and τ_2 are time constants of the model, α_d is the order of fractional differentiation at the premotor level, and $\alpha_\imath$ is the order of the fractional integrator term in Anastasio's model.

The relationship between $v(t)$ and $r(t)$ can be obtained by applying the inverse Laplace transform to equation (10.96). Let us denote

$$g(s) = R(s)/V(s),$$

where $g(s)$ is the Laplace transform of $G(t)$, and assume $\alpha_i > \alpha_d$. Writing

$$g(s) = \frac{\tau_2 s^{\alpha_d - \alpha_\imath + 1}}{s + \tau_1^{-1}} + \frac{s^{\alpha_d - \alpha_\imath}}{s + \tau_1^{-1}},$$

and using the Laplace transform of the Mittag-Leffler function (1.80) we obtain:

$$G(t) = \tau_2 t^{\alpha_i - \alpha_d - 1} E_{1,\,\alpha_i - \alpha_d}\left(-\frac{t}{\tau_1}\right) + t^{\alpha_i - \alpha_d} E_{1,\,\alpha_i - \alpha_d + 1}\left(-\frac{t}{\tau_1}\right). \quad (10.97)$$

Then

$$r(t) = \int_0^t G(t-\tau)v(\tau)d\tau. \quad (10.98)$$

T. J. Anastasio also suggested a more general hypothesis: since the muscle and joint tissues throughout the musculoskeletal system seem to behave as viscoelastic materials having fractional-order integration dynamics, then this could be compensated by the fractional-order differentiation dynamics of associated premotor neurons and motoneurons, and therefore the "fractional-order dynamics may be a property of the motor control system in general" [5].

10.10 Fractional Diffusion Equations

The modelling of diffusion in a specific type of porous medium (in fractal media) is one of the most significant applications of fractional-order derivatives. The order of the resulting equation is related to the so-called fractal dimension of the porous material.

For the description of transfer processes in fractals (in the sense of B. Mandelbrot [142]), A. Le Mehaute, A. de Guibert, M. Delaye, and Ch. Filippi [120] suggested the equation of the form

$$ {}_0D_t^{1/d-1} J(t) = LX(t), \quad (10.99)$$

where $J(t)$ is the macroscopic flow across the fractal interface, $X(t)$ is the local driving force, L is a constant, and d is the fractal dimension. The equation (10.99) has been then rigorously deduced by A. Le Mehaute and G. Crepy [121]. It is important that the fractional diffusion equation has been related to a dynamical process in fractal media: the order of the resulted equation depends on the fractal dimension of the fractal, which serves as a model of a porous material.

Further development led to two types of partial differential equations of fractional order.

The first type is a generalization of the fractional partial differential equation suggested by K. B. Oldham and J. Spanier as a replacement of

Fick's law [178, 179]. In this way, M. Giona and H. E. Roman constructed an equation, which in the simplest version takes on the form [78, 79, 224]:

$$ {}_0D_t^{1/d} P(r,t) = -A\Big(\frac{\partial P(r,t)}{\partial r} + \frac{\kappa}{r} P(r,t)\Big) \tag{10.100} $$

where $P(r,t)$ is the average probability density of random walks on fractals, A and κ are constant, and d is the anomalous diffusion exponent, which depends on the fractal dimension of the considered media [225].

The fractional-order diffusion equation, suggested by R. Metzler, W. G. Glöckle, and T. F. Nonnenmacher [150], is an example of the second type of fractional diffusion equation:

$$ {}_0D_t^{2/d_w} P(r,t) = \frac{1}{r^{d_s-1}} \frac{\partial}{\partial r}\Big(r^{d_s-1} \frac{\partial P(r,t)}{\partial r}\Big), \tag{10.101} $$

where d_w and d_s depend on the fractal dimension of the media.

Another example of the second type is the fractional diffusion equation in the form deduced by R. R. Nigmatullin [162, 164]. In the simplest case of spatially one-dimensional diffusion Nigmatullin's equation takes on the form

$$ {}_0D_t^{\alpha} u(x,t) = \frac{d^2 u(x,t)}{dx^2}. \tag{10.102} $$

Since the order α of the derivative with respect to time in equation (10.102) can be of arbitrary real order, including $\alpha = 1$ and $\alpha = 2$, it is called the *fractional diffusion-wave equation.* This name has been suggested by F. Mainardi [131, 135]. For $\alpha = 1$ equation (10.102) becomes the classical diffusion equation, and for $\alpha = 2$ it becomes the classical wave equation. For $0 < \alpha < 1$ we have so-called ultraslow diffusion, and values $1 < \alpha < 2$ correspond to so-called intermediate processes [89].

The solution of equation (10.102) for the spatially one-dimensional case is given in Chapter 4 (Example 4.4).

Equation (10.102), with the fractional derivative defined in the sense of the generalized functions approach (see Section 2.4.2), has been considered by W. Wyss [259]. Later W. R. Schneider and W. Wyss [235], and also T. F. Nonnenmacher and D. J. F. Nonnenmacher [171], suggested another approach to "fractionalization" and unification of the form of the classical diffusion and wave equations, which leads to partial integro-differential equations containing fractional integrals with respect to time [235]. The simplest form of such an equation in the case of the spatially one-dimentional problem is

$$ u(x,t) = u(x,0) + \lambda^2 \, {}_0D_t^{-\alpha} \frac{\partial^2 u(x,t)}{\partial x^2}, \tag{10.103} $$

and the solution of this equation is given in Chapter 4 (Example 4.5).

Equation (10.103) allows the use of the classical initial conditions in terms of integer-order derivatives. This is not so in the case of equation (10.102) with the Riemann–Liouville fractional derivative. However, the Schneider–Wyss fractional integrodifferential equation (10.103) is equivalent to the fractional differential equation (10.102), in which the fractional derivative is interpreted as the Caputo fractional derivative (see Section 2.4.1).

The fractional diffusion-wave equation (10.102) was intensively studied by F. Mainardi [131, 135, 133, 134, 137], and also by A. N. Kochubei [117], and A. M. A. El-Sayed [57].

A different type of equation has been proposed by J. D. Polack [210] for modelling wave propagation in certain media using fractional derivatives. The impact of fractional-derivative terms in Polack's equation on the spectrum and impulse response of a boundary-controlled-and-observed infinite-dimensional linear system has been studied by D. Matignon and B. d'Andréa-Novel [144].

10.11 Control Theory

Chapter 9 provides an example of the use of fractional derivatives in control theory.

The idea of using fractional-order controllers for the control of dynamical systems belongs to A. Oustaloup, who developed the so-called CRONE controller (CRONE is an abbreviation of *Commande Robuste d'Ordre Non Entier*), which is described in a series of his books on applications of fractional derivatives in control theory [183, 185, 186, 187]. A. Oustaloup demonstrated the advantage of the CRONE controller in comparison with the PID-controller. The $PI^{\lambda}D^{\mu}$-controller, described in Chapter 9, also shows better performance when used for the control of fractional-order systems than the classical PID-controller.

The work by R. L. Bagley and R. A. Calico [13], A. Makroglou, R. K. Miller and S. Skaar [140], M. Axtell and M. E. Bise [9], G. Kaloyanov and J. M. Dimitrova [110], D. Matignon [143], D. Matignon and B. d'Andréa-Novel [145, 146], also provide very interesting ideas for using fractional derivatives in control theory, as well as some methods of studying fractional-order control systems.

The use of fractional-order derivatives and integrals as boundary controls for integer-order infinite-dimensional systems has been recently studied by B. Mbodje and G. Montseny [147], and G. Montseny, J. Audounet, and D. Matignon [159].

The use of fractional-order derivatives and integrals in control theory leads to better results than integer-order approaches; in addition, it provides strong motivation for further development of control theory in generalizing classical methods of study and the interpretation of results.

10.12 Fitting of Experimental Data

In this section we demonstrate on the example of modelling the impact of hereditary effects in steel wires on the change of their mechanical properties that fractional derivatives can be successfully used as an instrument for fitting experimental measurement data. We do not consider noisy data, but concentrate on the presentation of the idea.

In a certain sense, the subject of this section is close to the system identification discussed in Section 9.3. However, in control theory the system identification is just a step to the efficient control of a real dynamical object. On the contrary, the fitting of experimental data in a general sense may also be used for modelling static objects, and it is often the final step in system modelling.

10.12.1 Disadvantages of Classical Regression Models

Let us start with polynomial regression. To determine the basic disadvantages of polynomial regression models which are frequently applied for estimation of reliability of steel wires, used in mining transport machines, it is necessary to recall the main features of the process of the change of properties of such a wire:

- during a certain period after installation of a wire an enhancement of its properties is observed;
- then the properties of a wire become worse and worse, until it breaks down;
- the period of enhancement is shorter than the period of decrease, and the general shape of the process curve is not symmetric.

Linear regression can give a rough estimate of the second phase (decrease of the performance of a wire), but it cannot describe the period of enhancement of the wire properties.

Parabolic regression gives a symmetric shape for the fitting curve, which does not correspond to the physical background of the considered process.

Higher-order polynomial regression models can give better interpolation within the time interval for which measurements are available, but they give a wrong picture if one tries to use them for the prediction of the change of wire properties.

In real industrial practice, parabolic regression is preferred in most cases in spite its physical inadequacy. As a consequence, this leads to an underestimation of the strength of a wire and to its premature replacement.

One may try to use another regression model, for example, exponential, logarithmic, combined, etc. However, all these types of regression curves, in fact, dictate a certain shape of the fitting curve, and the whole responsibility for the selected shape lies upon the researcher/engineer. All such approaches miss much of the necessary flexibility. Some considerations regarding the use of mathematical models for estimation and prediction of the state of steel wires are given in [21].

10.12.2 Fractional Derivative Approach

Perhaps it is possible to try to obtain a more or less rigorous mathematical model of the process of exploitation of a wire; however, the main problem is that each particular wire changes its properties due to certain very particular causes, which are too unique to be incorporated in a general model.

An alternative approach, which we introduce here, is based on the use of a fractional integral for the description of hereditary changes of mechanical properties of steel wires.

A set of experimental measurements

$$y_1, y_2, \ldots, y_n$$

is fitted with the help of the function $y(t)$ satisfying the following integral equation:

$$y(t) = \sum_{k=0}^{m-1} a_k t^k - a_m \, {}_0D_t^{-\alpha} y(t), \qquad (0 < \alpha \leq m), \tag{10.104}$$

and the constant α, a_k, $(k = 0, \ldots, m)$ must be determined. For the determination of these parameters we used the least squares method, although any other criterion can be used as well. Regarding the parameter

m it is worth mentioning that m is the smallest integer number which is not less than α, so once one knows α, m is also known.

The parameters in equation (10.104) allow obvious physical interpretation. Namely, a_k $(k = 0, \ldots, m-1)$ are initial values of the fitting function $y(t)$ and its first $(m-1)$ derivatives. The fractional-order integral in the right-hand side represents the cumulative impact of the previous history of loading on the present state of the wire, and the order of integration, α, determines the shape of the memory function of the wire material. By omitting the fractional-order integral we obtain the classical general polynomial regression model.

The problem is then reduced to the initial-value problem

$$ {}_0D_t^{\alpha} z(t) + a_m z(t) = -a_m \sum_{k=0}^{m-1} a_k t^k, \tag{10.105} $$

$$ z^{(k)}(0) = 0, \quad (k = 0, \ldots, m-1), $$

for the auxiliary unknown function $z(t)$, where

$$ z(t) = y(t) - \sum_{k=0}^{m-1} a_k t^k. \tag{10.106} $$

The fact that initial conditions are zero allows application of the fractional difference method, which is described in Chapter 8, for the numerical solution of the problem (10.105) for any fixed combination of parameters a_k $(k = 0, \ldots, m)$. After the solution $z(t)$ is computed, we can use the relationship (10.106) for performing the backward substitution, and evaluate the value of the least-squares criterion for the function $y(t)$. The optimal set of parameters a_k $(k = 0, \ldots, m)$ can be determined by using known optimization methods. In particular, in this case the simplex method for unconditional optimization was used, which is implemented as a one of the standard MATLAB functions.

10.12.3 Example: Wires at Nizna Slana Mines

This approach was applied to modelling the change of properties of wires of the transport equipment at the Nizna Slana mining enterprize. A set of 14 measurements made each 6 months during 7 years served as input.

We do now show the linear regression model, which cannot reflect the improvement of properties of wires during the initial period of exploitation.

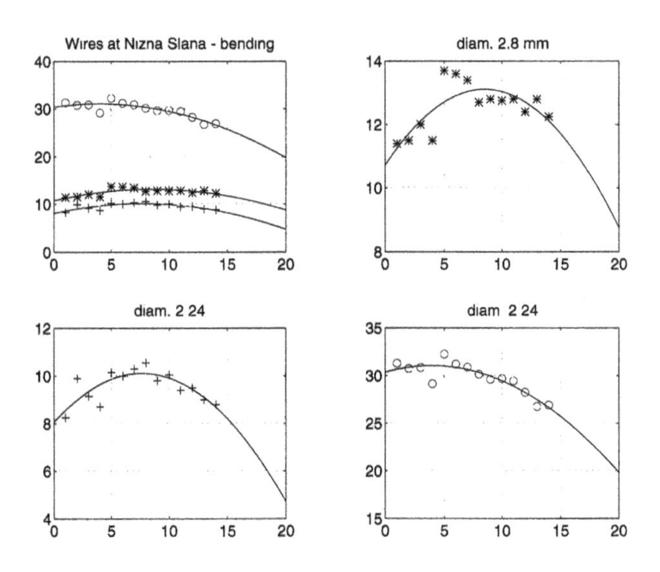

Figure 10.11: *Parabolic regression*

The parabolic regression model is shown in Fig. 10.11. The fitting curve is symmetric, and therefore it will not give a suitable prediction; the wire will be replaced ealier than necessary.

In Fig. 10.12 the third-order polynomial regression model is depicted. In contradiction with the physics of the considered process, the properties of one of the wires become better after a previous decrease of performance! The situation is much worse in the case of the fourth order polynomial regression model: according to this model, all three wires become better after a period of decrease of their performance (Fig. 10.13).

From this experience with polynomial regression models it follows that the most appropriate model is parabolic regression, in spite of its inaccuracy.

Finally, in Fig. 10.14 a comparison of the parabolic regression model

$$y(t) = -0.0330t^2 + 0.5619t + 10.7236$$

and the described fractional differential equation approach is given. The following parameters of equation (10.104) were computed:

$$\alpha = 1.32, \quad m = 2, \quad a_1 = 10.1955, \quad a_2 = 1.2760, \quad a_2 = 0.0457.$$

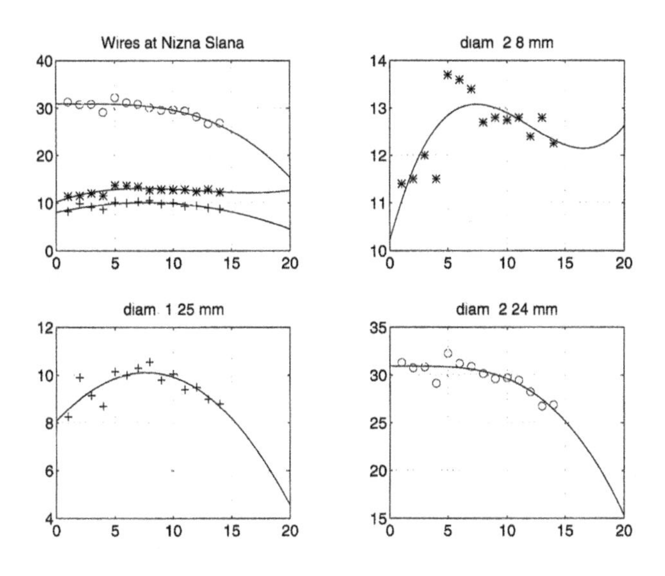

Figure 10.12: *Polynomial regression of third order.*

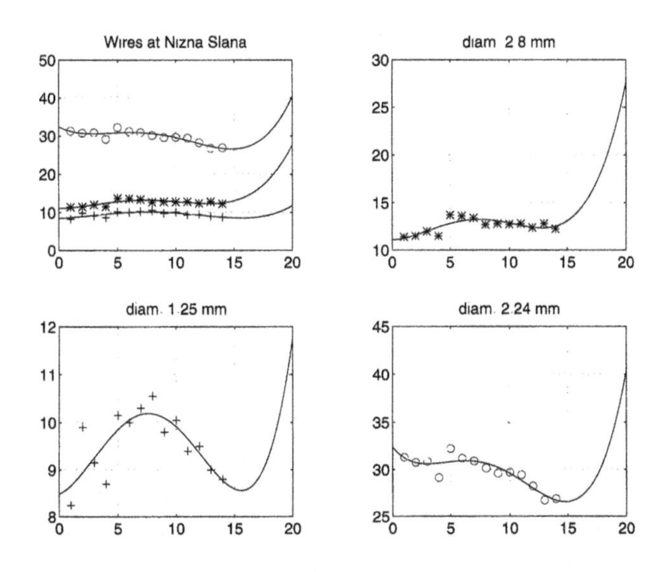

Figure 10.13: *Polynomial regression of fourth order.*

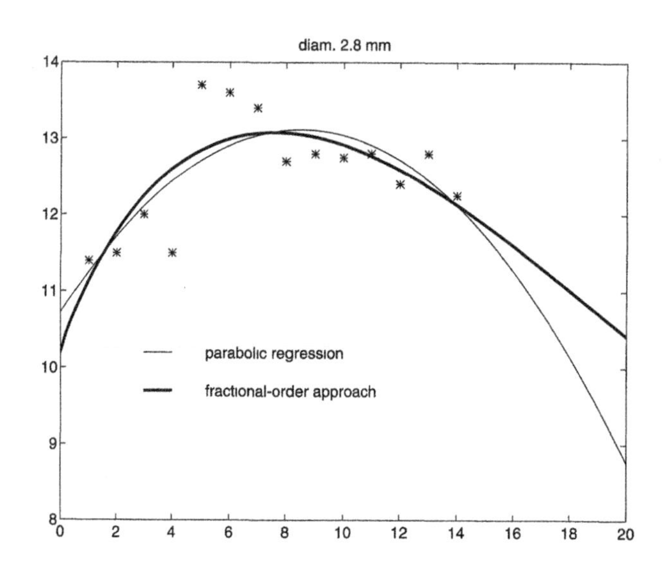

Figure 10.14: *Fractional order model versus parabolic regression.*

There is no surprise in the fact that the fractional-order model gives a lower value of the least squares criterion. More important is the fact that all significant characteristics of the shape of the process curve, which we mentioned in the beginning of this section, are preserved.

It is obvious that the order α of a fractional-order model will be different for different wires, because they work in different conditions. Therefore, it is necessary to apply the described approach in each case separately. However, it is not even a technical problem now, with modern computer facilities.

The outlined method is flexible. It allows continuous enhancement of the prediction of properties of wires after obtaining results of further measurements of mechanical properties of wires. It can be further generalized by introducing more fractional-order terms in equation (10.104) and/or replacing the set of functions t^k $(k = 0, 1, \ldots, n)$ with another suitable set of linearly independent functions.

Fitting of experimental data with the help of solutions of fractional differential equations is a promising approach which can be used in many experimental fields of science and engineering.

10.13 "Fractional-order" Physics?

In the previous sections we discussed some examples of the application of fractional derivatives in various fields of science and engineering.

The growing number of such applications indicates that there is a significant demand for better mathematical models of real objects, and that the fractional calculus provides one possible approach on the way to more adequate mathematical modelling of real objects and processes.

Among other results, some works must be mentioned, in which possible generalizations of important physical laws are suggested.

In 1991, S. Westerlund suggested using fractional derivatives for the description of propagation of plane electromagnetic waves in an isotropic and homogeneous, lossy dielectric. The equation suggested by S. Westerlund takes in the spatially one-dimensional case the following form:

$$\mu_0 \epsilon_0 \frac{\partial^2 E}{\partial t^2} + \mu_0 \epsilon_0 \chi_0 \; E^{(\nu)} + \frac{\partial^2 E}{\partial x^2} = 0, \qquad (10.107)$$

where E is the electric field, μ_0, ϵ_0, and χ_0 are constant, and ν $(1 < \nu < 2)$ is the order of differentiation of E with respect to time.

Later, in 1994, S. Westerlund [254] suggested replacing in the Maxwell equations the relationships $D = \epsilon E$ (E is the electric field, D is the electric field density) and $B = \mu H$ (B is the magnetic field, H is the magnetic field density) with their fractional-order generalizations

$$D = \epsilon E^{(\nu-1)}, \quad B = \mu H^{(\nu-1)}, \qquad (0 < \nu < 1), \qquad (10.108)$$

in which we see fractional-order integrals (since $\nu - 1 < 0$).

In the paper on electrochemically polarizable media [27], published in 1993, M. Caputo suggested the fractional-order version of the relationship between E (electric field) and D (electric flux density). In the spatially one-dimensional case this relationship has the form

$$\gamma D^{(\nu)} + \alpha D = \sigma E + \epsilon E^{(\nu)}, \qquad (10.109)$$

where γ, α, σ, and ϵ are constant, and ν denotes the (real) order of differentiation of D and E with respect to time. It is interesting to note that the relationship (10.109) is more general than (10.108) and has the same form as the four-parameter model of viscoelasticity (10.46), and is more general than (10.108).

Using some simplifying assumptions, M. Caputo reduced the Maxwell equations in the spatially one-dimensional case to the following system of two equations:

$$\frac{\partial^2 E}{\partial x^2} = -\mu \frac{\partial^2 D}{\partial t^2}; \qquad \gamma D^{(\nu)} + \alpha D = \sigma E + \epsilon E^{(\nu)}, \tag{10.110}$$

where μ is also constant. Using separation of variables and the Laplace transform of the Caputo fractional derivative (2.253) Caputo obtained a solution of the system (10.110) in terms of inverse Laplace transforms.

So, we see that the Maxwell equations have already been attacked, and we may expect further development in this direction.

Recently, in 1996–97, F. Riewe suggested a formulation of Lagrangian and Hamiltonian mechanics involving fractional derivatives [221, 222]. Lagrangians with fractional derivatives lead to equations of motion with non-conservative forces (such as friction, etc.). F. Riewe suggested a modified Hamilton principle, introduced two types of canonical transformations, and derived the Hamilton–Jacobi equation using fractional-order mechanics. In addition, he also proposed a fractional-order quantum-mechanical wave equation. He also suggested a generalized Euler–Lagrange equation, which involves fractional derivatives.

The formulas obtained by F. Riewe are two long to be included here even for illustration. However, it is worth mentioning that the appearance of Riewe's fractional mechanics was motivated by the well-known fact that the methods of classical mechanics deal only with conservative system, while almost all classical processes observed in the physical world are non-conservative, and exhibit irreversible dissipative effects.

A similar motivation, namely the wish to include dissipation, led S. Westerlund to a generalization of Newton's second law [254]. In this way, it is interesting to note that if F is an acting force and x is the displacement, then Hooke's model of elasticity ($F = kx$), Newton's model of a viscous fluid ($F = kx'$), and Newton's second law ($F = kx''$) can be considered as particular cases of a general relationship of the form

$$F = kx^{(\alpha)}, \tag{10.111}$$

in which α may be allowed to be any real number. In particular, S. Westerlund suggested that for $1 < \alpha < 2$ equation (10.111) can be considered as a generalization of Newton's second law, which better describes reality [254].

The mentioned recent work by M. Caputo, R. Riewe, and S. Westerlund, all trying to develop models describing dissipation, turn our attention to earlier work by G. W. Scott Blair [236, 237, 239], in which the use of fractional-order models of viscoelastic behaviour was interpreted as the introduction of "separate time scales for different materials" [236], based on the observation that "subjective judgements of time do not follow the Newtonian time scale" [239].

It is possible that in the future there will appear more "fractional-order" physical theories. We would like to end this section with the following two expressive quotations:

> "We may express our concepts in Newtonian terms if we find this convenient but, if we do so, we must realize that we have made a translation into a language which is foreign to the organism which we are studying." (G. W. Scott Blair, [238, p. 85])

> "... all systems need a fractional time derivative in the equations that describe them ... systems have memory of all earlier events. It is necessary to include this record of earlier events to predict the future ...
>
> The conclusion is obvious and unavoidable: *Dead matter has memory.* Expressed differently, we may say that Nature works with fractional time derivatives." (S. Westerlund, [253]).

Appendix: Tables of Fractional Derivatives

The short tables below contains Riemann–Liouville fractional derivatives of some functions which are frequently used in applications. In most cases, the order of differentiation, α, may be any real number, so replacing it with $-\alpha$ gives the Riemann–Liouville fractional integrals.

The tables can also be used for evaluating the Grünwald–Letnikov fractional derivatives, the Caputo fractional derivatives, and the Miller–Ross sequential fractional derivatives as well. In such cases, α should be taken between 0 and 1, and the Riemann–Liouville fractional derivative should be properly combined with integer- or fractional-order derivatives, in accordance with the considered definition.

1. Riemann–Liouville fractional derivatives with the lower terminal at 0

$f(t)$	${}_0D_t^{\alpha} f(t), \qquad (t > 0, \quad \alpha \in R)$
$H(t)$	$\dfrac{t^{-\alpha}}{\Gamma(1-\alpha)}$
$H(t-a)$	$\begin{cases} \dfrac{(t-a)^{-\alpha}}{\Gamma(1-\alpha)}, & (t > a) \\ 0, & (0 \le t \le a) \end{cases}$
$H(t-a)f(t)$	$\begin{cases} {}_aD_t^{\alpha} f(t), & (t > a) \\ 0, & (0 \le t \le a) \end{cases}$

$f(t)$	${}_0D_t^{\alpha} f(t), \qquad (t > 0, \quad \alpha \in R)$
$\delta(t)$	$\dfrac{t^{-\alpha-1}}{\Gamma(-\alpha)}$
$\delta^{(n)}(t)$	$\dfrac{t^{-\alpha-n-1}}{\Gamma(-\alpha-n)} \qquad (n \in N)$
$\delta^{(n)}(t-a)$	$\begin{cases} \dfrac{(t-a)^{-n-\alpha-1}}{\Gamma(-n-\alpha)}, & (t > a) \\ 0, & (0 \le t \le a) \end{cases} \qquad (n \in N)$
t^{ν}	$\dfrac{\Gamma(\nu+1)}{\Gamma(\nu+1-\alpha)} t^{\nu+\alpha} \qquad (\nu > -1)$
$e^{\lambda t}$	$t^{-\alpha} E_{1,1-\alpha}(\lambda t)$
$\cosh(\sqrt{\lambda} t)$	$t^{-\alpha} E_{2,1-\alpha}(\lambda t^2)$
$\dfrac{\sinh(\sqrt{\lambda} t)}{\sqrt{\lambda} t}$	$t^{1-\alpha} E_{2,2-\alpha}(\lambda t^2)$
$\ln(t)$	$\dfrac{t^{-\alpha}}{\Gamma(1-\alpha)} \Big(\ln(t) + \psi(1) - \psi(1-\alpha) \Big)$
$t^{\beta-1} \ln(t)$	$\dfrac{\Gamma(\beta) t^{\beta-\alpha-1}}{\Gamma(\beta-\alpha)} \Big(\ln(t) + \psi(\beta) - \psi(\beta-\alpha) \Big) \qquad (Re(\beta) > 0)$
$t^{\beta-1} E_{\mu,\beta}(\lambda t^{\mu})$	$t^{\beta-\alpha-1} E_{\mu,\beta-\alpha}(\lambda t^{\mu}) \qquad (\beta > 0, \quad \mu > 0)$
$t^{\beta-1}\, {}_2F_1(\mu,\, \nu;\, \beta;\, \lambda t)$	$\dfrac{\Gamma(\beta) t^{\beta-\alpha+1}}{\Gamma(\beta-\alpha)}\, {}_2F_1(\mu,\, \nu;\, \beta-\alpha;\, \lambda t) \qquad (Re(\beta) > 0)$
$P_m^{-\alpha,0}(2t-1)$	$\dfrac{m!\, t^{-\alpha}}{\Gamma(m-\alpha+1)} P_m^{0,-\alpha}(2t-1), \quad m = 1, 2, \ldots \qquad (0 < \alpha < 1, \quad 0 < t < 1)$

2. Riemann–Liouville fractional derivatives with the lower terminal at $-\infty$

$f(t)$	${}_{-\infty}D_t^{\alpha} f(t), \quad (t > 0, \quad \alpha \in R)$
$H(t-a)$	$\begin{cases} \dfrac{(t-a)^{-\alpha}}{\Gamma(1-\alpha)}, & (t > a) \\ 0, & (t \le a) \end{cases}$
$H(t-a)f(t)$	$\begin{cases} {}_aD_t^{\alpha} f(t), & (t > a) \\ 0, & (t \le a) \end{cases}$
$e^{\lambda t}$	$\lambda^{\alpha} e^{\lambda t}$ $\quad (\lambda > 0)$
$e^{\lambda t + \mu}$	$\lambda^{\alpha} e^{\lambda t + \mu}$ $\quad (\lambda > 0)$
$\sin \lambda t$	$\lambda^{\alpha} \sin\left(\lambda t + \frac{\pi\alpha}{2}\right)$ $\quad (\lambda > 0, \quad \alpha > -1)$
$\cos \lambda t$	$\lambda^{\alpha} \cos\left(\lambda t + \frac{\pi\alpha}{2}\right)$ $\quad (\lambda > 0, \quad \alpha > -1)$
$e^{\lambda t} \sin \mu t$	$r^{\alpha} e^{\lambda t} \sin(\mu t + \alpha\varphi)$ $\quad (r = \sqrt{\lambda^2 + \mu^2}, \quad \tan\varphi = \frac{\mu}{\lambda}, \quad \lambda > 0, \ \mu > 0)$
$e^{\lambda t} \cos \mu t$	$r^{\alpha} e^{\lambda t} \cos(\mu t + \alpha\varphi)$ $\quad (r = \sqrt{\lambda^2 + \mu^2}, \quad \tan\varphi = \frac{\mu}{\lambda}, \quad \lambda > 0, \ \mu > 0)$

Bibliography

[1] N. H. Abel, Resolution d'un problème de mécanique, *Oeuvres complètes de Niels Henrik Abel*, vol. 1, pp. 97–101.

[2] M. Abramowitz and I. A. Stegun, *Handbook of Mathematical Functions*, Nauka, Moscow, 1979 (references in the text are given to this Russian translation; original publication: Nat. Bureau of Standards, Appl. Math. Series, vol. 55, 1964)

[3] R. P. Agarwal. A propos d'une note de M. Pierre Humbert, *C. R. Séances Acad. Sci.*, vol. 236, no. 21, 1953, pp. 2031–2032.

[4] M. A. Al-Bassam, Some existence theorems on differential equations of generalized order, *Journal für Reine und Angewandte Mathematik*, vol. 218, 1965, pp. 70–78.

[5] T. J. Anastasio, The fractional-order dynamics of brainstem vestibulo-oculomotor neurons, *Biological Cybernetics*, vol. 72, 1994, pp. 69–79.

[6] C. P. Andrieux, L. Nadjo, and J. M. Savéant, Electrodimerization 1. One-electron irreversible dimerization. Diagnostic criteria and rate determination procedures for voltammetric studies, *J. Electroanal. Chem. and Interfacial Electrochem.*, vol. 26, 1970, pp. 147–186.

[7] N. Kh. Arutyunyan, Plane contact problem of the creep theory, *Prikl. Mat. Mekh.*, vol. 23, 1959, pp. 901–924 (in Russian).

[8] R. Askey, Inequalities via fractional integration, *Lect. Notes in Math.*, vol. 457, 1975, pp. 106–115.

[9] M. Axtell and M. E. Bise, Fractional calculus applications in control systems, *Proc. of the IEEE 1990 Nat. Aerospace and Electronics Conf., New York*, 1990, pp. 563–566.

[10] T. Ya. Azizov and I. S. Iokhvidov, *The Foundations of the Theory of Linear Operators in the Indefinite Metric Spaces*, Nauka, Moscow, 1986 (in Russian).

[11] Yu. I. Babenko, *Heat and Mass Transfer*, Khimiya, Leningrad, 1986 (in Russian).

[12] R. L. Bagley, Power law and fractional calculus model of viscoelasticity, *AIAA Journal*, vol. 27, no. 10, 1989, pp. 1412–1417.

[13] R. L. Bagley and R. A. Calico, Fractional order state equations for the control of viscoelastically damped structures, *J. Guidance*, vol. 14, no. 2, 1991, pp. 304–311.

[14] R. L. Bagley and P. J. Torvik, Fractional calculus – a different approach to the analysis of viscoelastically damped structures, *AIAA Journal* , vol. 21, no. 5, 1983, pp. 741–748

[15] R. L. Bagley and P. J. Torvik, A theoretical basis for the application of fractional calculus to viscoelasticity, *Journal of Rheology*, vol. 27, no. 3, 1983, pp. 201–210.

[16] R. L. Bagley and P. J. Torvik, On the appearence of the fractional derivative in the behavior of real materials, *J. Appl. Mech.*, vol. 51, 1984, pp. 294–298.

[17] R. L. Bagley and P. J. Torvik, Fractional calculus in the transient analysis of viscoelastically damped structures, *AIAA Journal*, vol. 23, no. 6, 1985, pp. 918–925.

[18] R. L. Bagley and P. J. Torvik, On the fractional calculus model of viscoelastic behavior, *Journal of Rheology*, vol. 30, 1986, pp. 133–155.

[19] H. Beyer and S. Kempfle, Definition of physically consistent damping laws with fractional derivatives, *Z. angew. Math. Mech.*, vol. 75, no. 8, 1995, pp. 623–635.

[20] H. W. Bode, *Network Analysis and Feedback Amplifier Design*, Tung Hwa Book Company, Shanghai, China, 1949.

[21] J. Boroska. Possibilities for using mathematical methods for evaluation and prediction of the state of transport steel wires. *Uhli*, no. 1, 1978, pp. 21–24 (in Slovak).

[22] F. Bella, P. F. Biagi, M. Caputo, G. Della Monica, A. Ermini, P. Manjgaladze, V. Sgrigna and D. Zilpimiani, Very slow-moving crustal strain disturbances, *Tectonophysics*, vol. 179, 1990, pp. 131–139.

[23] M. Caputo, Linear model of dissipation whose Q is almost frequency independent – II, *Geophys. J. R. Astr. Soc.*, vol. 13, 1967, pp. 529–539.

[24] M. Caputo, *Elasticità e Dissipazione*, Zanichelli, Bologna, 1969.

[25] M. Caputo, The rheology of an anelastic medium studied by means of the observation of the splitting of its eigenfrequencies, *J. Acoust. Soc. Am.*, vol. 86, no. 5, 1989, pp. 1984–1987.

[26] M. Caputo,. The splitting of the free oscillations of the Earth caused by the rheology, *Rend. Fis. Acc. Lincei*, ser. 9, vol. 1, 1990, pp. 119–125.

[27] M. Caputo, Free modes splitting and alterations of electrochemically polarizable media, *Rend. Fis. Acc. Lincei*, ser. 9, vol. 4, 1993, pp. 89–98.

[28] M. Caputo, *Lectures on Seismology and Rheological Tectonics*, Univ. degli studi di Roma "La Sapienza", 1992–1993.

[29] M. Caputo and F. Mainardi, Linear models of dissipation in anelastic solids, *Rivista del Nuevo Cimento (Serie II)*, vol. 1, no. 2, 1971, pp. 161–198.

[30] M. Caputo and F. Mainardi, A new dissipation model based on memory mechanism, *Pure and Applied Geophysics*, vol. 91, no. 8, 1971, pp. 134–147.

[31] G. E. Carlson, Investigation of fractional capacitor approximations by means of regular Newton processes, *Kansas State University Bulletin*, vol. 48, no. 1, special report no. 42, 1964.

[32] G. E. Carlson and C. A. Halijak, Simulation of the fractional derivative operator $\sqrt{s}$ and the fractional integral operator $1/\sqrt{s}$, *Kansas State University Bulletin*, vol. 45, no. 7, 1961, pp. 1–22.

[33] G. E. Carlson and C. A. Halijak, Approximation of fixed impedances, *IRE Transactions on Circuit Theory*, vol. CT-9, no. 3, 1962, pp. 302–303.

[34] G. E. Carlson and C. A. Halijak, Approximation of fractional capacitors $(1/s)^{1/n}$ by a regular Newton process, *IRE Transactions on Circuit Theory*, vol. CT-11, no. 2, 1964, pp. 210–213.

[35] A. Carpinteri and F. Mainardi (eds.), *Fractals and Fractional Calculus in Continuum Mechanics*, Springer Verlag, Vienna – New York, 1997.

[36] A. M. Chak, A generalization of the Mittag-Leffler function, *Mat. Vesnik*, vol. 19, no. 4, 1967, pp. 257–262.

[37] K. S. Cole, Electric conductance of biological systems, *Proc. Cold Spring Harbor Symp. Quant. Biol.*, Cold Spring Harbor, New York, 1933, pp. 107–116.

[38] G. Dahlquist, On accuracy and unconditional stability of linear multistep methods for second order differential equations, *BIT*, vol. 18, 1978, pp. 133–136.

[39] H. D. Davis, *The Theory of Linear Operators*, Principia Press, Bloomington, Indiana, 1936.

[40] K. Diethelm, An algorithm for the numerical solution of differential equations of fractional order, *Electronic Transactions on Numerical Analysis*, ISSN 1068-9613, vol. 5, March 1997, pp. 1–6.

[41] K. Diethelm, Numerical approximation of finite-part integrals with generalized compound quadrature formulae, *Hildesheimer Informatikberichte*, ISSN 0941-3014, no. 17/95, June 1995.

[42] G. Doetsch, *Anleitung zum Praktischen Gebrauch der Laplace-transformation*, Oldenbourg, Munich, 1956 (Russian translation: Fizmatgiz, Moscow, 1958).

[43] L. Dorcak, J. Prokop and I. Kostial, Investigation of the properties of fractional-order dynamical systems, *Proceedings of the 11th Int. Conf. on Process Control and Simulation ASRTP'94, Kosice-Zlata Idka, September 19-20, 1994*, pp. 58–66.

[44] S. Dugowson, *Les Différentielles Métaphysiques: Histoire et Philosophie de la Généralisation de l'Ordre de Dérivation*, thèse de Doctorat, University of Paris, 1994.

[45] M. M. Dzhrbashyan, *Integral Transforms and Representations of Functions in the Complex Domain*, Nauka, Moscow, 1966 (in Russian).

[46] M. M. Dzhrbashyan and A. B. Nersesyan, Criteria of expansibility of functions in Dirichlet series, *Izv. Akad. Nauk Arm. SSR, ser. fiz.-mat.*, vol. 11, no. 5, 1958, pp. 85–106.

[47] M. M. Dzhrbashyan and A. B. Nersesyan, On the use of some integro-differential operators, *Dokl. Akad. Nauk SSSR*, vol. 121, no. 2, 1958, pp. 210–213.

[48] M. M. Dzhrbashyan and A. B. Nersesyan, Expansions in spesial biorthogonal systems and boundary-value problems for differential equations of fractional order, *Dokl. Akad. Nauk SSSR*, vol. 132, no. 4, 1960, pp. 747–750.

[49] M. M. Dzhrbashyan and A. B. Nersesyan, Expansions in some biorthogonal systems and boundary-value problems for differential equations of fractional order, *Trudy Mosk. Mat. Ob.*, vol. 10, 1961, pp. 89–179.

[50] M. M. Dzhrbashyan and A. B. Nersesyan, Fractional derivatives and the Cauchy problem for differential equations of fractional order, *Izv. Akademii Nauk Arm. SSR*, vol. 3, no. 1, 1968, pp. 3–29.

[51] A. M. A. El-Sayed, Fractional differential equations, *Kyungpook Math. J.*, vol. 28, no. 2, 1988, pp. 119–122.

[52] A. M. A. El-Sayed, Fractional derivative and fractional differential equations, *Bull Fac. Sci.*, Alexandria Univ., vol. 28. 1988, pp. 18–22.

[53] A. M. A. El-Sayed, On the fractional differential equations, *Apll. Math. and Comput.*, vol. 49, 1992, pp. 2–3.

[54] A. M. A. El-Sayed, Linear differential equations of fractional order, *Apll. Math. and Comput.*, vol. 55, 1993, pp. 1–12.

[55] A. M. A. El-Sayed, Multivalued fractional differential equations, *Apll. Math. and Comput.*, vol. 80, 1994, pp. 1–11.

[56] A. M. A. El-Sayed, Fractional order evolution equations, *J. of Frac. Calculus*, vol. 7, May 1995, pp. 89–100.

[57] A. M. A. El-Sayed, Fractional order diffusion-wave equation, *Int. J. of Theor. Phys.*, vol. 35, 1996, pp. 311–322.

[58] M. Enelund, Å. Fenander and P. Olsson, Fractional integral formulation of constitutive equations of viscoelasticity, *AIAA Journal*, vol. 35, no. 8, 1997, pp. 1356–1362.

[59] M. Enelund and B. L. Josefson, Time-domain finite element analysis of viscoelastic structures with fractional derivatives constitutive equations, *AIAA Journal*, vol. 35, no. 10, 1997, pp. 1630–1637.

[60] N. Engheta, On fractional calculus and fractional multipoles in electromagnetism. *IEEE Trans. on Antennas and Propagations*, vol. 44, no. 4, 1996, pp. 554–566.

[61] N. Engheta, Electrostatic "fractional" image methods for perfectly conducting wedges and cones, *IEEE Trans. on Antennas and Propagations*, vol. 44, no. 2, 1996, pp. 1565–1574.

[62] A. Erdélyi (ed.), *Tables of Integral Transforms*, vol. 1, McGraw-Hill, New York, 1954.

[63] A. Erdélyi (ed.), *Higher Transcendental Functions*, vol. 1, McGraw-Hill, New York, 1955.

[64] A. Erdélyi (ed.), *Higher Transcendental Functions*, vol. 2, McGraw-Hill, New York, 1955.

[65] A. Erdélyi (ed.), *Higher Transcendental Functions*, vol. 3, McGraw-Hill, New York, 1955.

[66] Å. Fenander, Modal synthesis when modeling damping by use of fractional derivatives, *AIAA Journal*, vol. 34, no. 5, 1998, pp. 1051–1058.

[67] H. E. Fettis, On the numerical solution of equations of the Abel type, *Math. Comp.*, vol. 18, no. 84, 1964, pp. 491–496.

[68] G. M. Fikhtengoltz, *Course of Differential and Integral Calculus*, vol. 2, Nauka, Moscow, 1969.

[69] C. Fox, The G and H functions as symmetrical Fourier kernels, *Trans. Am. Math. Soc.*, vol. 98, 1961, pp. 395–429.

[70] Ch. Friedrich, Relaxation and retardation functions of the Maxwell model with fractional derivatives, *Rheologica Acta*, vol. 30, 1991, pp. 151–158.

[71] Ch. Friedrich, Rheological material functions for associating comb-shaped or H-shaped polymers: a fractional calculus approach, *Philosophical Magazine Letters*, vol. 66, no. 6, 1992, pp. 287–292.

[72] Ch. Friedrich, Mechanical stress relaxation in polymers: fractional integral model versus fractional differential model, *J. Non-Newtonian Fluid Mech.*, vol. 46, 1993, pp. 307–314.

[73] Ch. Friedrich, Linear viscoelastic behavior of branched plybutadiene: a fractional calculus approach, *Acta Polymer.*, vol. 46, 1995, pp. 385–390.

[74] Ch. Friedrich and H. Braun, Linear viscoelastic behavior of complex polymeric materials: a fractional mode representation, *Colloid and Polymer Science*, vol. 272, 1994, pp. 1536–1546.

[75] L. Gaul, S. Kempfle and P. Klein, Transientes Schwingungsverhalten bei der Dämpfungsbeschreibung mit nicht ganzzahligen Zeitableitungen, *Z. angew. Math. Mech.*, vol. 70, no. 4, 1990, pp. T139–T141.

[76] I. M. Gelfand and G. E. Shilov, *Generalized Functions*, vol. 1, Nauka, Moscow, 1959 (in Russian).

[77] A. N. Gerasimov, A generalization of linear laws of deformation and its application to inner friction problems, *Prikl. Mat. Mekh.*, vol. 12, 1948, pp. 251–259 (in Russian).

[78] M. Giona and H. E. Roman, A theory of transport phenomena in disordered systems, *Chemical Engineering Journal*, vol. 49, 1992, pp. 1–10.

[79] M. Giona, S. Gerbelli and H. E. Roman, Fractional diffusion equation and relaxation in complex viscoelastic materials, *Physica A*, vol. 191, 1992, pp. 449–453.

[80] W. G. Glöckle and T. F. Nonnenmacher, Fractional integral operators and Fox functions in the theory of viscoelasticity, *Macromolecules*, vol. 24, 1991, pp. 6426–6436.

[81] W. G. Glöckle and T. F. Nonnenmacher, Fractional relaxation and the time-temperature superposition principle, *Rheologica Acta*, vol. 33, 1994, pp. 337–343.

[82] R. Gorenflo, Abel integral equations: application-motivated solution concepts, *Methoden Verfahren Math. Phys.*, vol. 34, 1987, pp. 151–174.

[83] R. Gorenflo, Fractional calculus: some numerical methods, in Carpinteri and Mainardi [35].

[84] R. Gorenflo, Abel integral equations with special emphasis on applications, *Lectures in Mathematical Sciences*, vol. 13, University of Tokyo, 1996.

[85] R. Gorenflo and Y. Kovetz, Solution of an Abel type integral equation in the presence of noise by quadratic programming, *Numer. Math.*, vol. 8, 1966, pp. 392–406.

[86] R. Gorenflo and Yu. Luchko, *An Operational Method for Solving Generalized Abel Integral Equations of Second Kind*, preprint no. A-6/95, Department of Mathematics and Informatics, Free University of Berlin, 1995.

[87] R. Gorenflo, Yu. Luchko, and S. Rogosin, *Mittag-Leffler Type Functions: Notes on Growth Properties and Distribution of Zeros*, preprint no. A-97-04, Department of Mathematics and Informatics, Free University of Berlin, 1997.

[88] R. Gorenflo and F. Mainardi, Fractional calculus: integral and differential equations of fractional order, in Carpinteri and Mainardi [35].

[89] R. Gorenflo and R. Rutman, On ultraslow and on intermediate processes, in P. Rusev, I. Dimovski and V. Kiryakova (eds.), *Transform Methods and Special Functions*, SCT Publishers, Singapore, 1995.

[90] R. Gorenflo and S. Vessella, *Abel Integral Equations: Analysis and Applications*, Lecture Notes in Mathematics, vol. 1461, Springer-Verlag, Berlin, 1991.

[91] R. Gorenflo and Vu Kim Tuan, Singular value decomposition of fractional integration operators in L_2-spaces with weights, *J. Inverse and Ill-Posed Problems*, vol. 3, 1995, pp. 1–9.

[92] M. Goto and D. Ishii, Semidifferential electroanalysis, *J. Electroanal. Chem. and Interfacial Electrochem.*, vol. 61, 1975, pp. 361–365.

[93] M. Goto and K. B. Oldham, Semiintegral electroanalysis: shapes of neopolarograms, *Anal. Chem.*, vol. 45, no. 12, 1973, pp. 2043–2050.

[94] M. Goto and K. B. Oldham, Semiintegral electroanalysis: studies on the neopolarographic plateau, *Anal. Chem.*, vol. 46, no. 11, 1973, pp. 1522–1530.

[95] M. Goto and K. B. Oldham, Semiintegral electroanalysis: the shape of irreversible neopolarograms, *Anal. Chem.*, vol. 48, no. 12, 1976, pp. 1671–1676.

[96] M. Grenness and K. B. Oldham, Semiintegral electroanalysis: theory and verification, *Anal. Chem.*, vol. 44, 1972, pp. 1121–1129.

[97] B. Gross, On creep and relaxation, *J. Appl. Phys.*, vol. 18, 1947, pp. 212–221.

[98] A. K. Grünwald, Ueber "begrenzte" derivationen und deren anwendung, *Zeitschrift f. Mathematik u. Physik*, vol. 12, no. 6, pp. 441–480.

[99] J. Hadamard. *Lectures on Cauchy's Problem in Linear Partial Differential Equations.* Yale Univ. Press, New Haven, 1923.

[100] S. B. Hadid and Yu. Luchko, An operational method for solving fractional differential equations of an arbitrary real order, *Panam. Math. J.*, vol. 6, no. 1, 1996, pp. 57–73.

[101] T. T. Hartley and F. Mossayebi, Control of Chua's circuit, *J. Circuits, Syst., Comput.*, vol. 3, no. 1, 1993, pp. 173–194.

[102] T. T. Hartley, C. F. Lorenzo and H. K. Qammer, Chaos in fractional order Chua's system, *IEEE Trans. on Circuits nad Systems – I: Fundamental Theory and Applications*, vol. 42, no. 8, 1995, pp. 485–490.

[103] P. Hašek, *Tables for Thermal Devices*, VŠB Ostrava, 1984 (in Czech).

[104] X.-F. He, Dimensionality in optical spectra of solids: analysis by fractional calculus, *Solid State Comm.*, vol. 61, no. 1, 1987, pp. 53–55.

[105] P. Henrici, Fast Fourier methods in computational complex analysis, *SIAM Review*, vol. 21, no. 4, 1979, pp. 481–527.

[106] N. Heymans and J.-C. Bauwens, Fractal rheological models and fractional differential equations for viscoelastic behavior, *Rheologica Acta*, vol. 33, 1994, pp. 210–219.

[107] P. Humbert and R. P. Agarwal, Sur la fonction de Mittag-Leffler et quelques-unes de ses généralisations, *Bulletin des Sciences Mathématiques*, vol. 77, no. 10, 1953, pp. 180-185.

[108] P. Humbert and P. Delerue, Sur une extension à deux variables de la fonction de Mittag-Leffler, *C. R. Acad. Sci. Paris*, vol. 237, 1953, pp. 1059–1060.

[109] J. C. Imbeaux and J. M. Savéant, Convolutive potential sweep voltammetry I. Introduction, *J. Electroanal. Chem. and Interfacial Electrochem.*, vol. 44, 1973, pp. 169–187.

[110] G. Kaloyanov and J. M. Dimitrova, Theoretical-experimental determination of the area of applicability of a system "PI(I)-controller – object with non-integer astaticity", *Izv. Vysshykh Utchebnykh Zav. - Elektromekhanika*, 1992, no. 2, pp. 65–72 (in Russian).

[111] L. V. Kantorovich and G. P. Akilov, *Functional Analysis*, Nauka, Moscow, 1986 (in Russian).

[112] L. V. Kantorovich and V. I. Krylov, *Approximate Methods of the Higher Analysis*, Fizmatgiz, Moscow–Leningrad, 1962 (in Russian).

[113] T. Kaplan, L. J. Gray and S. H. Liu, Self-affine fractal model for a metal-electrolyte interface, *Phys. Review B*, vol. 35, no. 10, April 1987, pp. 5379–5381

[114] A. M. Keightley, J. C. Myland, K. B. Oldham and P. G. Symons, Reversible cyclic volammetry in the presense of product, *J. Electroanal. Chem.*, vol. 322, 1992, pp. 25–54.

[115] S. Kempfle and L. Gaul, Global solutions of fractional linear differential equations, *Proc. of ICIAM'95, Zeitschrift Angew. Math. Mech.*, vol. 76, suppl. 2, 1996, pp. 571–572.

[116] V. Kiryakova, *Generalized Fractional Calculus and Applications*, Pitman Research Notes in Math., no. 301, Longman, Harlow, 1994.

[117] A. N. Kochubei, Fractional order diffusion, *J. Diff. Equations*, vol. 26, 1990, pp. 485–492 (English translation from Russian).

[118] R. C. Koeller, Applications of fractional calculus to the theory of viscoelasticity, *Trans. ASME – J. Appl. Mech.*, vol. 51, 1984, pp. 299–307.

[119] G. A. Korn and T. M. Korn, *Mathematical Handbook*, 2nd ed., McGraw-Hill, New York, 1968.

[120] A. Le Mehaute, A. de Guibert, M. Delaye and Ch. Filippi, Note d'introduction de la cinétique des échanges d'énergies et de matières sur les interfaces fractales, *C. R. Acad. Sci. Paris*, vol. 294, ser. II, 1982, pp. 835–838.

[121] A. Le Mehaute and G. Crepy, Introduction to transfer and motion in fractal media: the geometry of kinetics, *Solid State Ionics*, 1983, no. 9 and 10, pp. 17–30.

[122] H. H. Lee and C.-S. Tsai, Analytical model of viscoelastic dampers for seismic mitigation of structures, *Computers and Structures*, vol. 50, no. 1, 1994, pp. 111–121.

[123] G. W. Leibniz, *Mathematische Schiften*, Georg Olms Verlagsbuchhandlung, Hildesheim, 1962.

[124] A. V. Letnikov, Theory of differentiation of an arbitrary order, *Mat. Sb.*, vol. 3, 1868, pp. 1–68 (in Russian).

[125] A. V. Letnikov, On the historical development of the theory of differentiation of an arbitrary order, *Mat. Sb.*, vol. 3, 1868, pp. 85–112 (in Russian).

[126] A. V. Letnikov, Treatment related to the theory of the integrals of the form $\int_a^x (x-u)^{p-1} f(u)du$, *Mat. Sb.*, vol. 7, 1872, pp. 5–205 (in Russian).

[127] Ch. Lubich, Discretized fractional calculus, *SIAM J. Math. Anal.*, vol. 17, no. 3, May 1986, pp. 704–719.

[128] Yu. F. Luchko and H. M. Srivastava, The exact solution of certain differential equations of fractional order by using operational calculus, *Computers Math. Applic.*, vol. 29, no. 8, 1995, pp. 73–85.

[129] N. Makris and M. C. Constantinou, Fractional-derivative Maxwell model for viscous dampers, *ASCE J. Structural Engineering*, vol. 117, no. 9, 1991, pp. 2708–2724.

[130] N. Makris, G. F. Dargush and M. C. Constantinou, Dynamic analysis of generalized viscoelastic fluids, *ASCE J. Eng. Mech.*, vol. 119, no. 8, 1993, pp. 1663–1679.

[131] F. Mainardi, On the initial value problem for the fractional diffusion–wave equation, in: S. Rionero and T. Ruggeri (eds.), *Waves and Stability in Continuous Media*, World Scientific, Singapore, 1994, pp. 246–251.

[132] F. Mainardi, Fractional relaxation in anelastic solids, *J. Alloys and Compounds*, vol. 211/212, 1994, pp. 534–538.

[133] F. Mainardi, Fractional diffusive waves in viscoelastic solids, in: J. L. Wegner and F. R. Norwood (eds.), *Nonlinear Waves in Solids*, ASME/AMR, Fairfield NJ, 1995, pp. 93–97.

[134] F. Mainardi, The time fractional diffusion–wave equation, *Radiofizika*, vol. 38, 1995, pp. 20–36.

[135] F. Mainardi, Fractional relaxation–oscillation and fractional diffusion-wave phenomena, *Chaos, Solitons and Fractals*, vol. 7, 1996, pp. 1461–1477.

[136] F. Mainardi, Applications of fractional calculus in mechanics, in: P. Rusev, I. Dimovski and V. Kiryakova (eds.), *Transform Methods and Special Functions, Varna'96*, SCT Publishers, Singapore, 1997.

[137] F. Mainardi, The fundamental solutions for the fractional diffusion-wave equation, *Appl. Math. Lett.*, vol. 9, no. 6, 1996, pp. 23–28.

[138] F. Mainardi, *Fractional calculus: some basic problems in continuum and statistical mechanics*, in Carpinteri and Mainardi [35].

[139] F. Mainardi and E. Bonetti, The application of real-order derivatives in linear viscoelasticity, *Rheologica Acta*, vol. 26 (suppl.), 1988, pp. 64–67.

[140] A. Makroglou, R. K. Miller and S. Skaar, Computational results for a feedback control for a rotating viscoelastic beam, *J. of Guidance, Control and Dynamics*, vol. 17, no. 1, 1994, pp. 84–90.

[141] N. N. Malinin, *Applied Theory of Plasticity and Creep*, Mashinostroenie, Moscow, 1975.

[142] B. Mandelbrot. *The Fractal Geometry of Nature.* Freeman, San Francisco, 1982.

[143] D. Matignon, Stability results on fractional differential equations with applications to control processing, *Computational Engineering in Systems Applications*, Lille, France, July 1996, IMACS, IEEE-SMC, vol. 2, pp. 963–968.

[144] D. Matignon and B. d'Andréa-Novel, Spectral and time-domain consequences of an integro-differential perturbation of the wave PDE, *3rd Int. Conf. on Math. and Numer. Aspects of Wave Propagation Phenomena*, Mandelieu, France, April 1995, INRIA, SIAM, pp. 769–771.

[145] D. Matignon and B. d'Andréa-Novel, Some results on controllability and observability of finite-dimensional fractional differential systems, *Computational Engineering in Systems Applications*, Lille, France, July 1996, IMACS, IEEE-SMC, vol. 2, pp. 952–956.

[146] D. Matignon and B. d'Andréa-Novel, Observer-based controllers for fractional differential systems, *36th IEEE Conference on Decision and Control*, San Diego, California, December 1997, IEEE-CSS, SIAM, pp. 4967–4972.

[147] B. Mbodje and G. Montseny, Boundary fractional derivative control of the wave equation, *IEEE Trans. Aut. Control*, vol. 40, 1995, pp. 378–382.

[148] A. C. McBride, *Fractional Calculus and Integral Transforms of Generalized Functions*, Res. Notes in Math., vol. 31, Pitman Press, San Francisco, 1979.

[149] S. I. Meshkov, *Viscoelastic Properties of Metals*, Metallurgia, Moscow, 1974.

[150] R. Metzler, W. G. Glöckle, and T. F. Nonnenmacher, Fractional model equation for anomalous diffusion, *Physica A*, vol. 211, 1994, pp. 13–24.

[151] R. Metzler, W. Schick, H.-G. Kilian, and T. F. Nonnenmacher, Relaxation in filled polymers: a fractional calculus approach, *J. Chem. Phys.*, vol. 103, 1995, pp. 7180–7186.

[152] M. W. Michalski, Derivatives of noninteger order and their applications, *Dissertationes Mathematicae*, CCCXXVIII, Inst. Math., Polish Acad. Sci., Warsaw, 1993.

[153] K. S. Miller and B. Ross, *An Introduction to the Fractional Calculus and Fractional Differential Equations*, John Wiley & Sons Inc., New York, 1993.

[154] G. N. Minerbo and M. E. Levy, Inversion of Abel's integral equation by means of orthogonal polynomials, *SIAM J. Numer. Anal.*, vol. 6, no. 4, 1969, pp. 598–616.

[155] G. M. Mittag-Leffler, Sur la nouvelle fonction $E_\alpha(x)$, *C. R. Acad. Sci. Paris*, vol. 137, 1903, pp. 554–558.

[156] G. M. Mittag-Leffler, Sopra la funzione $E_\alpha(x)$, *Rend. Acc. Lincei*, ser. 5, vol. 13, 1904, pp. 3–5.

[157] G. M. Mittag-Leffler, Sur la représentation analytique d'une branche uniforme d'une fonction monogène, *Acta Mathematica*, vol. 29, 1905, pp. 101–182.

[158] D. Mo, Y. Y. Lin, J. H. Tan, Z. X. Yu, G. Z. Zhou, K. C. Gong, G. P. Zhang and X.-F. He, Ellipsometric spectra and fractional derivative spectrum analysis of polyaniline films, *Thin Solid Films*, vol. 234, 1993, pp. 468–470.

[159] G. Montseny, J. Audounet and D. Matignon, Fractional integro-differential boundary control of the Euler-Bernoulli beam, *36th IEEE Conference on Decision and Control*, San Diego, California, December 1997, IEEE-CSS, SIAM, pp. 4973–4978.

[160] M. A. Naimark, *Linear Differential Operators*, Nauka, Moscow, 1969 (in Russian).

[161] M. Nakagawa and K. Sorimachi, Basic characteristics of a fractance device, *IEICE Trans. Fundamentals*, vol. E75-A, no. 12, Dec 1992, pp. 1814–1819.

[162] R. R. Nigmatullin, To the theoretical explanation of the "universal response", *Phys. Sta. Sol. (b)*, vol. 123, 1984, pp. 739–745.

[163] R. R. Nigmatullin, On the theory of relaxation for systems with "remnant memory", *Phys. Sta. Sol. (b)*, vol. 124, 1984, pp. 389–393.

[164] R. R. Nigmatullin, The realization of the generalized transfer equation in a medium with fractal geometry, *Phys. Sta. Sol. (b)*, vol. 133, 1986, pp. 425–430.

[165] R. R. Nigmatullin, Fractional integral and its physical interpretation, *Soviet J. Theor. and Math. Phys.*, vol. 90, no. 3, 1992, pp. 354–367.

[166] R. R. Nigmatullin and Ya. E. Ryabov, Cole–Davidson dielectric relaxation as a self-similar relaxation process, *Phys. Solid State*, vol. 39, no. 1, 1997, pp. 87–90 (translated from the Russian original published in: *Fyz. Tverd. Tela*, vol. 39, 1997, pp. 101–105).

[167] K. Nishimoto, *An Essence of Nishimoto's Fractional Calculus*, Descartes Press, Koriyama, 1991.

[168] T. F. Nonnenmacher, Fractional integral and differential equations for a class of Lévy-type probability densities, *J. of Physics A: Math. and Gen.*, vol. 23, 1990, pp. L697–L700.

[169] T. F. Nonnenmacher, Fractional relaxation equations for viscoelasticity and related phenomena, *Lect. Notes in Physics*, vol. 381, Springer-Verlag, Berlin, 1991, pp. 309–320.

[170] T. F. Nonnenmacher and W. G. Glöckle, A fractional model for mechanical stress relaxation, *Philosophical Magazine Letters*, vol. 64, no. 2, 1991, pp. 89–93.

[171] T. F. Nonnenmacher and D. J. F. Nonnenmacher, Towards the formulation of a non-linear fractional extended irreversible thermodynamics, *Acta Physica Hungarica*, vol. 66, 1989, pp. 145–154.

[172] P. G. Nutting, A new general law of deformation, *Journal of the Franklin Institute*, vol. 191, 1921, pp. 679–685.

[173] L. Nyikos and T.Pajkossy, *Electrochem. Acta*, vol. 30, 1985, pp. 1533.

[174] M. Ochmann and S. Makarov, Representation of the absorption of nonlinear waves by fractional derivatives, *J. Amer. Acoust. Soc.*, vol. 94, no. 6, 1993, pp. 3392–3399.

[175] K. B. Oldham, A signal-independent electroanalytical method, *Anal. Chem.*, vol. 44, no. 1, 1972, pp. 196–198.

[176] K. B. Oldham, Semiintegration of cyclic voltammograms, *J. Electroanal. Chem.*, vol. 72, 1976, pp. 371–378.

[177] K. B. Oldham, Interrelation of current and concentration at electrodes, *J. Appl. Electrochem.*, vol. 21, 1991, pp. 1068–1072.

[178] K. B. Oldham and J. Spanier, The replacement of Fick's law by a formulation involving semidifferentiation, *J. Electroanal. Chem. and Interfacial Electrochem.*, vol. 26, 1970, pp. 331–341.

[179] K. B. Oldham and J. Spanier, *The Fractional Calculus*, Academic Press, New York – London, 1974.

[180] K. B. Oldham and C. G. Zoski, Analogue insrumentation for processing polarographic data, *J. Electroanal. Chem.*, vol. 157, 1983, pp. 27–51.

[181] O. V. Onishchuk and G. Ya. Popov, On some problems of bending of plates with cracks and thin inclusions, *Izv. Akad. Nauk SSSR, Mekhanika Tverdogo Tela*, vol. 4, 1980, pp. 141–150 (in Russian).

[182] T. J. Osler, Open questions for research, *Lecture Notes in Mathematics*, vol. 457, 1975, pp. 376–381.

[183] A. Oustaloup, *Systèmes asservis linéaires d'ordre fractionnaire.* Masson, Paris, 1983.

[184] A. Oustaloup, From fractality to non integer derivation through recursivity, a property common to these two concepts: a fundamental idea for a new process control strategy, *Proceedings of the 12th IMACS World Congress*, Paris, July 18–22, 1988, vol. 3, pp. 203–208.

[185] A. Oustaloup, *La Commande CRONE*, Hermes, Paris, 1991.

[186] A. Oustaloup, *La Robustesse*, Hermes, Paris, 1994.

[187] A. Oustaloup, *La Dérivation Non Entière: Théorie, Synthèse et Applications*, Hermes, Paris, 1995.

[188] E. Pitcher and W. E. Sewell, Existence theorems for solutions of differential equations of non-integral order, *Bull. Amer. Math. Soc.*, vol. 44, no. 2, 1938, pp. 100-107; and a correction in: vol. 44, no. 9, 1938, p. 888.

[189] L. Pivka and V. Spany, Boundary surfaces and basin bifurcations in Chua's circuit, *J. Circuits, Syst., Comput.*, vol. 3, no. 2, 1993, pp. 441–470.

[190] Yu. I. Plotnikov, *Steady-state vibrations of plane and axesymmetric stamps on a viscoelastic foundation*, Ph.D. thesis, Moscow, 1979 (in Russian).

[191] I. Podlubny, *Discontinuous Harmonic and Biharmonic Problems for a Sector and a Strip*, Ph.D. thesis, Odessa State University, Odessa, 1989, (in Russian).

[192] I. Podlubny, Orthogonal with non-integrable weight function Jacobi polynomials and their application to singular integral equations in elasticity and heat conduction problems, in: *Computational and Applied Mathematics II* (eds.: W. F. Ames and P. J. van der Houwen), North-Holland, Amsterdam, 1992, pp. 207–216.

[193] I. Podlubny, A united form of solution of singular integral equations of the first kind with Cauchy's kernal, *Transactions of the Technical University of Kosice,* vol. 3, no. 4, 1993, pp. 379–383.

[194] I. Podlubny, Riesz potential and Riemann-Liouville fractional integrals and derivatives of Jacobi polynomials, *Appl. Math. Lett.*, vol. 10, no. 1, 1997, pp. 103–108.

[195] I. Podlubny and J. Misanek, The use of fractional derivatives for modelling the motion of a large thin plate in a viscous fluid, *Proceedings of the 9th Conference on Process Control, Tatranske Matliare, May 1993*, STU Bratislava, pp. 274–278.

[196] I. Podlubny and J. Misanek, The use of fractional derivatives for modelling adiabatic process of solution of gas in fluid, *Proceedings of the 9th Conference on Process Control, Tatranske Matliare, May 1993*, STU Bratislava, pp. 279–282.

[197] I. Podlubny and J. Misanek, The use of fractional derivatives for solution of heat conduction problems, *Proceedings of the 9th Conference on Process Control, Tatranske Matliare, May 1993*, STU Bratislava, pp. 270–273.

[198] I. Podlubny, Fractional derivatives: a new stage in process modelling and control, *4th International DAAAM Symposium, Brno, Czech Republic, September 16–18, 1993*, pp. 263–264.

[199] I. Podlubny and I. Kostial, Fractional derivative based process models and their applications, *4th International DAAAM Symposium, Brno, Czech Republic, September 16–18, 1993*, pp. 265–266.

[200] I. Podlubny, The use of derivatives of the fractional order for process modelling and simulation: the present state and perspectives, *Proceedings of Int. Sci. Conf. MICROCAD SYSTEM'93, Nov 9–10, Kosice*, p. 29 and p. 64.

[201] I. Podlubny, *The Laplace Transform Method for Linear Differential Equations of the Fractional Order*, Inst. Exp. Phys., Slovak Acad. Sci., no. UEF-02-94, 1994, Kosice.

[202] I. Podlubny, *Fractional-Order Systems and Fractional-Order Controllers*, Inst. Exp. Phys., Slovak Acad. Sci., no. UEF-03-94, 1994, Kosice.

[203] I. Podlubny, Numerical methods of the fractional calculus, *Transactions of the Technical University of Kosice*, vol. 4, no. 3–4, 1994, pp. 200–208.

[204] I. Podlubny, L. Dorcak and J. Misanek, Application of fractional-order derivatives to calculation of heat load intensity change in blast furnace walls, *Transactions of the Technical University of Kosice*, vol. 5, no. 2, 1995, pp. 137–144.

[205] I. Podlubny, Analytical solution of linear differential equations of the fractional order, *Proceedings of the 14th World Congress on Computation and Applied Mathematics, July 11–15, 1994, Atlanta,*

Georgia, USA, Late Papers volume (ed.: W. F. Ames), pp. 102–106.

[206] I. Podlubny, Numerical solution of initial value problems for ordinary fractional-order differential equations, *Proceedings of the 14th World Congress on Computation and Applied Mathematics, July 11–15, 1994, Atlanta, Georgia, USA, Late Papers volume* (ed.: W.F.Ames), pp. 107–111.

[207] I. Podlubny, Basic mathematical tools for the analysis of dynamic systems of non-integer order, *Proceedings of the 11th Int. Conf. on Process Control and Simulation, September 19–20, 1994, Kosice-Zlata Idka*, pp. 305–311.

[208] I. Podlubny, Solution of linear fractional differential equations with constant coefficients, in: P. Rusev, I. Dimovski, V. Kiryakova (eds.), *Transform Methods and Special Functions*, SCT Publishers, Singapore, 1995, pp. 217–228.

[209] I. Podlubny, Numerical solution of ordinary fractional differential equations by the fractional difference method, in: S. Elaydi, I. Gyori and G. Ladas, (eds.), *Advances in Difference Equations,* Gordon and Breach, Amsterdam, 1997, pp. 507–516.

[210] J. D. Polack, Time domain solution of Kirchhoff's equation for sound propagation in viscothermal gases: a diffusion process, *SFA J. Acoustique*, vol. 4, 1991, pp. 47–67.

[211] G. Ya. Popov, Some properties of classical polynomials and their application to contact problems, *Prikl. Math. Mekh.*, vol. 27, 1963, pp. 821–832 (in Russian).

[212] G. Ya. Popov, On the method of orthogonal polynomials in contact problems of the theory of elasticity, *Prikl. Mat. Mekh.,* vol. 33, no. 3, 1969, pp. 518–531.

[213] G. Ya. Popov. *Stress Concentration near Punches, Cuts, Thin Inclusions and Supporters.* Nauka, Moscow, 1982 (in Russian).

[214] T. Pritz, Analysis of four-parameter fractional derivative model of real solid materials, *Journal of Sound and Vibration*, vol. 195, no. 1, 1996, pp. 103–115.

[215] A. P. Prudnikov, Yu. A. Brychkov and O. I. Marichev, *Integrals and Series*, vol. 1, Nauka, Moscow, 1981.

[216] A. P. Prudnikov, Yu. A. Brychkov and O. I. Marichev, *Integrals and Series*, vol. 2, Nauka, Moscow, 1983.

[217] Yu. N. Rabotnov, Equilibrium of an elastic medium with aftereffect, *Prikl. Mat. Mekh.*, vol. 12, no. 1, 1948, pp. 53–62 (in Russian).

[218] Yu. N. Rabotnov, *Elements of Hereditary Solids Mechanics*, Nauka, Moscow, 1977 (in Russian).

[219] Yu. N. Rabotnov, *Creep of Structural Elements*, Nauka, Moscow, 1966 (in Russian).

[220] K. Rektorys, *Handbook of Applied Mathematics*, vols. I, II. SNTL, Prague, 1988 (in Czech).

[221] F. Riewe, Nonconservative Lagrangian and Hamiltonian mechanics, *Phys. Rev. E*, vol. 53, no. 2, 1996, pp. 1890–1899.

[222] F. Riewe, Mechanics with fractional derivatives, *Phys. Rev. E*, vol. 55, no. 3, 1997, pp. 3581–3592.

[223] L. Rogers, Operators and fractional derivatives for viscoelastic constituitive equations, *J. of Rheology*, vol. 27, 1983, pp. 351–372.

[224] H. E. Roman, Structure of random fractals and the probability distribution of random walks, *Phys. Rev. E*, vol. 51, no. 6, 1995, pp. 5422–5425.

[225] H. E. Roman and P. A. Alemany, Continuous-time random walks and the fractional diffusion equation, *J. Phys. A: Math. Gen.*, vol. 27, 1994, pp. 3407–3410.

[226] B. Ross, A brief history and exposition of the fundamental theory of the fractional calculus, *Lecture Notes in Mathematics,* vol. 457, Springer-Verlag, New York, 1975, pp. 1–36.

[227] B. Ross, Fractional calculus: an historical apologia for the development of a calculus using differentiation and antidifferentiation of non integral orders, *Mathematics Magazine*, vol. 50, no. 3, May 1977, pp. 115–122.

[228] Yu. A. Rossikhin and M. V. Shitikova, Applications of fractional calculus to dynamic problems of linear and nonlinear hereditary mechanics of solids, *Appl. Mech. Rev.*, vol. 50, no. 1, January 1997, pp. 15–67.

[229] Yu. A. Rossikhin and M. V. Shitikova, Application of fractional derivatives to the analysis of damped vibrations of viscoelastic single mass system, *Acta Mech.*, vol. 120, 1997, pp. 109–125.

[230] B. Rubin, *Fractional Integrals and Potentials*, Pitman Monographs and Surveys in Pure and Applied Mathematics, vol. 82, Longman, Harlow, 1996.

[231] R. S. Rutman, On the paper by R. R. Nigmatullin "Fractional integral and its physical interpretation", *Theor. Math. Phys.*, vol. 100, no. 3, 1994, pp. 1154–1156. (Translated from the Russian original published in: *Teoret. i Matem. Fyz.*, vol. 100, no. 3, 1994, pp. 476–478.)

[232] S. G. Samko, A. A. Kilbas and O. I. Maritchev, *Integrals and Derivatives of the Fractional Order and Some of Their Applications*, Nauka i Tekhnika, Minsk, 1987 (in Russian).

[233] H. Schiessel, R. Metzler, A. Blumen and T. F. Nonnenmacher, Generalized viscoelastic models: their fractional equations with solutions, *J. Phys. A: Math. Gen.*, vol. 28, 1995, pp. 6567–6584.

[234] R. F. Schmidt and G. Thews, *Human Physiology*, Springer-Verlag, Berlin – Heidelberg – New York, 1983.

[235] W. R. Schneider and W. Wyss, Fractional diffusion and wave equations, *J. Math. Phys.*, vol. 30, 1989, pp. 134–144.

[236] G. W. Scott Blair, The role of psychophysics in rheology, *J. of Colloid Sciences*, vol. 2, 1947, pp. 21–32.

[237] G. W. Scott Blair, Some aspects of the search for invariants, *British Journal for Philosophy in Science*, vol. 1, 1950, pp. 230–244.

[238] G. W. Scott Blair, *Measurements of Mind and Matter*, Dennis Dobson, London, 1950.

[239] G. W. Scott Blair, Psychoreology: links between the past and the present, *Journal of Texture Studies*, vol. 5, 1974, pp. 3–12.

[240] A. M. Sedletskii, Asymptotic formulas for zeros of a function of Mittag-Leffler type, *Analysis Mathematica*, vol. 20, 1994, pp. 117–132 (in Russian).

[241] G. L. Slonimsky, On the law of deformation of highly elastic polymeric bodies, *Dokl. Akad. Nauk SSSR*, vol. 140, no. 2, 1961, pp. 343–346 (in Russian).

[242] W. Smit and H. de Vries, Rheological models containing fractional derivatives, *Rheologica Acta*, vol. 9, 1970, pp. 525–534.

[243] H. M. Srivastava, On an extension of the Mittag-Leffler function, *Yokohama Math. J.*, vol. 16, no. 2, 1968, pp. 77–88.

[244] H. M. Srivastava, A certain family of sub-exponential series, *Int. J. Math. Educ. Sci. Technol.*, vol. 25, no. 2, 1994, pp. 211–216.

[245] H. M. Srivastava and R. G. Buschman, *Theory and Applications of Convolution Integral Equations*, Kluwer Academic Publishers, Dordrecht – Boston – London, 1992.

[246] Z. Sobotka, *Reology of Materials and Constructions*, Academia, Prague, 1981 (in Czech).

[247] G. Szegö, *Orthogonal Polynomials*, Amer. Math. Soc., New York, 1959 (Russian translation with additions: Moscow, Fizmatfiz, 1962).

[248] A. N. Tikhonov and V. Ya. Arsenin, *Methods of Solution of Ill-Posed Problems*, Nauka, Moscow, 1986 (in Russian).

[249] E. C. Titchmarsh, *Introduction to the Theory of Fourier Integrals*, Clarendon Press, Oxford, 1937 (Russian translation: GTTI, Moscow, 1948).

[250] A. I. Tseytlin, *Applied Methods of Solution of Boundary Value Problems in Civil Engineering*, Stroyizdat, Moscow, 1984 (in Russian).

[251] E. Vitasek, *Numerical Methods*, SNTL, Prague, 1987 (in Czech).

[252] V. Volterra. *Leçons sur la Théorie Mathématique de la Lutte pour la Vie.* Paris, Gauthier-Villars, 1931 (Russian translation: Moscow, Nauka, 1976).

[253] S. Westerlund. Dead matter has memory! *Physica Scripta*, vol. 43, 1991, pp. 174–179.

[254] S. Westerlund, *Causality*, report no. 940426, University of Kalmar, 1994.

[255] S. Westerlund and L. Ekstam, Capacitor theory, *IEEE Trans. on Dielectrics and Electrical Insulation*, vol. 1, no. 5, Oct 1994, pp. 826–839.

[256] A. Wiman, Über den fundamentalsatz in der teorie der funktionen $E_\alpha(x)$, *Acta Math.*, vol. 29, 1905, pp. 191–201.

[257] A. Wiman. Über die nulstellen der funktionen $E_\alpha(x)$. *Acta Math.*, vol. 29, 1905, pp. 217–234.

[258] E. M. Wright, On the coefficients of power series having exponential singularities, *J. London Math. Soc.*, 1933, vol. 8, pp. 71–79.

[259] W. Wyss, The fractional diffusion equation, *J. Math. Phys.*, vol. 27, no. 11, 1986, pp. 2782–2785.

Index